基于 Hadoop 的大数据平台构建

周静 谭亮 陈斌 刘贤 编著

西南交通大学出版社
·成都·

图书在版编目（CIP）数据

基于 Hadoop 的大数据平台构建 / 周静等编著. —成都：西南交通大学出版社，2018.11
ISBN 978-7-5643-6566-0

Ⅰ. ①基… Ⅱ. ①周… Ⅲ. ①数据处理软件 Ⅳ. ①TP274

中国版本图书馆 CIP 数据核字（2018）第 254837 号

基于 Hadoop 的大数据平台构建

周 静　谭 亮　陈 斌　刘 贤　编著

责 任 编 辑	穆　丰
封 面 设 计	何东琳设计工作室
出 版 发 行	西南交通大学出版社 （四川省成都市二环路北一段 111 号 西南交通大学创新大厦 21 楼）
发行部电话	028-87600564　028-87600533
邮 政 编 码	610031
网　　　址	http://www.xnjdcbs.com
印　　　刷	成都蓉军广告印务有限责任公司
成 品 尺 寸	170 mm × 230 mm
印　　　张	19.75
字　　　数	355 千
版　　　次	2018 年 11 月第 1 版
印　　　次	2018 年 11 月第 1 次
书　　　号	ISBN 978-7-5643-6566-0
定　　　价	68.00 元

图书如有印装质量问题　本社负责退换
版权所有　盗版必究　举报电话：028-87600562

前 言

近年来，大数据逐渐被人们所熟知，大数据技术也在人们的日常生活中得到了广泛的应用。随着计算机硬件性能、计算机软件技术、互联网技术的飞速发展，信息技术与社会各个方面的交汇融合引发了数据的爆发式增长，数据已然成为了国家的基础性战略资源，对全球的生产、流通、分配、消费等活动以及经济运行机制、社会生活方式、国家综合实力等方面产生着重大影响。而大数据技术正是对数量巨大、来源分散、格式多样的数据进行采集、存储、分析，然后从中发现新知识、创造新价值、提升新能力，成为了新一代的信息服务模式。

Apache Hadoop 作为大数据行业中的一位元老级成员，已经经过了十几年的发展，建立了成熟的生态圈，是大数据行业的重要技术标准之一。它是一个开源的大数据平台，提供了一种可以高效处理海量规模数据的方式，可以说是专门为大数据而生，为大数据的发展提供了巨大帮助。

本书以构建一个大数据存储与处理平台为例，以 Apache Hadoop 软件为基础，采用单节点平台到高可用平台再到平台相关工具的循序渐进的方式，使读者逐步了解大数据平台，并掌握基于 Apache Hadoop 的大数据平台的搭建和使用。全书的内容如下：

项目 1 以单节点模式的 Hadoop 整合平台的搭建为例，介绍大数据以及 Hadoop 平台。单节点模式是 Hadoop 平台最为简单的一种部署模式，多用于测试和开发环境，通过该部署模式能够更容易地了解 Hadoop 平台。

项目 2 以高可用模式的 Hadoop 整合平台的搭建为例，展示了更为完整的 Hadoop 平台的面貌。高可用模式是用于实际运行服务的 Hadoop 平台的完整部署模式，通过该部署模式能够进一步深入了解 Hadoop 平台的功能与特性。

项目 3 通过多个 Hadoop 平台的操作示例，介绍了 HDFS 文件系统、MapReduce 框架、HBase 数据库的常见使用和维护方法。

项目 4 介绍了基于 Linux 操作系统的 MySQL 数据库平台的搭建，包括单节点和集群两种部署模式。MySQL 数据库作为最为常用的关系型数据库之一，被 Hadoop 生态圈中的很多组件所使用，是完整的基于 Hadoop 的大数据平台中不可或缺的一部分。

项目 5 介绍了 Hadoop 生态圈中的数据仓库平台 Hive 的搭建，并以操作示例展示了其基本使用方法。数据仓库是用于支持大数据决策分析的重要工具，而 Hive 所提供的类 SQL 查询功能更是让传统的关系型数据库的使用者也能轻松地使用 Hadoop 平台来进行数据的分析和处理。

项目 6 介绍了 Hadoop 生态圈中的 ETL 工具 Sqoop 的安装，并通过示例展示了如何在关系型数据库和 Hadoop 平台之间转换数据。ETL 工具是不同数据存储平台之间进行数据转换的重要手段，而由于 Hadoop 平台不善于处理随机数据存储的特性，使其必然需要从其他存储业务数据的数据库中导入数据，这就使得 Sqoop 工具显得尤为重要。

本书在编写过程中参阅了国内外同行的相关著作和文献，谨向各位作者致以深深谢意！由于作者水平有限，书中难免存在错误与疏漏之处，敬请广大读者批评指正。

作　者

2018 年 6 月

目 录

项目 1　搭建单节点 Hadoop 整合平台 ··· 1
　任务 1.1　认识大数据 ··· 1
　任务 1.2　了解 Hadoop ··· 9
　任务 1.3　安装 JDK ··· 17
　任务 1.4　安装 Zookeeper 伪分布模式 ··· 24
　任务 1.5　安装 Hadoop 伪分布模式 ··· 42
　任务 1.6　安装 HBase 伪分布模式 ··· 80

项目 2　搭建高可用 Hadoop 整合平台 ··· 95
　任务 2.1　集群网络属性配置 ··· 96
　任务 2.2　集群下 JDK 的便捷安装 ··· 105
　任务 2.3　安装 Zookeeper 完全分布模式 ··· 106
　任务 2.4　安装 Hadoop 完全分布模式 ··· 112
　任务 2.5　安装 Hadoop 高可用模式 ··· 122
　任务 2.6　安装 HBase 完全分布模式和高可用模式 ··· 151

项目 3　Hadoop 整合平台的使用与管理 ··· 160
　任务 3.1　Hadoop 的使用和管理 ··· 160
　任务 3.2　HBase 的使用和管理 ··· 191

项目 4　基于 Linux 的 MySQL 数据库平台的搭建 ··· 208
　任务 4.1　安装基于 Linux 的 MySQL 单机模式 ··· 208
　任务 4.2　安装 MySQL 集群模式 ··· 222

项目 5　Hive 数据仓库的搭建和使用 ……………………………………… 245
任务 5.1　Hive 数据仓库的搭建 ………………………………………… 245
任务 5.2　Hive 数据仓库的使用和管理 ………………………………… 267

项目 6　使用 ETL 工具 Sqoop 转换数据 ……………………………… 281
任务 6.1　Sqoop 工具的安装 …………………………………………… 281
任务 6.2　使用 Sqoop 工具进行数据转换 ……………………………… 295

参考文献 ………………………………………………………………………… 309

项目 1　搭建单节点 Hadoop 整合平台

近年来,"大数据"一词已经越来越被人们所熟知,它引领着新一轮的信息技术革命,人类也步入了一个崭新的大数据时代。随着大数据的概念、技术、应用不断地深入到社会中的各个方面,它在迅速而深刻地改变着我们工作方式和生活方式。

Hadoop 作为大数据行业中的一位元老级成员,可以说是专门为大数据而生的。它提供了一种可以高效处理海量规模数据的方式,为大数据的发展提供了巨大帮助。这里以单节点 Hadoop 整合平台的搭建为例,让大家对大数据以及 Hadoop 平台有一个初步的了解。Hadoop 整合平台指的是 Hadoop 软件自身,其中包括了 HDFS 分布式文件系统和 MapReduce 分布式计算框架,同时搭配上 HBase 分布式数据库,以及 HBase 所需要使用到的 Zookeeper 分布式协调服务。

任务 1.1　认识大数据

随着计算机硬件性能的发展、互联网传输速度的提升、智能便携移动终端设备的出现和快速普及,人类所产生的数字信息量也越来越大,近几年来甚至是以爆发式的速度增长。以 2017 年的天猫"双十一"购物狂欢节活动为例,仅一天的成交额就达到了 1 682 亿元,产生的交易记录达到 14.8 亿条,产生的物流订单超过 8 亿单。而在最近几年,这样会产生海量规模数据的应用数不胜数。2017 年,新浪微博的每日活跃用户数量达到 1.65 亿,而微博内容的总条数已累积超过千亿。YouTube 在 2017 年中平均每天视频播放的总时长超过了 10 亿小时,这个时长的视频如果由一个人来看的话需要约 10 万年的时间,这足够使用光速旅行从银河系的一端飞到另一端,并且 YouTube 上的新视频还在以每分钟 300 小时的数量增长。而 2017 年春节期间,使用微信收发的红包总数量达到了 460 亿个,同时通过微信的音视频通话功能进行新年问候的总时长也是达到了 21 亿分钟。全球最大的搜索引擎谷歌(Google)每分钟的搜索量可达 278 万次,全球最大的社交网站脸书(Facebook)每分钟的点赞数量超过 400 万次。预计到了 2020 年，全球的数据总量将会达到 44 ZB，这换算成我们更为熟悉的单位钛字节 TB 和吉字节 GB，那就是 440 亿 TB 和 44 万亿 GB。

2013 年被定义为世界大数据元年，它是世界步入大数据时代的一个标志。想要深入了解大数据，可以从理论（Theory）、技术（Technology）、实践（Utilization）三个方面来进行，如图 1-1-1 所示。

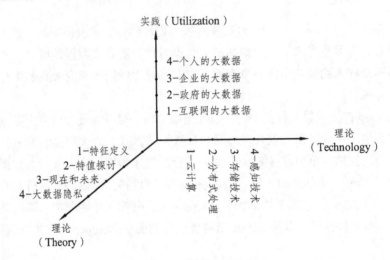

图 1-1-1　深入理解大数据

1.1.1　从理论上了解大数据

最早提出大数据时代的麦肯锡公司在其有关大数据研究的报告中将大数据定义为大小超出常规的数据库工具获取、存储、管理、分析能力的数据集。而 IBM 公司将大数据的特征归纳为 5 个"V"，分别为 Volume、Velocity、Variety、Value、Veracity。

Volume 代表的是大容量，即数据体量巨大。大数据的计量单位符号一般都是 PB（1 024 TB）或 EB（1 024 PB）级别，甚至现在在朝着 ZB（1 025 EB）级别发展。

Velocity 代表的是高速度，即处理速度快。由于数据量巨大，只有达到了一定的处理速度，才能实时地从数据中发掘出有用的信息。

Variety 代表的是多样性，即数据类型的多样性。这里的数据类型包括了视频、音频、图片、地理信息、日志等多种多样的类型，不再只是传统关系型数据库中的结构化数据。

Value 代表的是价值，即以低成本创造高价值。现今随着互联网和计算机硬件的飞速发展，数据的产生和收集变得越来越简单，数据的获取和存放的成本也越来越低。而当很多不全面、不连续，甚至看似无关联的数据，在达到一定规模时，便可以从中产生出不可估量的价值。

Veracity 代表的是真实，指数据的质量真实可靠。只有从用户和业务系统产生的真实数据才是有价值的，才能从中挖掘出有用的信息。

当下，大数据应用价值已在各行各业凸显。大数据帮助政府实现市场经济调控、公共卫生安全防范、灾难预警、社会舆论监督、犯罪预防；大数据帮助城市实现智慧交通、智慧楼宇、智慧公共设施；大数据帮助医疗机构建立患者的疾病风险跟踪机制，帮助医药企业提升药品的临床使用效果，帮助艾滋病研究机构为患者提供定制的药物；大数据帮助航空公司节省运营成本，帮助电信企业实现售后服务质量提升，帮助保险企业识别欺诈骗保行为，帮助快递公司检测分析运输车辆的故障险情以便提前预警维修，帮助电力公司有效识别预警即将发生故障的设备。不管大数据的核心价值是不是预测，基于大数据形成决策的模式已经为不少的企业带来了盈利和声誉。

而大数据时代也有其面临的问题，其中最主要的就是侵犯隐私问题。当在不同网站上注册了个人信息之后，可能这些信息就已经被自动扩散出去了。当人们莫名其妙地接收到各种邮件、电话、短信的滋扰时，不会想到自己的电话号码、邮箱、生日、购买记录、收入水平、家庭住址、亲朋好友等私人信息早就被各种商业机构非法存储或出售给其他任何有需要的企业和个人。即使用户在某个地方删除了相关信息，也许这些信息已经被其他人转载或保存，更有可能已经被存为网页快照，早已提供给任意用户搜索。因此在大数据背景下，很多人都在积极抵制无底线的数字化，这种大数据和个体之间的博弈还会一直持续下去。

当很多互联网企业意识到隐私对于用户的重要性时，为了继续得到用户的信任，采取了很多办法。如谷歌承诺保留用户的搜索记录时间为 9 个月，而有的浏览器厂商提供了无痕冲浪模式，还有社交网站拒绝公共搜索引擎的爬虫，并将提供出去的数据全部采取匿名方式处理等。被誉为大数据商业应用第一人的维克托·迈尔·舍恩伯格在《大数据时代》一书中提出了一些如何有效保护大数据背景下隐私权的建议，如减少个人信息的数字化、建立完善隐私权保护法、增强数字隐私权基础设施建设等。

1.1.2 从技术上了解大数据

大数据常和云计算联系在一起，因为实时的大型数据集的分析需要分布式处理框架来向数十、数百甚至上万的计算机分配工作，它的特色在于对海量数据的挖掘。如今在谷歌、亚马逊、Facebook 等一批互联网企业引领下，创建了一种行之有效的模式，即云计算提供基础架构平台，大数据应用可以运行在这个平台之上，如图 1-1-2 所示。业内这样认为两者的关系：没有大数据的信息积淀，云计算的处理能力再强大，也难以找到用武之地；而没有云计算的处理能力，大数据的信息积淀再丰富，也同样没有用武之地。大数据和云计算两者之间有效结合之后，便可以提供更多基于海量业务数据的创新型服务，并通过云计算技术的不断发展降低大数据业务的创新成本。

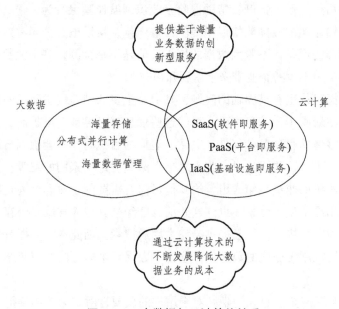

图 1-1-2　大数据与云计算的关系

大数据与云计算的最显著区别表现在两个方面。一方面是概念上的不同，云计算改变了整个 IT（Information Technology）行业，而大数据则是改变了构建在 IT 行业基础上的各种业务，而其中大数据又必须以云计算作为基础架构，才能得以顺畅运行。另一方面是受众群体不同，云计算属于 CIO（首席信息官）等技术相关职位人员所关心的技术层面，是一个进阶的 IT 行业解决方案；而大数据则是属于 CEO（首席执行官）等管理相关人员所关心的业务层面的产品。

大数据可以抽象地分为大数据存储和大数据分析两个部分，其中大数据存储的目的是为了支撑大数据分析业务。目前来看，大数据存储和大数据分析已经发展成为两个截然不同的计算机技术领域。大数据存储致力于研发可以扩展至 PB 甚至 EB 级别的数据存储平台，而大数据分析关注在最短时间内处理大量不同类型的数据集的计算平台。

　　大数据的技术最核心、最重要的部分就是分布式处理技术。分布式处理系统可以将处于不同地点、具备不同功能、拥有不同数据的多台计算机用通信网络连接起来，然后在控制系统的统一管理下，协调完成信息处理任务。本书所要介绍的 Hadoop 就是大数据分布式处理系统的典型代表。它所包含的 MapReduce 软件框架，能以一种可靠、高效、可伸缩的方式对大数据进行分布式处理。同时 MapRedcue 还利用 HDFS 分布式文件系统来存储数据，采用将数据移动到计算的方式进一步提高大数据分布式处理的效率。

　　另外，大数据的采集与感知技术的发展也是紧密联系在一起的。以传感器技术、指纹识别技术、RFID 技术、坐标定位技术等为基础的计算机感知能力的提升同样是物联网发展的基石。全世界的工厂自动化生产线、汽车、电表等设备上有着无数的数据传感器，随时测量和传递着有关位置、运动、震动、温度乃至空气中化学物质变化等信息，都会产生海量的数据。而随着智能移动终端设备的发展和普及，感知技术可谓迎来了发展的高峰期。除了地理位置信息被广泛地应用之外，一些新的感知手段也开始出现，如手机中内嵌的指纹传感器，即将面世的可以检测空气污染及危险化学药品的带有嗅觉传感器的智能手机，感知用户当前情绪的手机智能技术等。这些事物被逐渐感知和捕获的过程，其实就是世界被逐渐数据化、信息化的过程。

1.1.3 从实践上了解大数据

　　大数据的主要来源是互联网。互联网上的数据每年增长 50%，几乎每两年便会翻一番，而目前世界上 90%以上的数据都是最近十年左右才产生的。据预测，到 2020 年，全球将总共拥有 35 ZB 的数据量。互联网是大数据发展的前哨阵地，随着 Web 2.0 时代的发展，人们已经越来越习惯于将自己的生活通过网络进行数据化，方便分享与记录与回忆。

　　互联网上的大数据很难清晰地界定分类界限，大致可以分为以下五类：

　　（1）用户行为数据。这类数据主要用于精准广告投放、内容推荐、行为习惯分析、喜好分析、产品优化等。

（2）用户消费数据。这类数据主要用于精准营销、信用记录分析、活动促销、理财推荐等。

（3）用户地理位置数据。这类数据主要用于在线或离线的推广、线上到线下的推广、商家推荐、交友推荐等。

（4）互联网金融数据。这类数据主要用于点对点网络借款、小额贷款、支付、信用分析、供应链金融等。

（5）用户社交数据。这类数据主要用于潮流趋势分析、流行元素分析、受欢迎程度分析、舆论监控分析、社会问题分析等。

近几年来，各国政府也越来越重视大数据的发展。美国政府曾投资 2 亿美元致力于拉动大数据相关产业的发展，将大数据战略上升到国家意志层面。美国政府将大数据定义为"未来的新石油"，并表示一个国家拥有数据的规模、活性及解释运用的能力将成为综合国力的重要组成部分。未来，对数据的占有和控制甚至将成为国家核心资产。在国内，政府各个部门都掌握着构成社会基础的原始数据，如气象数据、金融数据、信用数据、电力数据、煤气数据、自来水数据、道路交通数据、客运数据、安全刑事案件数据等。这些数据在每个政府部门里面看起来都是单一的、静态的，但是如果政府可以将这些数据关联起来，并对这些数据进行有效的关联分析和统一管理，这些数据必定将产生无法估量的价值。

而现在的城市也已经逐渐在走向智能化，如智能电网、智慧交通、智慧医疗、智慧环保、智慧城市等。这些都是依托于大数据实现的，可以说大数据是智能化的核心能源。从国内整体的投资规模来看，自 2012 年至今，全国开始进行智慧城市建设的城市超过了 290 个，通信网络和数据平台等基础设施的建设投资规模接近 5 000 亿元。大数据为智慧城市的各个领域提供角色支持。在城市规划方面，通过对城市地理、气象等自然信息，以及经济、社会、文化、人口等人文社会信息的挖掘，可以为城市规划提供决策依据，强化城市管理服务的科学性和前瞻性；在交通管理方面，通过对道路交通信息的实时挖掘，能有效地缓解交通拥堵，并快速响应突发状况，为城市交通的良性运转提供科学的决策依据；在舆情监控方面，通过网络关键词搜索及语义智能分析，能提高舆情分析的及时性、全面性，快速全面地掌握社情民意，提高公共服务能力，快速高效地应对网络突发的公共事件，打击违法犯罪；在安防和防灾领域，通过大数据的挖掘，可以及时发现人为和自然灾害、恐怖事件等，提高应急处理能力和安全防范能力。另外作为国家的管理者，政府还应该将国家所掌控的数据逐步开放，提供给更多有能力的组织、机构、个人来分析和利用，加快大数据对社会发展的促进作用。

项目 1　搭建单节点 Hadoop 整合平台

对于企业来说，其管理者最关注的还是报表曲线的背后所具有的信息，以及通过这些报表曲线应该做出怎样的决策，这些也都需要通过数据来传递和支撑。在理想的状态下，大数据可以改变公司的影响力，它可以节省开支、增加利润、取悦买家、增加用户忠诚度、转化潜在客户、增加吸引力、提高竞争力、开拓市场。对于企业的大数据，随着数据逐渐成为企业的一种资产，数据产业会向传统企业的供应链的模式发展，最终形成数据供应链。这主要表现在两个方面。第一，外部数据的重要性日益超过内部数据，互联网时代单一企业的内部数据与整个互联网数据相比只是沧海一粟。第二，能够提供包括数据供应、数据整合、数据加工、数据应用等多环节服务的公司会有明显的综合竞争优势。

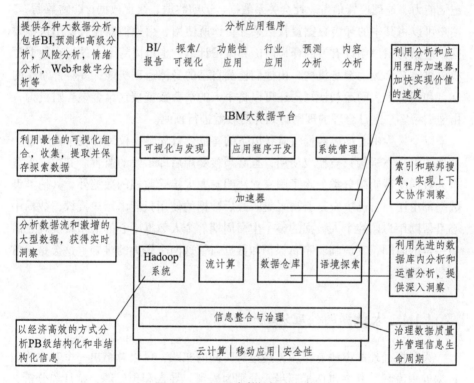

图 1-1-3　IBM 大数据平台架构

从 IT 产业的发展来看，第一代 IT 巨头大多是提供 ToB（面向企业）服务的，如 IBM、微软、Oracle、SAP、HP 这类传统 IT 企业。而第二代 IT 巨头大多是提供 ToC（面向用户）服务的，如雅虎、Google、亚马逊、Facebook 这类互联网企业。大数据到来之前，这两类公司彼此之间基本上是井水不犯河水，相互之间的竞争关系较少。但大数据时代来临之后，这两类公司开始了直接的竞争。

出现这种现象的主要原因是：在互联网巨头的带动下，传统 IT 巨头的客户普遍开始从事电子商务业务，正是由于客户进入了互联网，所以传统 IT 巨头们也被拖入了互联网领域。以 IBM 为例，在上一个十年中成功抛弃了计算机硬件开发与生产，成功转向了软件和服务，并提出了大数据平台架构，如图 1-1-3 所示。

大数据对于个人而言现在还只是需要数据被采集以及享受最终产生的结果，并且这种采集和结果都还是分布在不同的平台和应用之上的。未来，每个用户将可以在互联网上注册个人的数据中心，以存储个人的大数据信息；同时用户可以确定哪些个人数据可被采集，并通过可穿戴设备或植入芯片等感知技术来捕获个人的大数据信息，如牙齿数据、心律数据、体温数据、视力数据、记忆能力、地理位置信息、社会关系数据、运动数据、饮食数据、购物数据等。用户可以将其中的牙齿数据授权给某牙科诊所使用，由他们监控和使用这些数据，进而为用户制订有效的牙齿防治和维护计划；也可以将个人的运动数据授权提供给某个运动健身机构，由他们检测自己的身体运动机能，并有针对性的制订和调整个人的运动计划；还可以将个人的消费数据授权给金融理财机构，由他们帮客户制订合理的理财计划并对收益进行预测。

以个人为中心的大数据将具备以下的特征。首先数据仅留存在个人中心，其他的第三方机构只被授权使用，且必须接受用后即销毁的监管。其次，采集个人数据应该明确分类，除了国家立法明确要求接受监控的数据外，其他类型数据都由用户自己决定是否被采集。最后数据的使用只能由用户授权，数据中心可帮助用户监控个人数据的整个生命周期。个人数据中心的愿望要实现还需要一定的时间，其面临着隐私保护和数据监管这两个最大的难题，这会是异常激烈的博弈。

1.1.4 大数据的处理流程

大数据处理的方法有很多，经过长时间的实践，可以总结出一个基本的大数据处理流程，其中包括了四步，分别为采集、导入和预处理、统计和分析、挖掘。

大数据的采集是指利用多个数据库来接收来自客户端（Web、App、传感器）的数据，并且用户可以通过这些数据库来进行简单的查询和处理工作。大数据采集过程中的主要问题就是高并发数，因为同时有可能会有成千上万的用户来进行访问和操作，这需要在采集端部署大量数据库才能支持，并且如何在这些数据库之间进行负载均衡和数据同步也是需要深入思考的问题。

采集端本身会有很多数据库，如果要对这些海量数据进行有效的分析，需要将这些来自前端的数据导入到一个集中的大型分布式数据库或分布式存储集群中，并且可以在导入的基础上进行一些简单的清晰和预处理工作，这就是导入和预处理阶段。这个过程的主要问题是导入的数据量大，每秒的导入数据量经常会达到百兆甚至千兆级别。

统计和分析阶段主要是利用分布式数据库，或者分布式计算集群来对存储于其内部的海量数据进行普通的分析和分类汇总等，以满足大多数常见的分析需求。这个阶段的主要问题是分析涉及的数据量大，对系统资源，特别是 I/O 资源的消耗极大。

数据的挖掘与统计与分析过程有所不同，数据的挖掘一般没有什么预先设定好的主题，主要是在现有数据上面进行基于各种算法的计算，从而起到预测的效果，以便实现一些高级别的数据分析需求。该阶段的主要问题是用于挖掘的算法一般都比较复杂，并且计算涉及的数据量和计算量都很大。

任务 1.2 了解 Hadoop

Hadoop 是一个能够对大量数据进行分布式存储和处理的软件框架，能够以一种可靠、高效、可伸缩的方式进行数据处理。Hadoop 通过维护多个数据和工作的副本来提供可靠性，以并行处理的方式来提供高效性，平台规模能够动态调整，可以处理 PB 级别的数据。并且 Hadoop 是属于 Apache 基金会下的开源软件，可以方便地获取，任何人都可以使用。

Hadoop 的出现和 Google 有着千丝万缕的联系。Hadoop 技术思想的根本就是来自这家著名的 IT 企业。Google 最为大家所熟知的就是它的搜索引擎，而它所使用的大数据平台的构建一直遵循着低成本和"Power of Many"的思想，那就是不使用超级计算机、大型机、专业存储等价格昂贵且升级困难的设备，转而使用大量的普通 PC（个人计算机）服务器。通过软件来将这些大量的普通 PC 服务器连接起来，从而达到高性能的存储和计算要求，同时还要能够自由变更规模来达到理论上的无性能瓶颈。而 Google 将其用于实现其搜索引擎的大数据平台的相关技术先后以公开论文的方式发表，分别是 2003 年的《The Google File System》，2004 年的《MapReduce: Simplified Data Processing on Large Clusters》，2006 年的《Bigtable: A Distributed Storage System for Structured Data》，分别对应了分布式存储技术、分布式计算技术、非关系型分布式数据库技术。这三篇公开论文直接导致了之后 Hadoop 的诞生，其分布式存储技术对应了 Hadoop 中的 HDFS 文件系统，分布式计算技术对应了 Hadoop 中同名的 MapRedcue 计算框架，非关系型分布式数据库技术对应了 Hadoop 中的 Hbase 数据库。可以说 Hadoop 就是 Google 的大数据平台的一个开源实现版本。

1.2.1 Hadoop 的发展历程

Hadoop 是由 Apache Lucene 的创始人 Doug Cutting 创建的，而 Lucene 是一个被广泛使用的文本搜索系统库。Lucene 项目中有一个名为 Apache Nutch 的子项目，这是一个网络搜索引擎，而 Hadoop 就起源于该项目。

该项目中有一个开放源代码的全文检索引擎工具包，提供了一个全文检索引擎的架构，包括完整的查询引擎和索引引擎，以及部分文本分析引擎（英文与德文），目的是方便软件开发人员在目标系统中实现全文检索功能或是以此为基础构建完整的全文检索引擎。2005 年，Nutch 的开发者在 Nutch 上发布了一个可工作的 MapReduce 应用，并将所有主要的 Nutch 算法移植到了使用 MapReduce 的 NDFS 上运行。之后 Doug Cutting 将 Nutch 的这一部分作为独立项目引入 Apache 软件基金会，并命名为 Hadoop。如图 1-2-1 所示，其中展示了 Hadoop 初期的大致发展过程。

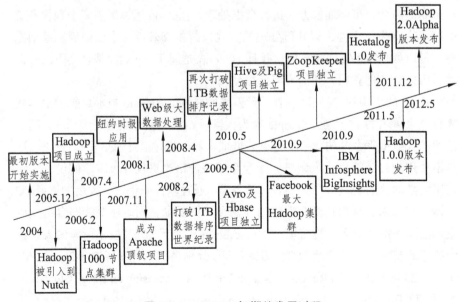

图 1-2-1　Hadoop 初期的发展过程

Hadoop 发展至今，先后经历过几次较大的版本更新。在这些版本更新中，软件的功能、架构、组件、代码、API（应用程序接口）都发生了不同程度的变化，甚至有些版本更新的变化非常大。所以这些不同版本的 Hadoop 平台相互之间兼容性较差，甚至有一些版本之间完全不兼容，而不同版本分支的 Hadoop 平台所对应的相关组件的版本也会有所不同，若选择的 Hadoop 相关组件的版本与 Hadoop 的版本分支不对应，则会出现兼容性问题，甚至组件完全无法使用。所以在选择 Hadoop 的版本时，也要注意选择对应该版本的相关组件的版本。

1. 0.20.X 系列

该版本分支是由最原始的 Hadoop 版本发展而来，基本保留了最早 Hadoop 的原貌。后来被重命名为 1.X 版本系列。

2. 0.21.0/0.22.X 系列

该版本分支将整个 Hadoop 项目分割成了三个独立的模块，分别是 Common、HDFS、MapReduce。HDFS 和 MapReduce 都对 Common 模块有依赖性，但是 MapReduce 对 HDFS 并没有依赖性。这样，MapReduce 便可以更容易地运行其他分布式文件系统。同时，模块之间也可以独立进行开发。在该版本分支中 HDFS 模块的变化最大，主要是将 SecondaryNameNode 剔除，转而使用 CheckpointNode 取代，同时添加了一个名为 BackupNode 的角色作为 NameNode 的冷备份。而 0.22.X 是在 0.21.0 的基础上修复一些 Bug 并进行部分优化产生。

3. 0.23.X 系列

该版本分支是为了克服 Hadoop 在扩展性和框架通用性方面的不足而提出来的。它实际上是一个全新的平台，包含了分布式文件系统 HDFS Federation 和资源管理框架 YARN 两个部分，可对接入的各种计算框架（如 MapReduce、Spark 等）进行统一管理。它的发行版自带 MapReduce 库，而该库集成了迄今为止所有的 MapReduce 新特性。

4. 2.X 系列

同 0.23.X 系列的版本分支一样，2.X 系列也属于新一代 Hadoop。与 0.23.X 系列相比，2.X 系列增加了 NameNode HA 和 Wire-compatibility 等新特性。

5. 3.X 系列

作为最近才发布的最新的 Hadoop 版本，该版本分支在 2.X 系列的版本分支上做出了很多重大改进，特别是在运行性能方面。该版本分支中引入了一些重要的功能和优化，包括 HDFS 的可擦除编码、多 Namenode 支持、MapReduce Native Task 优化、YARN 基于 cgroup 的内存和磁盘 I/O 隔离、YARN container resizing 等。该版本分支的 Hadoop 平台能够支持 Storm、JStorm、Flink 实时计算，总体性能上比现在流行的 Spark 快 10 倍。

1.2.2 Hadoop 生态圈成员

Hadoop 经过多年的发展，如今已经成为 Apache 的顶级项目，除了最初的 HDFS 分布式文件系统、MapReduce 分布式计算框架、HBase 分布式非关系型数据库之外，还新增了众多的子项目，以及很多需要使用 Hadoop 的 HDFS、MapRedcue 等组件作为基础的周边项目，这些子项目和周边项目使得 Hadoop 平台的功能越来越完善。如图 1-2-2 所示，这些子项目和周边项目共同构成了 Hadoop 的生态圈。

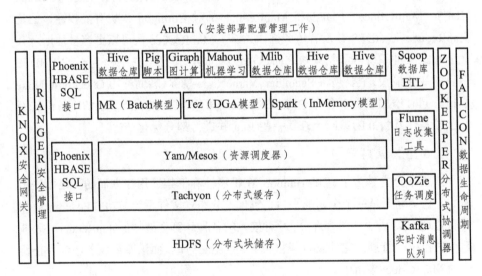

图 1-2-2　Hadoop 生态图

1. HDFS（Hadoop Distributed File System，Hadoop 分布式文件系统）

Hadoop 的基础组件之一，源自 Google 于 2003 年 10 月发表的关于 GFS 的论文《The Google File System》，可以看作是 GFS 的克隆版本或开源实现版本。HDFS 是 Hadoop 体系中数据存储管理的基础。它是一个高度容错的系统，能够检测和应对硬件故障，用于在低成本的通用硬件上运行。HDFS 简化了文件的一致性模型，通过流式数据访问，提供高吞吐量应用程序数据访问功能，适合带有大型数据集的应用程序。

2. MapReduce（分布式计算框架）

Hadoop 的基础组件之一，源自 Google 于 2004 年 12 月发表的论文《MapReduce：Simplified Data Processing on Large Clusters》，可以看作是 Google MapReduce 的克隆版本或开源实现版本。MapReduce 是一种分布式计算模型，用于进行大数据量的计算。它屏蔽了分布式计算框架细节，将计算抽象成 Map 和 Reduce 两部分，其中 Map 对数据集上的独立元素进行指定的操作生成中间结果，而 Reduce 则对中间结果进行整合得到最终结果。MapReduce 非常适合在大量计算机组成的分布式并行环境里进行数据处理。

3. HBase（分布式列存储数据库）

源自 Google 于 2006 年 11 月发表的论文《Bigtable：A Distributed Storage System for Structured Data》，可以看作是 Google BigTable 的克隆版本或开源实现版本。HBase 是一个建立在 HDFS 之上，面向列的针对结构化数据的可伸缩、高可靠、高性能、分布式的动态模式数据库。HBase 采用了和 BigTable 一样的数据模型，即增强的稀疏排序映射表（Key/Value），同时提供了对大规模数据的随机、实时的读写访问。同时，HBase 中保存的数据可以使用 MapReduce 来处理，它将数据存储和并行计算完美地结合在一起。

4. Zookeeper（分布式协作服务）

源自 Google 于 2006 年 11 月发表的论文《The Chubby Lock Service for Loosely-Coupled Distributed Systems》，可以看作是 Chubby 的克隆版本或开源实现版本。Zookeeper 解决了分布式环境下的数据管理问题，为分布式应用提供一致性服务，其中包括命名空间、状态同步、集群管理、配置维护等。它将复杂易出错的关键服务进行封装，将简单易用的接口和性能高效、功能稳定的系统提供给用户。Hadoop 的许多组件依赖于 Zookeeper，它运行在计算机集群上面，用于管理 Hadoop 操作。

5. Hive（数据仓库）

由 Facebook 开发并在之后开源，最初用于解决海量结构化的日志数据统计问题。Hive 可以将结构化的数据文件映射为一张数据库表，并且定义了一种类似于 SQL 的查询语言 HQL，它将 SQL 转化为 MapReduce 任务在 Hadoop 上执行，通常用于离线分析。Hive 使得不熟悉 MapReduce 开发的用户也能够编写数据查询语句快速方便的检索数据，而不必专门去学习如何编写 MapReduce 程序。

6. Pig（ad-hoc 脚本）

由 Yahoo 开发并在之后开源，设计动机是提供一种基于 MapReduce 的 ad-hoc 数据分析工具。Pig 定义了一种数据流语言 Pig Latin，它是 MapReduce 编程的复杂性抽象。Pig 平台包括了运行环境和用于分析 Hadoop 数据集的脚本语言（Pig Latin），其编译器将脚本语言翻译成 MapReduce 程序序列，然后转换为 MapReduce 任务在 Hadoop 上执行。通常用于进行离线分析。

7. Sqoop（数据 ETL/同步工具）

Sqoop 是 SQL-to-Hadoop 的缩写，主要用于在传统数据库和 Hadoop 之间传输数据。数据的导入和导出本质上是 Mapreduce 程序，充分利用了 MapReduce 的并行化和容错性。Sqoop 利用数据库技术描述数据架构，用于在传统关系型数据库、数据仓库、Hadoop 之间转移数据。

8. Flume（日志收集工具）

Cloudera 开发并在之后开源的海量日志收集系统，具有分布式、高可靠、高容错、易定制、可扩展的特点。它将数据从产生、传输、处理并最终写入目标路径的过程抽象为数据流，在具体的数据流中，数据源支持在 Flume 中定制数据发送方，从而支持收集各种不同协议下的数据。同时，Flume 数据流还提供对日志数据进行简单处理的能力，如过滤、格式转换等。此外，Flume 还具有能够将日志写往各种定制的数据目标的能力，同时也可以用于收集其他类型的数据。

9. Mahout（数据挖掘算法库）

Mahout 起源于 2008 年，最初是 Apache Lucene 的一个子项目。Mahout 的主要目标是创建一些可扩展的机器学习领域经典算法的实现，旨在帮助开发人员更加方便快捷地创建智能应用程序。Mahout 现在已经包含了聚类、分类、推荐引擎（协同过滤）、频繁集挖掘等广泛使用的数据挖掘方法，并通过 Hadoop 大大提升了算法可处理的数据量和处理性能。除了算法，Mahout 还包含了数据的输入输出工具，以及与其他存储系统集成的数据挖掘支持架构。

10. Oozie（工作流调度器）

Oozie 是一个可扩展的工作体系，集成于 Hadoop 的堆栈，用于协调多个 MapReduce 作业的执行，可以将多个 MapReduce 作业组合到一个逻辑工作单元中形成工作流，从而达到完成更大型任务的目的。它能够管理一个复杂的系统，基于数据定时和数据出现等外部事件来执行。Oozie 工作流是放置在控制依赖

DAG（Direct Acyclic Graph，有向无环图）中的一组动作，其中指定了动作执行的顺序，同时使用 hPDL（一种 XML 流程定义语言）来描述这个图。

11. Yarn（分布式资源管理器）

Yarn 是下一代的 MapReduce，即 MapReduce v2，是在第一代 MapReduce 的基础上演变而来的，主要是为了解决原始 Hadoop 的扩展性较差、不支持多计算框架等问题而提出的。Yarn 是下一代的 Hadoop 计算平台，是一个通用的运行时框架，用户可以编写自己的计算框架，在该运行环境中运行。

12. Mesos（分布式资源管理器）

Mesos 诞生于 UC Berkeley 的一个研究项目，著名的 Twitter 便是使用 Mesos 来管理集群资源。Mesos 与 Yarn 类似，是一个对资源进行统一管理和调度的平台，而且同样支持 MapReduce、Steaming 等多种运算框架。

13. Tachyon（分布式内存文件系统）

Tachyon 诞生于 UC Berkeley 的 AMP Lab，是以内存为中心的分布式文件系统，拥有高性能和容错能力，能够为集群框架（如 Spark、MapReduce）提供可靠的内存级的文件共享服务。

14. Tez（DAG 计算模型）

Tez 是 Apache 最新开源的支持 DAG 作业的计算框架，它直接源于 MapReduce 框架，核心思想是将 Map 和 Reduce 两个操作进一步进行拆分。它将 Map 进一步拆分成 Input、Processor、Sort、Merge、Output，将 Reduce 进一步拆分成 Input、Shuffle、Sort、Merge、Processor、Output。这些分解后的原子操作可以进行灵活的任意组合，进而产生新的操作，这些操作经过一些控制程序的组装之后，可以形成一个大的 DAG 作业。

15. Spark（内存 DAG 计算模型）

Spark 被标榜为"快如闪电的集群计算"，是目前最活跃的 Apache 项目，最早是 UC Berkeley AMP Lab 所开发并在之后开源的类 Hadoop MapReduce 的通用并行计算框架。Spark 提供了一个更快、更通用的数据处理平台。和 Hadoop 相比，Spark 可以让程序在磁盘上运行时速度提升 10 倍，在内存中运行时速度提升 100 倍。Spark 是对 Hadoop 的补充，可并行运行在 Hadoop 的 HDFS 文件系统中，能更好地适用于数据挖掘、机器学习等需要迭代的 MapReduce 算法。

16. Giraph（图计算模型）

Giraph 是一个基于 Hadoop 平台的可伸缩的分布式迭代图处理系统，采用了 Google 于 2010 年发表的论文《Pregel： A System for Large-Scale Graph Processing》中的原理。Giraph 最早由 Yahoo 开发，后来捐赠给了 Apache 软件基金会成为开源项目，之后得到了 Facebook 的支持，获得了多方面的改进。

17. GraphX（图计算模型）

GraphX 最早是 UC Berkeley AMP Lab 的一个分布式图计算框架项目，目前整合在 Spark 运行框架中，为其提供 BSP 大规模并行图计算能力。

18. MLib（机器学习库）

Spark MLlib 是一个机器学习库，它提供了各种各样的算法，这些算法用来在集群上实现分类、回归、聚类、协同过滤等。

19. Streaming（流计算模型）

Spark Streaming 支持对流数据的实时处理，以微批的方式对实时数据进行计算。

20. Kafka（分布式消息队列）

Kafka 是 Linkedin 开发并在之后开源的消息系统，它主要用于处理活跃的流式数据。活跃的流式数据在 Web 网站应用中很常见，这些数据包括网站访问量、用户访问的内容、搜索记录等。这些数据通常以日志的形式被记录下来，然后每隔一段时间进行一次统计处理。

21. Phoenix（HBase SQL 接口）

Phoenix 是 HBase 的 SQL 驱动，它能够使 Hbase 支持通过 JDBC 的方式进行访问，并将 SQL 查询转换成 Hbase 的扫描和相应的动作。

22. Ranger（安全管理工具）

Ranger 是一个 Hadoop 集群的权限框架，可以操作、监控、管理复杂的数据权限。它提供了一个集中的管理机制，能够管理基于 Yarn 的 Hadoop 生态圈的所有数据权限。

23. Knox（Hadoop 安全网关）

Knox 是一个访问 Hadoop 集群的 REST API 网关，它为所有 REST 访问提供了一个简单的访问接口点，能够实现 3A 认证（Authentication、Authorization、Auditing）和 SSO（单点登录）等。

24. Falcon（数据生命周期管理工具）

Falcon 是一个面向 Hadoop 的数据处理和管理平台，其设计目的是用于数据移动、数据管道协调、生命周期管理、数据发现等。它使得终端用户可以快速地将他们的数据及其相关的处理和管理任务"上载（onboard）"到 Hadoop 集群。

25. Ambari（安装部署配置管理工具）

Ambari 的作用是创建、管理、监视 Hadoop 集群，支持 HDFS、MapReduce、Hive、Pig、HBase、Zookeeper、Sqoop 等大部分常见的 Hadoop 组件，是为了让 Hadoop 以及相关的大数据软件更容易使用的一个 Web 工具。

Hadoop 生态圈的成员除了上面介绍到的之外还有很多，而且还在不断增加中。但在搭建 Hadoop 平台时，并非所有组件全部都要安装，只需要根据具体需求来选择合适的组件即可。除了 Hadoop 的基础组件 HDFS 和 MapReduce 是必须的，其他的组件都是可选的。其中常用的有 HBase、Hive、Pig、Ambari、Sqoop 等。

任务 1.3　安装 JDK

1.3.1　JDK

JDK（Java Development Kit），是 Java 语言的软件开发工具包（SDK，Software Development Kit），是整个 Java 开发的核心。主流的 JDK 由 Sun 公司发布（已被 Oracle 收购），但作为开放源代码的项目，很多公司或组织，如 IBM 公司、BEA 公司、GNU 组织等，都开发并发布了自己的 JDK。

JDK 本身是一个开发环境，其中包含了 Java 工具（Java Tools）、Java 核心类库（Java API）、Java 运行环境（JRE）、Java 虚拟机（JVM）等所有相关的内容。而 JRE（Java Runtime Environment）则只是一个运行环境，其中包含了 Java 程序在运行时所需 Java 核心类库的一个子集以及 Java 虚拟机。所以 JDK 是在编写 Java 程序的时候需要使用，而运行 Java 程序时需要使用的是 JRE。不过由于 JRE 是包含在 JDK 之中的，所以在安装了完整的 JDK 软件之后，即可进行 Java 程序的开发，也可以运行已有的 Java 程序。

JDK 有三个大的版本分类，其中最常用的是标准版（Standard Edition），即

J2SE，现今改名为了 Java SE。大部分 Java 程序都是使用的该版本 JDK，我们搭建 Hadoop 的平台所使用的也是该版本。另一个是主要用于企业级应用开发的企业版（Enterprise Edition），即 J2EE，现今改名为 Jakarta EE。该版本为企业级应用的数据库连接、事务处理、邮件服务等业务功能提供了专门的技术架构和模块组件，以此来简化企业级应用的开发过程。最后一个是用于开发移动设备和嵌入式设备上的应用程序的精简版（Micro Edition），即 J2ME，现今改名为 Java ME，随着近几年智能移动设备和物联网的快速发展，该版本也成为一个非常常用的 JDK 版本。

Hadoop 平台的相关组件或工具基本都是使用 Java 语言实现，所以在搭建 Hadoop 平台之前需要先安装并配置好 JDK。不过由于 JDK 中包含了很多与运行 Java 程序无关的内容，会占用大量的磁盘空间，而 Hadoop 平台中只需要运行 Java 程序，不需要开发 Java 程序，即便要进行 Hadoop 项目的开发，一般也是另行搭建开发环境连接到 Hadoop 平台进行开发。所以如果磁盘空间不足，可以只安装并配置 JRE 即可。

1.3.2 JDK 安装规划

1. 环境要求

本地磁盘剩余空间 400 MB 以上；
已安装 CentOS 7 1611 64 位操作系统。

2. 软件版本

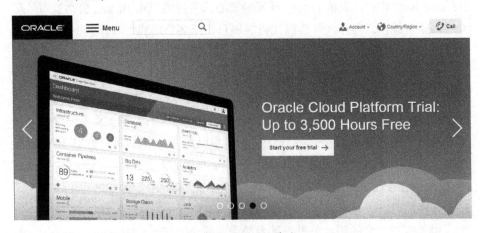

图 1-3-1 Oracle 官方

选用 JDK 的 1.8.0_131 版本，并对应 64 位的操作系统选用该版本 JDK 的 64 位版本，其软件包名为 jdk-8u131-linux-x64.tar.gz，该软件包可以在 Oracle 的官方网站（http：//www.oracle.com）的 Oracle Technology Network → Java → Java SE 页面（http：//www.oracle.com/technetwork/java/javase/downloads/jdk8-downloads-2133151.html）获取，如图 1-3-1，图 1-3-2 所示。

图 1-3-2　Java SE 页面

1.3.3　配置 SUDO 用户组

在 Hadoop 平台的安装和配置过程中，我们都是使用的为 Hadoop 平台创建的专用用户"hadoop"。但安装和配置过程中的某些步骤，可能需要使用到操作系统的 root 用户的权限才能进行操作，如配置操作系统的网络属性，设置操作系统的防火墙策略，修改操作系统的主机名，配置 IP 地址与主机名的映射关系，挂载和卸载存储设备，卸载操作系统中已安装的软件等。这些步骤可以通过使用"su"命令切换到 root 用户进行操作，也可以通过配置 SUDO 用户组来使得指定的非 root 用户也可以使用相关命令来完成操作步骤。

这里以挂载和卸载存储设备操作中的"mount"和"umount"两个命令为例，通过配置 SUDO 用户组，使得 Hadoop 的专用用户 hadoop 也能够像 root 用户一样使用这两个命令来挂载和卸载存储设备。

配置 SUDO 用户组需要对配置文件"sudoers"进行编辑，该文件位于操作系统的配置文件目录"/etc"下，如图 1-3-3 所示。该文件即便是操作系统中拥有最高权限的 root 用户，使用普通文本编辑器 vi/vim 打开也只能进行查看，而不能进行编辑。如图 1-3-4 所示，root 用户使用 vi/vim 文本编辑器打开该文件，左下角显示的是"readonly"。而要对该文件进行编辑，需要使用其专有的编辑命令"visudo"。

```
[root@localhost ~]# ls -l /etc/sudo*
-rw-r-----. 1 root root 1786 Nov  4  2016 /etc/sudo.conf
-r--r-----. 1 root root 3947 Jul 17 11:13 /etc/sudoers
-rw-r-----. 1 root root 3181 Nov  4  2016 /etc/sudo-ldap.conf
```

图 1-3-3　文件"sudoers"路径

```
## Sudoers allows particular users to run various commands as
## the root user, without needing the root password.
##
## Examples are provided at the bottom of the file for collections
## of related commands, which can then be delegated out to particular
## users or groups.
##
## This file must be edited with the 'visudo' command.

## Host Aliases
## Groups of machines. You may prefer to use hostnames (perhaps using
## wildcards for entire domains) or IP addresses instead.
# Host_Alias    FILESERVERS = fs1, fs2
# Host_Alias    MAILSERVERS = smtp, smtp2

## User Aliases
## These aren't often necessary, as you can use regular groups
## (ie, from files, LDAP, NIS, etc) in this file - just use %groupname
## rather than USERALIAS
# User_Alias ADMINS = jsmith, mikem

## Command Aliases
## These are groups of related commands...

## Networking
# Cmnd_Alias NETWORKING = /sbin/route, /sbin/ifconfig, /bin/ping, /sbin/dhclient, /usr/bin/net, /sbi
n/iptables, /usr/bin/rfcomm, /usr/bin/wvdial, /sbin/iwconfig, /sbin/mii-tool

## Installation and management of software
# Cmnd_Alias SOFTWARE = /bin/rpm, /usr/bin/up2date, /usr/bin/yum

## Services
# Cmnd_Alias SERVICES = /sbin/service, /sbin/chkconfig, /usr/bin/systemctl start, /usr/bin/systemctl
 stop, /usr/bin/systemctl reload, /usr/bin/systemctl restart, /usr/bin/systemctl status, /usr/bin/sy
stemctl enable, /usr/bin/systemctl disable
"/etc/sudoers" [readonly] 111L, 3907C
```

图 1-3-4　查看"sudoers"文件

（1）使用"root"用户登录操作系统。

（2）使用命令"visudo"打开配置文件"sudoers"进行编辑，编辑操作与在 vi/vim 文本编辑器内对文本进行编辑的操作相同。

（3）找到配置项"root   ALL=（ALL）   ALL"的所在行，如图 1-3-5 所示，并在其下面添加一行，该行的内容如下：

hadoop   ALL=（ALL）   /bin/mount, /bin/umount

命令解释：第一列指定需要加入到 SUDO 用户组的用户的用户名。

命令解释：第二列指定加入的用户所能具备的权限和用户身份。

命令解释：第三列指定加入的用户所能使用的命令，多个命令之间以逗号隔开，这里需要注意的是必须书写该命令在操作系统中的完整绝对路径，而不仅仅是命令的名称。

```
81 ## Next comes the main part: which users can run what software on
82 ## which machines (the sudoers file can be shared between multiple
83 ## systems).
84 ## Syntax:
85 ##
86 ##      user        MACHINE=COMMANDS
87 ##
88 ## The COMMANDS section may have other options added to it.
89 ##
90 ## Allow root to run any commands anywhere
91 root    ALL=(ALL)       ALL
92 hadoop  ALL=(ALL)       /bin/mount,/bin/umount
```

图 1-3-5　添加内容

（4）编辑完成之后保存更改内容并退出编辑器，操作与 vi/vim 文本编辑器中保存并退出的操作相同，退出时编辑器会对配置文件中的语法格式是否正确进行简单的自动检查。若编写内容的语法格式有错误则会有相应提示信息，如图 1-3-6 所示，并且保存并退出操作会失败。

```
"/etc/sudoers.tmp" 112L, 3946C written
visudo: >>> /etc/sudoers: syntax error near line 92 <<<
What now?
```

图 1-3-6　错误提示

（5）在"sudoers"文件编辑完成之后，指定的用户便可以如图 1-3-7 所示，通过在原有命令行之前加上"sudo"命令，组成"sudo　命令行"的命令格式的方式，来运行原本自身没有权限使用的指定命令。只不过在每次用户登录操作系统之后第一次使用 SUDO 方式来运行命令时，会提示输入当前登录用户的密码，如图 1-3-8 所示。

```
[hadoop@localhost ~]$ sudo mount /dev/sdb4 /mnt
```

图 1-3-7　"sudo　命令行"命令格式

```
We trust you have received the usual lecture from the local System
Administrator. It usually boils down to these three things:

    #1) Respect the privacy of others.
    #2) Think before you type.
    #3) With great power comes great responsibility.

[sudo] password for hadoop:
```

图 1-3-8　提示输入当前密码

1.3.4 卸载系统原有 JDK

对于已经用作于其他用途的主机，或者是安装了桌面版 CentOS 操作系统的主机，会有已经安装有其他版本的 JDK 或相关软件包的情况。在这种情况下，为了避免与将要使用的 JDK 软件之间可能出现的软件冲突的情况，建议先将系统中已有的 JDK 以及相关软件包进行卸载。JDK 以及相关软件包的包名中一般都包含了"jdk"或"java"关键字，可以通过 RPM 的查询命令进行查看，然后将之卸载。

对于新安装的最简化 CentOS 操作系统，一般没有安装任何 JDK 以及相关软件包，可以选择忽略该步骤。

（1）使用"root"用户登录操作系统或使用 SUDO 方式执行后面步骤中的命令。

（2）分别使用命令"rpm -qa | grep jdk"和"rpm -qa | grep java"搜索当前系统中所有已经安装的 JDK 以及相关的软件包，如图 1-3-9 所示。

```
[root@localhost ~]# rpm -qa | grep java
tzdata-java-2016j-1.el6.noarch
java-1.7.0-openjdk-1.7.0.131-2.6.9.0.el6_8.x86_64
java-1.6.0-openjdk-1.6.0.41-1.13.13.1.el6_8.x86_64
```

图 1-3-9　显示系统中已经安装的 JDK 及相关软件包

（3）使用命令"rpm -e --nodeps 软件包名"依次删除所有搜索到的当前系统中已经安装的 JDK 及相关软件包。

1.3.5 安装配置 JDK

这里安装 JDK 所使用的是"tar.gz"安装包，也就是俗称的绿色安装包，直接解包解压之后再进行一些简单的配置便可以使用，而不需要使用 RPM 软件包管理工具来进行安装。后面的 Hadoop、HBase、MySQL、Hive、Sqoop 等软件的安装和配置也都是使用的绿色安装包。

（1）使用"hadoop"用户登录操作系统，本章后面的所有安装和配置操作，在没有特殊说明的情况下，都默认使用"hadoop"用户登录系统进行。

（2）创建"java"目录用于存放 JDK 相关文件，该目录可自行选择创建位置，创建完成后将当前工作目录切换到该目录。

（3）使用命令"tar -xzf JDK 安装包路径"将软件安装包解压解包到"java"目录下，解压解包出来的目录名称为"jdk1.8.0_131"，如图 1-3-10 所示。

```
[hadoop@localhost java]$ ls -l
total 0
drwxr-xr-x. 8 hadoop hadoop 255 Mar 15  2017 jdk1.8.0_131
```

图 1-3-10　解压解包

（4）配置 JDK 相关的环境变量，需要修改用户的配置文件".bash_profile"。该文件位于 JAVA_HOME 目录下，是一个隐藏文件。对该文件进行编辑，如图 1-3-11 所示，在文件末尾添加以下内容：

\# java environment
JAVA_HOME=（JDK 软件目录路径）
CLASSPATH=.：$JAVA_HOME/lib/tools.jar：$JAVA_HOME/lib/dt.jar
PATH=$JAVA_HOME/bin：$PATH
export　JAVA_HOME　CLASSPATH　PATH

注意事项：

JDK 软件目录即 JDK 软件包解压解包出来的"jdk1.8.0_131"目录，这里需要书写该目录及其所在的完整绝对路径。

命令"export"后面的变量名必须严格按照前面的定义顺序书写，否则会导致后面的环境变量不正确，在后面配置其他软件的环境变量时也需要注意。

环境变量"PATH"的值要确保输入正确，若输入错误可能会导致除 Shell 内部命令之外的所有命令都无法直接使用，在后面配置其他软件的环境变量时同样需要注意。

```
# java environment
JAVA_HOME=/home/hadoop/java/jdk1.8.0_131
CLASSPATH=.:$JAVA_HOME/lib/tools.jar:$JAVA_HOME/lib/dt.jar
PATH=$JAVA_HOME/bin:$PATH
export JAVA_HOME CLASSPATH PATH
```

图 1-3-11　对文件添加内容

（5）使用命令"source　~/.bash_profile"使新配置的环境变量立即生效。需要注意的是如果使用的是桌面版的 CentOS 操作系统，该方式只能让环境变量在当前打开的命令行终端中生效。若重新打开一个新的命令行终端，将无法获得新添加的环境变量，也无法获得修改的环境变量的最新值。所以对于桌面版的 CentOS 操作系统，需要采用登出当前用户然后再次登入的方式来让配置的环境变量生效。

（6）使用命令"echo　$变量名"可以查看新添加和修改的环境变量的值是否正确，如图 1-3-12 所示。

```
[hadoop@localhost ~]$ echo $JAVA_HOME
/home/hadoop/java/jdk1.8.0_131
```

图 1-3-12　使用 echo 命令

（7）使用命令"java -version"或"javac -version"验证 JDK 的安装配置是否成功。如图 1-3-13 所示，若显示有 JDK 的版本信息，并且与所使用的 JDK 软件版本相符，则表示 JDK 已经安装配置成功，可以正常使用。

```
[hadoop@localhost ~]$ java -version
java version "1.8.0_131"
Java(TM) SE Runtime Environment (build 1.8.0_131-b11)
Java HotSpot(TM) 64-Bit Server VM (build 25.131-b11, mixed mode)
```

图 1-3-13　验证 JDK 安装是否成功

任务 1.4　安装 Zookeeper 伪分布模式

1.4.1　Zookeeper

ZooKeeper 是一个分布式的、开放源码的分布式应用程序协调服务，是 Google 的 Chubby 一个开源的实现，提供的功能包括：配置维护、域名服务、分布式同步、组服务等，还是 Hadoop 和 HBase 的重要依赖组件。它一般用于存储配置、发布、订阅等类型的少量信息服务，不适合用来存储大量信息。其应用场景包括 Hadoop 集群平台、Storm 集群平台、分布式数据库同步系统、RPC 服务框架、消息中间件等。

1.4.1.1　Paxos 算法

Zookeeper 为分布式应用提供了一致性服务，为保障分布式环境中数据的一致性，其节点之间的一致性算法采用的是一种基于 Paxos 实现的算法。

Paxos 算法解决的问题是分布式系统如何就某个值（决议）达成一致。此算法基于一个典型的场景：在一个分布式数据库系统中，如果各个节点的初始状态一致，那么每个节点在执行相同的操作序列之后，它们最后都将得到一个一致的状态。而为了保证每个节点都执行相同的操作序列，需要在每一条操作指令上执行一致性算法，以保证每个节点获得的指令都是一致的。一个通用的一致性算法可以在多个场景中应用，是分布式计算中的重要问题。

在分布式系统中，节点通信一般有共享内存（Shared memory）和消息传递

（Messages passing）两种模型，而 Paxos 算法就是一种基于消息传递模型的一致性算法。Paxos 算法采用选举机制，即少数服从多数的思想。简单来说就是只要在 $2N+1$ 个节点中，有 $N+1$ 个或是以上的节点表示同意某个决定（指令），则认为该系统达到了一致，不会再改变。所以在使用基于 Paxos 算法实现的应用中，拥有奇数个节点是最佳的选择，因为在 $2N$ 个节点中，同样需要至少有 $N+1$ 个节点表示同意某个决定（指令），才能认为该系统达到了一致。

不仅是在分布式系统中，凡是多个过程需要达成某种一致的场合都可以使用 Paxos 算法。该算法适用于以下的几种情况：

（1）一台机器中多个进程或线程间达成数据一致。

（2）分布式文件系统或者分布式数据库中多客户端并发读写数据。

（3）分布式存储中多个副本响应读写请求。

Zookeeper 默认使用 Fast Paxos 作为一致性算法，它以 Paxos 算法为基础，在其上进行了一些优化和改进。因为 Paxos 算法本身存在活锁的问题，即当有多个申请者交错提交请求时，有可能因为互相排斥导致没有一个申请者能够提交成功。而 Fast Paxos 在这方面做出了优化，它通过选举产生一个 Leader（领导者）来作为申请者提交请求。

1.4.1.2　Zookeeper 文件系统

在 Zookeeper 中每个子目录被称作为 znode，和文件系统中的目录类似，能够自由地对 znode 进行增加、删除，在一个 znode 下还可以增加、删除子 znode，如图 1-4-1 所示。唯一的不同是数据是存储于每个 znode 自身，而不是存储于 znode 下。

Zookeeper 中共有 4 种类型的 znode：

（1）持久化目录节点（PERSISTENT），该类节点在客户端与 Zookeeper 断开连接后依然存在。

（2）持久化顺序编号目录节点（PERSISTENT_SEQUENTIAL），该类节点与上一个类型相似，在客户端与 Zookeeper 断开连接后依旧存在，只是 Zookeeper 会给该类型节点的名称进行顺序编号。

（3）临时目录节点（EPHEMERAL），该类节点在客户端与 Zookeeper 断开连接后会被自动删除。

（4）临时顺序编号目录节点（EPHEMERAL_SEQUENTIAL），该类型的目录节点与上一个类型相似，在客户端与 Zookeeper 断开连接后会被自动删除，只是 Zookeeper 会给该类型节点的名称进行顺序编号。

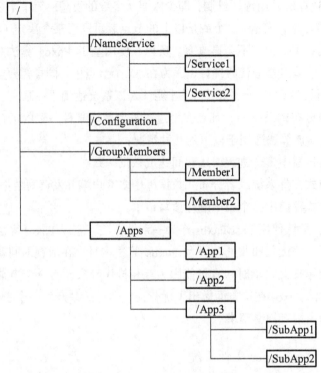

图 1-4-1　Zookeeper 目录结构

1.4.1.3　Zookeeper 的服务

1. 命名服务

在 Zookeeper 的文件系统里创建一个拥有唯一 Path 的目录节点，在无法确定上游程序的机器部署情况时，可以通过与下游程序约定好这个固定的 Path，然后通过这个 Path 实现相互之间机器部署情况的探索和发现。

2. 配置管理

大部分服务器平台或应用平台的程序都会有大量需要配置的内容，如果程序分散部署在多台机器上，在配置变更时要逐个改变每台机器上的配置就会变得非常困难。这种情况下可以将配置全部保存到 Zookeeper 上的某个目录节点中，然后所有相关程序都对这个目录节点进行监听，如图 1-4-2 所示。一旦配置信息发生了变化，每个程序就会收到来自 Zookeeper 的通知，然后所有程序便可以从 Zookeeper 获取新的配置信息。

3. 集群管理

集群管理主要是做两件事，一个是检测是否有机器退出或加入集群，另一个是选举集群中的 Master 节点。其结构如图 1-4-3 所示。

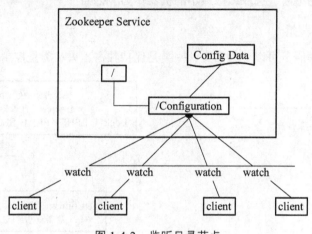

图 1-4-2　监听目录节点

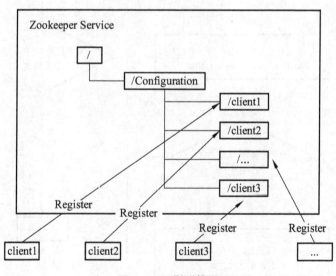

图 1-4-3　集群管理

对于第一个，集群中所有机器会在父目录节点下创建临时目录节点，然后 Zookeeper 会监听父目录节点的子目录节点变化消息。一旦有机器因为挂掉等原因退出集群，则该机器与 Zookeeper 之间的连接就会断开，对应其所创建的临时目录节点也会被删除，然后所有其他机器都会收到 Zookeeper 发出的相应通知。

而集群中有新机器加入的情况也类似，新加入的机器会在父目录节点下创建自己的临时目录节点，然后 Zookeeper 会发出通知信息告知其他机器。

对于第二点，则需要所有机器在父目录节点下创建临时顺序编号目录节点。然后在每次选举时，选取编号最小的机器作为 Master。

4. 分布式锁

分布式锁服务可以分为两类，一类是保持独占，另一类是控制时序。

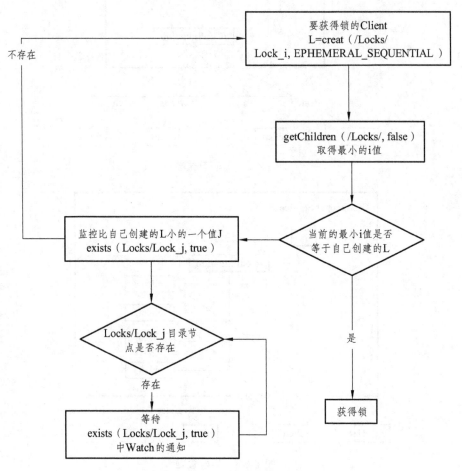

图 1-4-4　分布式锁服务

对于第一类，可以将 Zookeeper 上的一个 znode 看作是一把锁，所有请求都同时进行同一名称的锁目录节点 "/distribute_lock" 的创建，最终成功创建的那个请求便拥有了这把锁。并且在使用完成之后删除掉自身所创建的目录节点

"/distribute_lock"便可以释放该锁供其他请求继续使用。

对于第二类，会预先创建好锁目录节点"/distribute_lock"，而所有请求在该目录节点下面创建自身的临时顺序编号目录节点，并且和选举 Master 一样，编号拥有最小编号的目录节点的请求获得锁，并且在用完之后删除该目录节点，并将锁交给拥有下一个编号的目录节点的请求。如图 1-4-4 所示。

5. 队列管理

队列的类型也有两种，一种是同步队列，一种是 FIFO 队列。对于同步队列，只有当队列的所有成员都聚齐时，该队列才可用，否则将会一直等待所有成员到达。对于该类型队列，Zookeeper 会在指定目录节点下创建临时目录节点，监听节点数目是否达到要求的数目。而对于 FIFO 队列，Zookeeper 采用类似于分布式锁服务中的控制时序方式，入列时赋予一个编号，出列时按照编号执行。

1.4.1.4 Zookeeper 整体结构

领导者（Leader）：Leader 节点负责进行投票的发起和决议，更新系统状态。

学习者（Learner）- 跟随者（Follower）：Follower 节点用于接收客户端请求并向客户端返回结果，在选举过程中参与投票。

学习者（Learner）- 观察者（Observer）：Observer 节点可以接收客户端连接，将写请求转发给 Leader 节点。但不参与投票过程，只同步 Leader 节点的状态。其主要作用是扩展系统，提高读取速度。

客户端（Client）：即发起请求方。

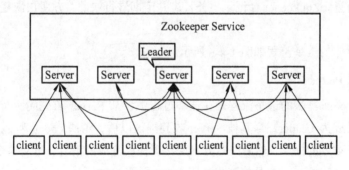

图 1-4-5 Zookeeper 整体结构

1.4.1.5 Zookeeper 的主节点选举

当 Leader 崩溃或者 Leader 失去与大多数 Follower 之间的联系时，Zookeeper

会进入恢复模式。在恢复模式中，需要重新选举出一个新的 Leader，让所有的 Server 都恢复到一个正确的状态。Zookeeper 的选举算法有两种，一种是 Basic Paxos 算法，另外一种是 Fast Paxos 算法，其中系统默认的选举算法为 Fast Paxos。

1. Basic Paxos

（1）选举进程由当前 Server 发起，当前 Server 一般是新加入的或是与 Leader 失去联系的 Server，其主要功能是对投票结果进行统计，并选出推荐的 Server。

（2）选举进程首先向所有 Server 发起一次询问，其中包括自己，并将自己作为 Leader 进行推荐。

（3）选举进程收到回复后，首先验证是否是自己发起的询问，即验证 zxid 是否一致；然后获取对方的 id（myid），并存储到当前询问对象列表中；最后获取对方提议的 Leader 的相关信息，包括 id 和 zxid，并将这些信息存储到当次选举的投票记录表中。

（4）收到所有 Server 的回复以后，计算出 zxid 最大的 Server，并将该 Server 的相关信息设置为下一次要投票的 Server，即下一次的推荐 Leader。

（5）若当前推荐的 Leader 获得 $N/2 + 1$ 的 Server 票数，设置当前推荐的 Leader 为获胜的 Server，并根据获胜的 Server 的相关信息设置状态。否则继续之前的过程，直到 Leader 最终被选举出来。要使推荐的 Leader 获得多数 Server 的支持，则集群中存活的 Server 的总数不得少于 $N/2 + 1$。

（6）在集群中每个 Server 启动之后都会重复以上的选举流程。在恢复模式下，如果是刚从崩溃状态恢复的或者刚启动的 Server 还会从磁盘快照中恢复数据和会话信息，Zookeeper 会记录事务日志并定期进行快照，方便在恢复时进行状态恢复。

（7）具体的选举流程如图 1-4-6 所示。

2. Fast Paxos

Fast Paxos 算法的选举流程是在选举过程中，某 Server 首先向所有 Server 提议自己要成为 Leader，当其他 Server 收到提议以后，解决 epoch 和 zxid 的冲突，并接受对方的提议，然后向对方发送接受提议完成的消息，重复这个流程，直到最后选举出 Leader。具体流程如图 1-4-7 所示。

项目 1 搭建单节点 Hadoop 整合平台

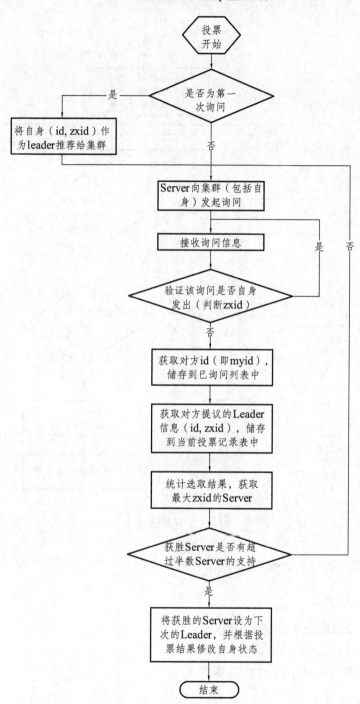

图 1-4-6 Basic Paxos 选举流程

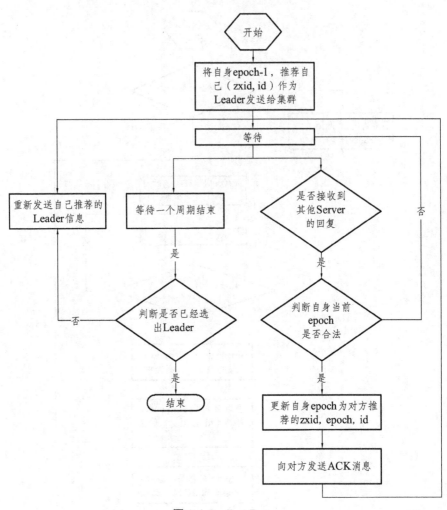

图 1-4-7 Fast Paxos

1.4.2 Zookeeper 安装规划

1. 环境要求

本地磁盘剩余空间 100 MB 以上；

已安装 CentOS 7 1611 64 位操作系统；

已安装 JDK 的 1.8.0_131 版本。

2. 软件版本

选用 Zookeeper 的 3.4.9 版本，其软件包名为 zookeeper-3.4.9.tar.gz，该软件

包可以在 Zookeeper 位于 Apache 的官方网站（http://zookeeper.apache.org）的 Releases 页面获取。

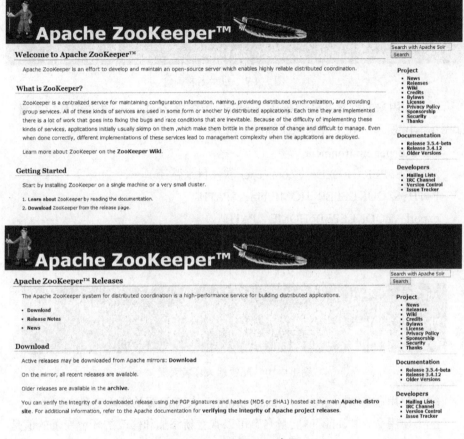

图 1-4-8　Zookeeper 官网

3. 服务规划

Zookeeper 采用了 Paxos 投票算法，要求至少 3 个及以上的服务节点数量。

Zookeeper 的伪分布模式实际上是在单台主机上使用多个不同的进程来模拟出多个不同的服务节点。

1.4.3　Zookeeper 基础环境配置

（1）创建"zookeeper"目录用于存放 Zookeeper 相关文件，该目录可自行选择创建位置，创建完成后将当前工作目录切换到该目录。

（2）使用命令"tar -xzf Zookeeper 安装包路径"将软件包解压解包到"zookeeper"目录下，解压解包出来的目录名称为"zookeeper-3.4.9"，如图 1-4-9 所示。

```
[hadoop@localhost zookeeper]$ ls -l
total 4
drwxr-xr-x. 10 hadoop hadoop 4096 Aug 23  2016 zookeeper-3.4.9
```

图 1-4-9　目录名称

（3）在用户的配置文件".bash_profile"中配置 Zookeeper 相关的环境变量，如图 1-4-10 所示，在文件末尾添加以下内容：

# zookeeper environment
ZOOKEEPER_HOME=（Zookeeper 软件目录路径）
PATH=$ZOOKEEPER_HOME/bin：$PATH
export　ZOOKEEPER_HOME　PATH

Zookeeper 软件目录即 Zookeeper 软件包解压解包出来的"zookeeper-3.4.9"目录，这里需要书写该目录及其所在的完整绝对路径。

```
# zookeeper environment
ZOOKEEPER_HOME=/home/hadoop/zookeeper/zookeeper-3.4.9
PATH=$ZOOKEEPER_HOME/bin:$PATH
export ZOOKEEPER_HOME PATH
```

图 1-4-10　配置相关环境变量

（4）使用命令"source　~/.bash_profile"使新配置的环境变量立即生效。

（5）使用命令"echo　$变量名"可以查看新添加和修改的环境变量的值是否正确，如图 1-4-11 所示。

```
[hadoop@localhost ~]$ echo $ZOOKEEPER_HOME
/home/hadoop/zookeeper/zookeeper-3.4.9
```

图 1-4-11　查看新添加和修改的环境变量

1.4.4　Zookeeper 伪分布模式配置

（1）进入 Zookeeper 相关文件的目录，如图 1-4-12 所示，对应单节点模式的三个节点进程分别创建"zookeeper1""zookeeper2""zookeeper3"三个节点进程的专用目录。

```
[hadoop@localhost zookeeper]$ ls -l
total 4
drwxrwxr-x. 2 hadoop hadoop     6 Apr 22 22:03 zookeeper1
drwxrwxr-x. 2 hadoop hadoop     6 Apr 22 22:03 zookeeper2
drwxrwxr-x. 2 hadoop hadoop     6 Apr 22 22:03 zookeeper3
```

图 1-4-12　创建三个节点进程的专用目录

（2）分别在三个节点进程的目录中创建数据文件的存放目录"data"和日志文件的存放目录"logs"，如图 1-4-13 所示。

```
[hadoop@localhost zookeeper1]$ ls -l
total 0
drwxrwxr-x. 2 hadoop hadoop 6 Apr 22 22:04 data
drwxrwxr-x. 2 hadoop hadoop 6 Apr 22 22:04 logs
```

图 1-4-13　创建目录

（3）分别在三个节点进程的数据文件存放目录"data"中创建标识文件"myid"，在标识文件中写入对应节点进程的编号，如图 1-4-14 所示，节点进程"zookeeper1"的标识文件中写入的内容为"1"。

```
[hadoop@localhost zookeeper1]$ cat data/myid
1
```

图 1-4-14　创建标识文件

（4）Zookeeper 的配置文件位于其软件目录中的"conf"目录下，进入该目录，复制并重命名配置文件模板"zoo_sample.cfg"，分别生成三个节点进程的对应配置文件"zoo1.cfg""zoo2.cfg""zoo3.cfg"，如图 1-4-15 所示。

```
[hadoop@localhost conf]$ pwd
/home/haduup/zookeeper/zonkeeper-3.4.9/conf
[hadoop@localhost conf]$ ls -l
total 24
-rw-rw-r--. 1 hadoop hadoop  535 Aug 23  2016 configuration.xsl
-rw-rw-r--. 1 hadoop hadoop 2161 Aug 23  2016 log4j.properties
-rw-rw-r--. 1 hadoop hadoop  922 Apr 22 23:37 zoo1.cfg
-rw-rw-r--. 1 hadoop hadoop  922 Apr 22 23:38 zoo2.cfg
-rw-rw-r--. 1 hadoop hadoop  922 Apr 22 23:38 zoo3.cfg
-rw-rw-r--. 1 hadoop hadoop  922 Aug 23  2016 zoo_sample.cfg
```

图 1-4-15　生成对应配置文件

（5）编辑生成的节点进程的对应配置文件。找到配置项"dataDir"，将其值改为节点进程对应数据文件的存放目录"data"的路径，如图 1-4-16 所示。

```
 9 # the directory where the snapshot is stored.
10 # do not use /tmp for storage, /tmp here is just
11 # example sakes.
12 dataDir=/home/hadoop/zookeeper/zookeeper1/data
```

图1-4-16　修改配置项"dataDir"

然后找到配置项"clientPort"。由于单节点模式是在一台机器上模拟出三个不同的节点进程，所以每个节点进程所使用的端口不能相同。如图1-4-17所示，第一个节点进程"zookeeper1"可以使用默认的端口"2181"，然后分配第二个节点进程"zookeeper2"使用的端口为"2182"，以及第三个节点进程"zookeeper3"使用的端口为"2183"。同样，下面的服务器列表配置中每个服务器也需要使用完全不同的端口号。

```
13 # the port at which the clients will connect
14 clientPort=2181
```

图1-4-17　节点进程对应不同端口号

最后如图1-4-18所示，在配置文件末尾添加如下内容：

dataLogDir=（节点进程对应日志文件的存放目录"logs"的路径）

server.1=localhost：2888：3888

server.2=localhost：2889：3889

server.3=localhost：2890：3890

```
dataLogDir=/home/hadoop/zookeeper/zookeeper1/logs
server.1=localhost:2888:3888
server.2=localhost:2889:3889
server.3=localhost:2890:3890
```

图1-4-18　修改配置文件

配置项说明：

dataDir——指定Zookeeper的数据的存放目录，默认情况下Zookeeper也会将事务日志文件存放在该目录。

clientPort——指定客户端连接Zookeeper服务器所使用的端口号，Zookeeper服务器会监听该端口来获取客户端的访问请求。

dataLogDir——指定Zookeeper的事务日志文件的存放目录，该文件为二进制文件，需要使用Zookeeper自带的JAR包中的程序才能查看。

server.*——指定Zookeeper平台的服务器列表，包括服务器的主机名或IP地址，以及Zookeeper服务所使用的LF通信端口号和选举端口号，其中"*"

代表服务器的编号，需要与"myid"文件中的编号相对应。

三个节点进程的对应配置文件"zoo1.cfg""zoo2.cfg""zoo3.cfg"都需要进行编辑设置，配置时注意节点进程要使用各自的相关目录的路径。

（6）进入 Zookeeper 的脚本文件所在目录"bin"，对 Zookeeper 的脚本文件"zkServer.sh"进行编辑，如图 1-4-19 所示，找到"ZOO_LOG_DIR"关键字所在行，在其上面一行添加如下脚本代码：

ZOO_LOG_DIR=

"$（$GREP　"^[[：space：]]*dataLogDir"　"$ZOOCFG"　|　sed　-e 's/.*=//'）"

```
125 ZOO_LOG_DIR="$($GREP "^[[:space:]]*dataLogDir" "$ZOOCFG" | sed -e 's/.*=//')"
126 if [ ! -w "$ZOO_LOG_DIR" ] ; then
127 mkdir -p "$ZOO_LOG_DIR"
128 fi
```

图 1-4-19　修改脚本文件

1.4.5　Zookeeper 伪分布模式启动和验证

（1）使用命令"zkServer.sh　start　节点进程对应配置文件路径"依次启动 Zookeeper 伪分布模式的三个节点进程服务，若启动过程没有报错，并如图 1-4-20 所示，显示"STARTED"信息，则表示节点进程服务启动完成。

```
[hadoop@localhost ~]$ zkServer.sh start zookeeper/zookeeper-3.4.9/conf/zoo1.cfg
ZooKeeper JMX enabled by default
Using config: zookeeper/zookeeper-3.4.9/conf/zoo1.cfg
Starting zookeeper ... STARTED
```

图 1-4-20　启动三个节点进程服务

（2）使用 JDK 的命令"jps"查看 Java 相关的进程信息，如图 1-4-21 所示，若存在三个名为"QuorumPeerMain"的进程，则表示 Zookeeper 伪分布模式启动成功。

```
[hadoop@localhost ~]$ jps
4240 QuorumPeerMain
4336 Jps
4293 QuorumPeerMain
4060 QuorumPeerMain
```

图 1-4-21　查看 Java 进程信息

（3）使用命令"zkServer.sh status 节点进程对应配置文件路径"可以查看 Zookeeper 节点进程服务的状态，如图 1-4-22 所示，从显示的状态信息中可以看到哪个节点进程服务是 Leader，哪个节点进程服务是 Follower。

```
[hadoop@localhost ~]$ zkServer.sh status zookeeper/zookeeper-3.4.9/conf/zoo1.cfg
ZooKeeper JMX enabled by default
Using config: zookeeper/zookeeper-3.4.9/conf/zoo1.cfg
Mode: follower
[hadoop@localhost ~]$ zkServer.sh status zookeeper/zookeeper-3.4.9/conf/zoo2.cfg
ZooKeeper JMX enabled by default
Using config: zookeeper/zookeeper-3.4.9/conf/zoo2.cfg
Mode: leader
[hadoop@localhost ~]$ zkServer.sh status zookeeper/zookeeper-3.4.9/conf/zoo3.cfg
ZooKeeper JMX enabled by default
Using config: zookeeper/zookeeper-3.4.9/conf/zoo3.cfg
Mode: follower
```

图 1-4-22 查看 Zookeeper 节点进程

（4）使用命令"zkCli.sh"可以利用 Zookeeper 的客户端工具来连接 Zookeeper 服务。如图 1-4-23 所示，若显示"CONNECTED"并正常进入 Zookeeper 客户端程序的控制台界面，则表示 Zookeeper 服务连接正常。

```
[hadoop@localhost ~]$ zkCli.sh
Connecting to localhost:2181
2018-04-24 03:58:30,571 [myid:] - INFO  [main:Environment@100] - Client environment:zookeeper.versio
n=3.4.9-1757313, built on 08/23/2016 06:50 GMT
2018-04-24 03:58:30,578 [myid:] - INFO  [main:Environment@100] - Client environment:host.name=localh
ost
2018-04-24 03:58:30,584 [myid:] - INFO  [main:Environment@100] - Client environment:java.version=1.8
.0_131
2018-04-24 03:58:30,590 [myid:] - INFO  [main:Environment@100] - Client environment:java.vendor=Orac
le Corporation
2018-04-24 03:58:30,594 [myid:] - INFO  [main:Environment@100] - Client environment:java.home=/home/
hadoop/java/jdk1.8.0_131/jre
2018-04-24 03:58:30,597 [myid:] - INFO  [main:Environment@100] - Client environment:java.class.path=
/home/hadoop/zookeeper/zookeeper-3.4.9/bin/../build/classes:/home/hadoop/zookeeper/zookeeper-3.4.9/b
in/../build/lib/*.jar:/home/hadoop/zookeeper/zookeeper-3.4.9/bin/../lib/slf4j-log4j12-1.6.1.jar:/hom
e/hadoop/zookeeper/zookeeper-3.4.9/bin/../lib/slf4j-api-1.6.1.jar:/home/hadoop/zookeeper/zookeeper-3
.4.9/bin/../lib/netty-3.10.5.Final.jar:/home/hadoop/zookeeper/zookeeper-3.4.9/bin/../lib/log4j-1.2.1
6.jar:/home/hadoop/zookeeper/zookeeper-3.4.9/bin/../lib/jline-0.9.94.jar:/home/hadoop/zookeeper/zook
eeper-3.4.9/bin/../zookeeper-3.4.9.jar:/home/hadoop/zookeeper/zookeeper-3.4.9/bin/../src/java/lib/*.
jar:/home/hadoop/zookeeper/zookeeper-3.4.9/bin/../conf:.:/home/hadoop/java/jdk1.8.0_131/lib/tools.ja
r:/home/hadoop/java/jdk1.8.0_131/lib/dt.jar
2018-04-24 03:58:30,600 [myid:] - INFO  [main:Environment@100] - Client environment:java.library.pat
h=/usr/java/packages/lib/amd64:/usr/lib64:/lib64:/lib:/usr/lib
2018-04-24 03:58:30,610 [myid:] - INFO  [main:Environment@100] - Client environment:java.io.tmpdir=/
tmp
2018-04-24 03:58:30,613 [myid:] - INFO  [main:Environment@100] - Client environment:java.compiler=<N
A>
2018-04-24 03:58:30,615 [myid:] - INFO  [main:Environment@100] - Client environment:os.name=Linux
2018-04-24 03:58:30,618 [myid:] - INFO  [main:Environment@100] - Client environment:os.arch=amd64
2018-04-24 03:58:30,619 [myid:] - INFO  [main:Environment@100] - Client environment:os.version=3.10.
0-514.el7.x86_64
2018-04-24 03:58:30,622 [myid:] - INFO  [main:Environment@100] - Client environment:user.name=hadoop
2018-04-24 03:58:30,623 [myid:] - INFO  [main:Environment@100] - Client environment:user.home=/home/
hadoop
2018-04-24 03:58:30,626 [myid:] - INFO  [main:Environment@100] - Client environment:user.dir=/home/h
adoop
```

```
2018-04-24 03:58:30,630 [myid:] - INFO  [main:ZooKeeper@438] - Initiating client connection, connect
String=localhost:2181 sessionTimeout=30000 watcher=org.apache.zookeeper.ZooKeeperMain$MyWatcher@2770
50dc
2018-04-24 03:58:30,686 [myid:] - INFO  [main-SendThread(localhost:2181):ClientCnxn$SendThread@1032]
 - Opening socket connection to server localhost/0:0:0:0:0:0:0:1:2181. Will not attempt to authentic
ate using SASL (unknown error)
Welcome to ZooKeeper!
JLine support is enabled
2018-04-24 03:58:30,949 [myid:] - INFO  [main-SendThread(localhost:2181):ClientCnxn$SendThread@876]
 - Socket connection established to localhost/0:0:0:0:0:0:0:1:2181, initiating session
2018-04-24 03:58:31,006 [myid:] - INFO  [main-SendThread(localhost:2181):ClientCnxn$SendThread@1299]
 - Session establishment complete on server localhost/0:0:0:0:0:0:0:1:2181, sessionid = 0x162f69a8fc
c0002, negotiated timeout = 30000

WATCHER::

WatchedEvent state:SyncConnected type:None path:null
[zk: localhost:2181(CONNECTED) 0]
```

图 1-4-23　连接 Zookeeper 服务

（5）在 Zookeeper 控制台界面中使用命令"quit"可以退出客户端程序，返回操作系统的控制台界面，如图 1-4-24 所示。

```
[zk: localhost:2181(CONNECTED) 0] quit
Quitting...
2018-04-24 04:09:07,423 [myid:] - INFO  [main:ZooKeeper@684] - Session: 0x162f69a8fcc0002 closed
2018-04-24 04:09:07,429 [myid:] - INFO  [main-EventThread:ClientCnxn$EventThread@519] - EventThread
shut down for session: 0x162f69a8fcc0002
```

图 1-4-24　退出客户端程序

（6）使用命令"zkServer.sh　stop　节点进程对应配置文件路径"可以停止指定的 Zookeeper 节点进程服务，如图 1-4-25 所示。

```
[hadoop@localhost ~]$ zkServer.sh stop zookeeper/zookeeper-3.4.9/conf/zoo1.cfg
ZooKeeper JMX enabled by default
Using config: zookeeper/zookeeper-3.4.9/conf/zoo1.cfg
Stopping zookeeper ... STOPPED
```

图 1-4-25　停止指定的 Zookeeper 节点进程服务

1.4.6　安装与配置的常见问题及解决方法

常见问题如下：

Zookeeper 节点进程服务启动失败。

Java 进程信息中没有"QuorumPeerMain"进程。

无法查看 Zookeeper 节点进程服务的状态或查看进程状态时报错。

Zookeeper 的命令行工具连接 Zookeeper 服务失败。

出现这些问题时一般在控制台中不会打印出详细的问题描述和报错信息。但导致这些问题的原因大多都是由于配置过程中输入错误或遗漏某个步骤造成的。可以先检查配置文件中的配置项的名称和值是否书写正确，比如数据文件和日志文件的存放目录路径是否正确，或者三个节点进程的配置文件中的端口号是否冲突，以及三个节点进程的标识文件"myid"中的值是否正确等。

当然也可以通过查看 Zookeeper 的日志文件来确定问题描述和报错信息，日志文件的文件名为"zookeeper.out"。Zookeeper 默认将日志文件输出到当前目录，若在配置文件中配置了配置项"dataLogDir"，并在脚本文件"zkServer.sh"中修改了变量"ZOO_LOG_DIR"的值，则会输出到指定的日志文件目录。

1.4.7 Zookeeper 的客户端基本操作

（1）查看当前 Zookeeper 中所包含的内容，对应命令为"ls /"，如图 1-4-26 所示。

```
[zk: localhost:2181(CONNECTED) 0] ls /
[zookeeper]
```

图 1-4-26 查看当前 Zookeeper 内容

（2）在 Zookeeper 中创建一个新的目录节点"mynode"，并设定目录节点中存放的数据为"TestData"，对应命令为"create /节点名 数据字符串"，如图 1-4-27 所示。

```
[zk: localhost:2181(CONNECTED) 1] create /mynode TestData
Created /mynode
[zk: localhost:2181(CONNECTED) 2] ls /
[mynode, zookeeper]
```

图 1-4-27 创建新的目录节点

（3）读取新创建的目录节点"mynode"中存放的数据，查看其中是否包含了创建时设定的数据"TestData"，对应命令为"get /节点名"，如图 1-4-28 所示。

```
[zk: localhost:2181(CONNECTED) 3] get /mynode
TestData
cZxid = 0x100000006
ctime = Wed Jul 18 10:34:35 EDT 2018
mZxid = 0x100000006
mtime = Wed Jul 18 10:34:35 EDT 2018
pZxid = 0x100000006
cversion = 0
dataVersion = 0
aclVersion = 0
ephemeralOwner = 0x0
dataLength = 8
numChildren = 0
```

图 1-4-28 读取新创建的目录节点数据

(4）在目录节点"mynode"下创建一个子目录节点"childnode"，并设定该子目录节点中存放的数据为"NewData"，对应命令为"create /节点名/子节点名　数据字符串"，如图 1-4-29 所示。然后查看目录节点"mynode"所包含的子目录节点的内容，对应命令为"ls /节点名"，如图 1-4-30 所示。并读取子目录节点中存放的数据，对应命令为"get /节点名/子节点名"，如图 1-4-31 所示。

```
[zk: localhost:2181(CONNECTED) 4] create /mynode/childnode NewData
Created /mynode/childnode
```

图 1-4-29　创建子目录节点

```
[zk: localhost:2181(CONNECTED) 5] ls /mynode
[childnode]
```

图 1-4-30　查看目录节点

```
[zk: localhost:2181(CONNECTED) 6] get /mynode/childnode
NewData
cZxid = 0x100000007
ctime = Wed Jul 18 10:45:26 EDT 2018
mZxid = 0x100000007
mtime = Wed Jul 18 10:45:26 EDT 2018
pZxid = 0x100000007
cversion = 0
dataVersion = 0
aclVersion = 0
ephemeralOwner = 0x0
dataLength = 7
numChildren = 0
```

图 1-4-31　读取子目录节点

（5）修改目录节点"mynode"中存放的数据为"ChangeData"，对应命令为"set /节点名　数据字符串"，如图 1-4-32 所示。然后再次读取该目录节点中存放的数据，查看数据是否修改成功，如图 1-4-33 所示。

```
[zk: localhost:2181(CONNECTED) 7] set /mynode ChangeData
cZxid = 0x100000006
ctime = Wed Jul 18 10:34:35 EDT 2018
mZxid = 0x100000008
mtime = Wed Jul 18 10:56:05 EDT 2018
pZxid = 0x100000007
cversion = 1
dataVersion = 1
aclVersion = 0
ephemeralOwner = 0x0
dataLength = 10
numChildren = 1
```

图 1-4-32　修改目录节点数据

```
[zk: localhost:2181(CONNECTED) 8] get /mynode
ChangeData
cZxid = 0x100000006
ctime = Wed Jul 18 10:34:35 EDT 2018
mZxid = 0x100000008
mtime = Wed Jul 18 10:56:05 EDT 2018
pZxid = 0x100000007
cversion = 1
dataVersion = 1
aclVersion = 0
ephemeralOwner = 0x0
dataLength = 10
numChildren = 1
```

图 1-4-33　查看是否修改成功

（6）删除手动创建的目录节点"mynode"及其子目录节点"childnode"，对应命令为"rmr　/节点名"，如图 1-4-34 所示。另一个命令"delete"也可以删除目录节点，但需要注意的是该命令无法删除包含有子目录节点的目录节点，只有先删除目录节点的所有的子目录节点之后，才能使用该命令来删除目录节点。如图 1-4-35 所示，在目录节点存在子目录节点的时候使用"delete"命令删除会有错误提示信息。

```
[zk: localhost:2181(CONNECTED) 11] rmr /mynode
[zk: localhost:2181(CONNECTED) 12] ls /
[zookeeper]
```

图 1-4-34　删除目录节点及子目录节点

```
[zk: localhost:2181(CONNECTED) 9] delete /mynode
Node not empty: /mynode
```

图 1-4-35　"delete"命令删除错误

任务 1.5　安装 Hadoop 伪分布模式

1.5.1　HDFS

1.5.1.1　分布式文件系统

分布式文件系统（Distributed File System）是指文件系统管理的物理存储资源不一定直接连接在本地节点上，而是通过计算机网络与节点相连。一般采用客户机服务器模式,并基于操作系统的本地文件系统构建。

分布式文件系统原本将固定于某个地点的某个文件系统能扩展到任意多个地点的多个文件系统,使用众多的节点组成一个文件系统网络。每个节点可以分布在不同的地点,通过网络进行节点间的通信和数据传输。在使用分布式文件系统时,无须关心数据存储在哪个节点之上,只需要像使用本地文件系统一样管理文件系统中的数据。

分布式文件系统相较于传统的本地文件系统有以下的优势:

突破了传统文件系统的容量限制,能够以更低廉的成本以及更方便快捷的方式扩充存储空间。

突破了传统文件系统的吞吐量限制,能够为多用户、多应用的服务提供高性能的并行读写。

可以提供冗余备份和容错机制。

为分布式计算提供基础。

1.5.1.2 HDFS

HDFS(Hadoop Distributed File System)是 Hadoop 项目的核心子项目,为 Hadoop 项目中的分布式计算模型 MapReduce 提供了数据存储管理的基础。其实现原理源自 Google 在 2003 年 10 月份发表的有关于 GFS(Google File System)的一篇论文,可以认为 HDFS 是 GFS 的一个开源实现版本。HDFS 基于流数据访问模式和超大文件处理的需求而开发,可以运行于廉价的商用服务器上,为海量数据和超大数据集应用提供了稳定可靠的低成本存储解决方案。

HDFS 主要具备以下优势和特点:

1. 高容错性

在 HDFS 中,数据会自动保存有多个副本。在某一个副本损坏或丢失之后,可以通过其他完整的副本来自动进行恢复,恢复过程由 HDFS 的内部机制自动实现。并且可以通过增加副本数量的方式来进一步提高容错性。

2. 适合分布式计算

HDFS 通过将数据位置暴露给计算框架,将计算移动到数据端,从而实现高效的分布式计算模式。

3. 适合大数据处理

HDFS 采用顺序的数据流访问方式,能够处理 TB、PB 甚至 EB 级别的数据。

同时支持百万规模以上的文件数量，以及 10 000 个节点数量的横向扩展。

4. 简单的一致性模型

HDFS 采用一次写入、多次读取的访问模式，文件中的数据一旦被写入便不能被修改，仅能在文件末尾追加新的数据，这样便能够很容易地保证各数据副本之间的一致性。

5. 基于廉价的普通硬件

HDFS 对 X86 架构的硬件有着良好的支持。同时由于其提供的高容错性，使得即便使用廉价的低端商用服务器，甚至是 PC 级别的硬件，也不用担心因为硬件故障而导致的数据丢失等问题。

1.5.1.3 HDFS 的块

在传统的存储介质中，块是读写的最小数据单位，比如磁盘中的扇区。传统文件系统中数据的读写就是基于存储块进行操作，但为了节约文件分配表所占用的空间，一般会对物理磁盘的存储块进行整合，大部分时候默认的存储块大小为 4 096 字节。

HDFS 中也使用了块的概念，但是默认大小为 64 M 字节，并且可以针对每个文件单独进行配置。从默认块大小可以看出，HDFS 并不适合处理小文件，这也正对应了其应用于大数据处理的特性。HDFS 将一个文件分为一个或数个块来存储，每个块都是一个独立的存储单位，并且有一个自己的全局 ID。不过 HDFS 中的块与传统文件系统不同的是，如果实际数据没有达到指定的块大小，则并不实际占用磁盘空间。如一个文件是 200 MB，则它会被分为 4 个块，大小分别为 64 MB、64 MB、64 MB、8 MB。

使用块的优点在于当一个文件大于集群中任意一个磁盘的剩余存储空间的时候，文件系统可以充分利用集群中所有磁盘的剩余空间来存储该文件。同时管理块可以让底层的存储子系统可以直接使用本地文件系统，使得分布式文件系统的底层实现变得相对简单。还有就是块本身相对于原本的文件体积更小，更加适合于进行备份，从而为容错性和高可用性的实现提供了便利。

HDFS 中每个块默认会有 3 个备份，并且可以针对每个文件单独进行配置，甚至可以在 HDFS 服务运行过程中动态修改。某个块的所有备份都享有同一个 ID，这样文件系统便无须再记录哪些块是同一份数据。HDFS 中某个块的所有备份在集群中如何进行存放，需要对可靠性、写入带宽、读取带宽等方面进行权衡,大部分时候备份副本之间需要遵循尽量远离的基本原则分布在集群之中,比如位于不同的机器、位于不同的机架等。Hadoop 中对 HDFS 中块的备份副本的存放有着自己的一套副本策略,如图 1-5-1 所示。

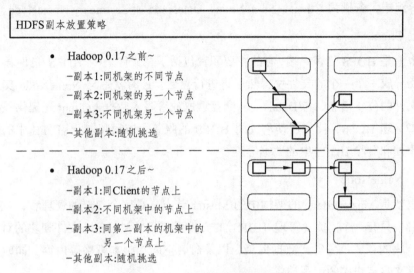

图 1-5-1 Hadoop 备份副本策略

1.5.1.4 HDFS 的元数据

HDFS 中的元数据包含了以下三大类信息:

(1)文件系统目录树信息,包括文件名、目录名、文件和目录的从属关系、文件或目录的大小、创建时间、最后访问时间、权限信息等。

(2)文件和块的对应关系,记录了文件由哪些块组成。

(3)块的存放位置,包括块的 ID、所存放机器的主机名或地址、所存放的路径等。

HDFS 对元数据和实际数据采取了分开存储的方式,元数据存储在一台或数台指定的服务器上,而实际数据则是储存在集群中的其他机器的本地文件系统中。

HDFS 中使用两个文件来持久化保存元数据信息。一个是映像文件 fsimage,其中保存了文件系统目录树信息以及文件和块的对应关系。另一个是事务日志文件 edits,其中保存了文件系统的更改记录。当客户端对文件进行写操作(包括新建或移动)的时候,首先将该操作记入 edits 文件,成功之后再更改内存中加载的元数据信息,并不会立刻更改磁盘中的 fsimage 文件的内容。元数据信息中的块的位置信息并不会做持久化保存。

1.5.1.5 HDFS 的架构

HDFS 采用了 Master/Slave 的架构来存储数据，如图 1-5-2 所示，主要由四个部分组成，分别为 Client、NameNode、DataNode 和 Secondary NameNode。

1. Client

也就是 HDFS 的客户端，它的主要职责包括了在文件上传 HDFS 的时候将文件切分成一个一个的文件块，然后再进行存储。还有就是与 NameNode 交互获取文件的位置信息，与 DataNode 交互读取或者写入数据。Client 还提供一些命令来管理 HDFS，比如启动和关闭 HDFS 的服务、访问和查看 HDFS 中的数据、添加和删除 HDFS 中的数据等。

2. NameNode

也就是 Master/Slave 的架构中的 Master 节点，是整个集群的管理者，在整个集群中只有一个。负责管理 HDFS 的文件系统目录树信息、文件和块的对应关系、块的存放位置等元数据信息，以及命名空间、副本策略等内容，同时负责处理来自客户端的读写请求。

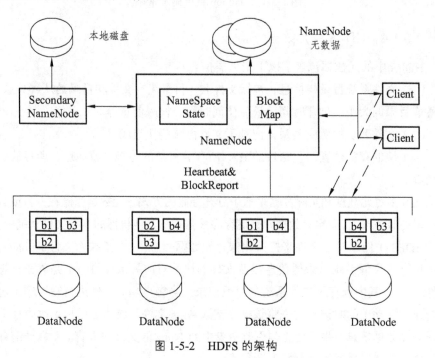

图 1-5-2　HDFS 的架构

3. DataNode

也就是 Master/Slave 的架构中的 Slave 节点，在整个集群中有多个。接收来自 NameNode 下达的命令，然后执行实际的操作。负责数据块的实际存储和读写操作的执行。每个数据块会在 DataNode 的本地文件系统中产生两个文件，一个是实际的数据文件，另一个是数据块的附加信息文件，其中包括校验和、创建时间等信息。

4. SecondaryNameNode

SecondaryNameNode 并非是 NameNode 的备份。当 NameNode 挂掉的时候，它并不能替换 NameNode 来提供服务。SecondaryNameNode 的主要职责是辅助 NameNode，分担一些工作量。其主要职责是定期合并 fsimage 和 edits 文件，并推送给 NameNode。在紧急情况下，也可以辅助恢复 NameNode。

1.5.1.6 HDFS 的写操作

HDFS 的文件写入操作如图 1-5-3 所示，主要包括以下几个步骤：

（1）客户端通过调用 DistributedFileSystem 对象的 create()方法，申请创建一个新的文件。

（2）DistributedFileSystem 的 create()方法通过 RPC（远程过程调用）在 NameNode 中创建一个没有 Block 关联的新文件。新文件创建前，NameNode 会进行各种校验，比如文件是否存在、客户端是否有创建权限等。如果校验通过，NameNode 就会将新文件添加到元数据记录中，否则就会抛出 IO 异常信息。

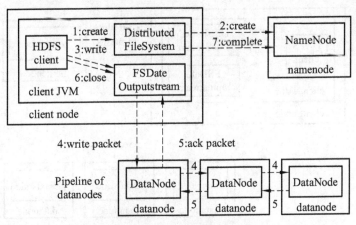

图 1-5-3　HDFS 文件写入操作

(3)前两步结束后会返回一个被封装成 DFSOutputStream 对象的 FSDataOutputStream 对象,DFSOutputStream 会自动协调与 NameNode 和 DataNode 之间的通信和操作。然后客户端开始写数据到 DFSOutputStream,DFSOutputStream 会自动把数据切分成一个个小的 Packet,然后排成数据队列 Data Queue。这时用户数据是被缓存在客户端本地,而每次本地缓存收集到足够一个 HDFS 设定的 Block 大小的时候,客户端便会向 NameNode 申请并写入一个新的 Block。

(4)DataStreamer 负责接收并处理 Data Queue,它会先问询 NameNode 新的 Block 最适合存储的在哪几个 DataNode 中,如果设定的 Block 副本数量为 3,那么就需要找到 3 个最适合存储的 DataNode,然后把它们排成一个 Pipeline。DataStreamer 把 Packet 按照 Data Queue 中的顺序输出到管道的第一个 DataNode 中,然后第一个 DataNode 又把 Packet 输出到第二个 DataNode 中,以此类推。

(5)DFSOutputStream 中还有一个队列叫作 Ack Queue,也是由 Packet 组成,这个队列等待 DataNode 的接收响应,当 Pipeline 中的所有 DataNode 都表示已经收到指定 Packet 的时候,Ack Queue 就会把队列中对应的 Packet 移除掉。

(6)DFSOutputStream 继续接收数据缓存,若客户端写数据完成,调用 close() 方法关闭写入流。同时 DataStreamer 继续处理 Data Queue 中的 Packet,直到 DataStreamer 移除掉 Ack Queue 中的最后一个 Packet 之后,通知 DataNode 把文件标示为已完成。

1.5.1.7 HDFS 的读操作

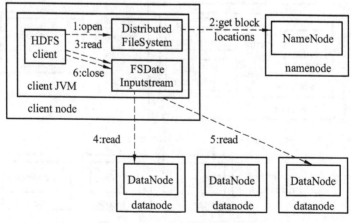

图 1-5-4 HDFS 文件读取操作

HDFS 的文件读取操作如图 1-5-4 所示，主要包括以下几个步骤：

（1）首先调用 DistributedFileSystem 对象的 open()方法，获取一个该对象的实例。

（2）DistributedFileSystem 通过 RPC（远程过程调用）获取文件的第一批 Block 的 Location，同一个 Block 会按照设定的副本数量返回多个 Location，这些获取的 Location 会按照 Hadoop 的拓扑结构进行排序，距离客户端近的会排在前面。

（3）和写操作的时候相似，前两步结束后会返回一个被封装成 DFSInputStream 对象的 FSDataInputStream 对象，DFSInputStream 可以方便地管理客户端与 DataNode 和 NameNode 之间的数据流。然后客户端调用 read()方法，此时 DFSInputStream 便会从多个 Location 中找出离客户端最近的 DataNode 并进行连接。

（4）数据从 DataNode 源源不断地读取并流向客户端。

（5）如果第一个 Block 的数据完成读取，指向第一个 Block 的 DataNode 连接就会关闭，然后接着读取下一个 Block 块。这些操作对客户端而言是透明的，从客户端的角度来看只是读一个持续不断的数据流。

（6）如果第一批 Block 全部都读取完成，DFSInputStream 就会去连接 NameNode 获取下一批 Block 的 Location，然后继续读取。如果所有的 Block 全部读取完成，这时就会关闭掉所有的数据流。

1.5.1.8 HDFS 的追加操作

HDFS 的文件追加操作大部分都与写操作的过程一样，主要有区别的就是刚开始的几个步骤。首先客户端与 NameNode 通信，获得需要进行追加操作的指定文件的写保护锁，以及文件最后一个 Block 的 Locaotion。这个过程与读取文件有些相似，获得的 Location 会按照设定的副本数量返回多个，并按照 Hadoop 的拓扑结构进行排序，距离客户端近的会排在前面。然后客户端选择一个 Location 作为主写入节点，并对所有 Location 上的该数据块进行加锁。接着开始写入需要追加的数据，这个过程就和写操作的过程一样了。

1.5.1.9 HDFS 的中元数据的载入和更新

NameNode 在启动时载入并更新元数据的过程如下：

（1）通过 fsimage 文件读取元数据信息并载入到本地内存之中。

（2）依次执行 edits 文件中的写操作记录，在内存中生成最新的元数据信息。

（3）生成新的 edits 文件用于在 NameNode 完成启动之后记录新的写操作。

（4）生成新的 fsimage 文件，将本地内存中最新的元数据信息保存到文件之中。

（5）收集所有 DataNode 汇报的数据块的位置信息，记录到本地内存的元数据信息之中。

而在 NameNode 运行过程中，所有写操作首先记录到 edits 文件当中。然后在写操作成功完成之后，更新本地内存中的元数据信息。同时收集 DataNode 汇报的新创建的数据块和数据块的所有副本的位置信息，记录到本地内存中的元数据信息之中。至于 fsimage 文件的更新，则需要使用到 SecondaryNameNode。

SecondaryNameNode 的主要工作就是阶段性的合并 fsimage 和 edits 两个文件，并以此来控制 edits 文件的大小在合理的范围内。这样做可以缩短 HDFS 集群在启动时 NameNode 重建 fsimage 文件所需要花费的时间。一般情况下，Secondary Namenode 运行在不同于 NameNode 的主机之上，并且由于在合并 fsimage 和 edits 两个文件时，同样需要将 fsimage 文件的内容预先加载到本地内存当中，然后再依次执行 edits 文件中的写操作，所以它的内存需求也是和 NameNode 一样的。

SecondaryNameNode 根据配置好的策略决定多长时间做一次合并，合并 fsimage 和 edits 两个文件的具体过程如图 1-5-5 所示。

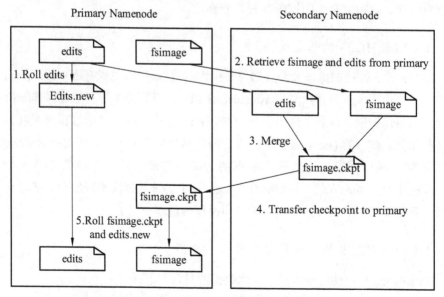

图 1-5-5　合并 fsimage 和 edits 两个文件

（1）SecondaryNameNode 通知 NameNode 现在准备提交 edits 文件，

NameNode 会创建一个新的操作日志文件 edits.new，然后将新的操作写入 edits.new 文件中。

（2）SecondaryNameNode 通过 HTTP GET 方式从 NameNode 获取 fsimage 与 edits 两个文件。

（3）SecondaryNameNode 将 fsimage 文件中的元数据信息载入到本地内存中，并执行 edits 文件中的所有操作，然后根据本地内存中合并完成之后的新的元数据信息生成新的映像文件 fsimage.ckpt。

（4）SecondaryNameNode 通过 HTTP POST 方式将新生成的映像文件 fsimage.ckpt 发送给 NameNode。

（5）NameNode 将新的映像文件 fsimage.ckpt 与新的操作日志文件 edits.new 文件分别重命名为 fsimage 与 edits，替换掉原有的两个文件。然后更新 fstime 文件中的内容，记录此次 CheckPoint 的时间。

1.5.1.10 HDFS 的可靠性策略

1. 冗余副本策略

所有 HDFS 的数据块默认都具备多个副本，并且可以在配置文件中手动设置副本数量。DataNode 在启动时会遍历本地文件系统，产生一份 HDFS 数据块和本地文件系统中文件的对应关系列表（BlockReport）并汇报给 NameNode。

2. 机架策略

集群中的机器一般会放在不同的机架之上，而机架间网络带宽要一般又比机架内网络带宽要小。HDFS 提供了对机架策略的支持，一般多个数据块副本会在当前机架存放一个，然后在其他机架上再存放别的副本，这样可以防止在机架功能失效时丢失数据，也可以提高带宽的利用率。

3. 心跳机制

NameNode 会周期性地从 DataNode 接收心跳信号和数据块报告。对于没有按时发送心跳信号的 DataNode 会被标记为宕机，不会再给它发送任何 I/O 请求。同时 NameNode 还会根据接收到的数据块报告来验证当前本地内存中的元数据信息。如果因为 DataNode 失效而造成数据块的副本数量下降到低于预先设置的阈值，NameNode 会检测出这些数据块，并在合适的时机进行重新复制。引发数据块重新复制的原因还包括了数据块副本本身损坏、磁盘读写错误、设置的副本数量被增大等。

4. 安全模式

NameNode 启动时会先经过一个安全模式阶段，在安全模式阶段中不会产生数据写操作。在此阶段中 NameNode 收集来自各个 DataNode 报告的数据块信息，当数据块的副本数量达到最小副本数量以上时，会被认为是安全的，而在一定比例的数据块被确定为安全后，再经过一些时间，安全模式便会结束。当检测到副本数量不足的数据块时，该数据块块会被复制直到达到最小副本数量。

5. 校验和

在文件创建时，每个数据块都会产生一个校验和信息。校验和信息会作为一个单独的隐藏文件保存在命名空间下。客户端在获取数据块数据时可以通过检查校验和信息来发现数据块是否已经损坏。如果正在读取的数据块损坏，则可以继续读取该数据块的其他副本。

6. 回收站

删除文件时并不会直接进行删除，而是先将其放入回收站。回收站里的文件可以快速进行恢复。并且可以对回收站设置一个时间阈值，当回收站里的文件存放时间超过该阈值时，就会被彻底删除，并且释放占用的数据块。

7. 元数据保护

映像文件 fsimage 和事务日志 edits 是整个 HDFS 的核心数据，可以配置其拥有多个副本。但需要注意的是多个元数据文件的副本会降低 NameNode 的处理速度，但可以增加整个系统的安全性。

8. 快照机制

存储在某个时间点时整个系统的状态，并且在需要时使数据重新返回到该时间点的状态。

1.5.2 MapReduce

MapReduce 最早是由 Google 公司研究并提出的一种面向大规模数据处理的并行计算模型和方法。Google 公司设计 MapReduce 的初衷主要是为了解决其搜索引擎中大规模网页数据的并行化处理问题。Google 公司发明了 MapReduce 之后首先用其重新改写了其搜索引擎中的 Web 文档索引处理系统。但由于 MapReduce 可以普遍应用于很多大规模数据的计算问题，因此自发明

MapReduce 以后，Google 公司内部进一步将其广泛应用于很多大规模数据处理问题。到目前为止，Google 公司内有上万个各种不同的算法问题和程序都使用 MapReduce 进行处理。

而 Hadoop 中的 MapReduce 的基本原理和设计思想也是源自 Google 在 2004 年发表的有关于 MapReduce 的一篇论文，可以认为 Hadoop 的 MapReduce 就是 Google 的 MapReduce 的一个开源实现版本，并且沿用了其原有的名称。

MapReduce 是一个基于集群的高性能并行计算平台（Cluster Infrastructure），它允许使用刀片机、机架式服务器甚至普通 PC 机来构成一个包含数千个节点的分布式并行计算集群。并且 MapReduce 还是一个并行程序的开发与运行框架（Software Framework），它提供了一个庞大但设计精良的并行计算软件构架，能自动完成计算任务的并行化处理，自动划分计算数据和计算任务，并在集群的节点上自动分配和执行子任务，同时收集计算结果。MapReduce 将数据分布存储、数据通信、容错处理等并行计算中的很多复杂细节交由系统负责处理，软件开发人员不再需要掌握分布式并行编程的很多细节，大大减少了软件开发人员的负担。同时 MapReduce 也是一个并行程序的设计模型与方法（Programming Model & Methodology），它借助于函数式语言中的设计思想，提供了一种简便的并行程序设计方法，将复杂的、运行于大规模集群上的并行计算过程高度地抽象为 Map 和 Reduce 两个函数来实现基本的并行计算任务，并提供了完整的并行编程接口来完成大规模数据处理。

MapReduce 的设计思想基于"计算向数据靠拢"的理念，而不是传统的"数据向计算靠拢"的理念，这大大减少了移动数据所需要消耗大量的网络带宽，但同时也使得 MapReduce 的实现必须完全依赖于一个良好的分布式文件系统。所以 Hadoop 中的 HDFS 和 MapReduce 虽然是独立的两个组件，但却是密不可分的。

1.5.2.1 数据的并行处理

假设有一个巨大的数据集，需要对其中每一个元素进行开立方运算。那么这是对于数据集中的每一个元素，我们需要进行的处理是相同的，并且元素与元素之间不存在任何数据依赖关系，那么就可以用不同的划分方法将这个数据集其划分为多个子数据集，交由多个处理器或计算机同时进行处理，这就是数据的并行处理。

而数据的并行处理的过程如图 1-5-6 所示，首先将数据集中的元素划分为多个子数据集。然后如图 1-5-7 所示，将不同子数据集交由不同的计算机来进行处理，并且在所有计算机完成对自身所负责的子数据集的处理之后，返回结果子集，然后再由一台专门的计算机将所有子数据集运算得到的结果子集进行合并，得到最终所需要的结果集。

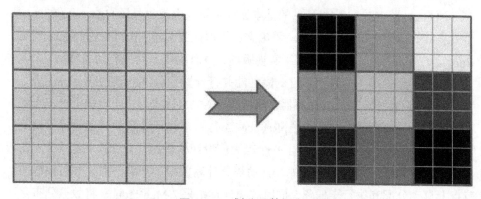

图 1-5-6　划分子数据集

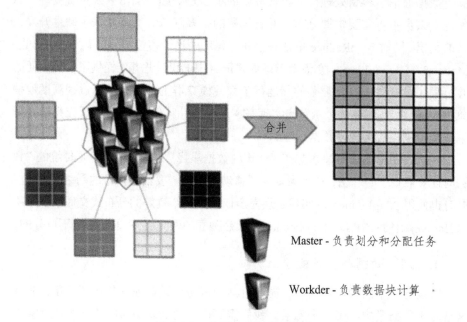

图 1-5-7　处理与合并子数据集

如图 1-5-8 所示，并行计算的整个过程并不复杂，其中最主要的问题在于如何确定哪些数据处理任务可以使用并行计算。一般来说一个数据集若是可以分为具有同样计算过程的数据块，并且这些数据块之间不存在数据依赖关系，那么就可以使用并行计算，而且使用并行计算是提高这类数据处理速度的最好办法。而对于不可分拆的计算任务，或是相互之间存在依赖关系的数据，则无法使用并行计算进行处理，如大家所熟知的 Fibonacci 函数就是一个典型的无法使用并行计算进行处理的例子。

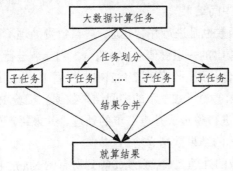

图 1-5-8　并行计算过程

1.5.2.2　MapReduce 的程序模型

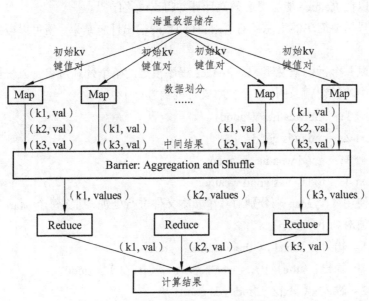

图 1-5-9　MapReduce 程序模型

MapReduce 程序模型中，最重要的两个角色就是 Map 和 Reduce。其中 Map 的职责是对大量数据记录和元素进行重复的处理，在其中计算或提取出感兴趣的内容。而 Reduce 的职责则是收集整理 Map 处理所产生的中间结果，然后整合产生最终的输出结果。

如图 1-5-9 所示，各个 Map 函数对所划分的数据并行处理，从不同的输入数据产生不同的中间结果输出。Map 函数键接收以键值对（k1；v1）的形式表示的文档数据记录，一般为文本文件中的行数据或数据表格中的行数据。然后 Map 函数将对这些键值对的内容进行处理，并以另一种键值对[（k2；v2）]的形式输出处理所得到的中间结果。

而各个 Reduce 函数也是进行并行处理，各自负责处理不同的中间结果数据集合。由 Map 函数输出的一组键值对[（k2；v2）]会被进行合并处理，将同样的键所对应的不同值合并到一个值列表[v2]中，然后 Reduce 函数接收合并之后的以键值对（k2；[v2]）的形式表示的中间结果数据。接着 Reduce 函数对这些传入的中间结果数据进行整理或是进一步的处理，并最终产生以某种形式的键值对[（k3；v3）]表示的结果数据并输出。

进行 Reduce 函数的处理之前，必须先等到所有的 Map 函数处理完成，所以 Map 函数和 Reduce 函数之间需要有一个同步障（Barrier）。这个阶段同时也能负责对 Map 函数所产生的中间结果数据进行收集和整理（Aggregation & Shuffle）的处理，以便 Reduce 能够更有效地计算并得出最终结果。

最后只需要简单的汇总所有 Reduce 函数的输出计算结果，便可以得到最终计算结果。

下面以一个示例来说明 MapReduce 程序模型的整个处理过程，以统计 4 个文本文件中的的词频为例，4 个文本文件的内容如下：

➢ Text 1： the weather is good
➢ Text 2： today is good
➢ Text 3： good weather is good
➢ Text 4： today has good weather

这里使用 4 个 Map 来分别对 4 个文本文件进行处理，那么每个 Map 的输入和输出键值对如下所示：

Map 1 - 输入：（text1,"the weather is good"）
        输出：（the, 1），（weather, 1），（is, 1），（good, 1）
Map 2 - 输入：（text2,"today is good"）
        输出：（today, 1），（is, 1），（good, 1）
Map 3 - 输入：（text3,"good weather is good"）
        输出：（good, 1），（weather, 1），（is, 1），（good, 1）
Map 4 - 输入：（text3,"today has good weather"）
        输出：（today, 1），（has, 1），（good, 1），（weather, 1）

然后使用 3 个 Reduce 来对 Map 输出的中间结果键值对进行处理，并输出最

终的结果键值对,那么每个 Reduce 的输入和输出键值对如下所示:

Reduce 1 - 输入:(good,1),(good,1),(good,1),(good,1),(good,1)
　　　　　　输出:(good,5)
Reduce 2 - 输入:(has,1),(is,1),(is,1),(is,1)
　　　　　　输出:(has,1),(is,3)
Reduce 3 - 输入:(the,1),(today,1),(today,1),(weather,1),
　　　　　　　　(weather,1),(weather,1)
　　　　　　输出:(the,1),(today,2),(weather,3)

最后整合所有 Reduce 所输出的结果键值对,形成最终的结果如下:
good:5,is:3,has:1,the:1,today:2,weather:3

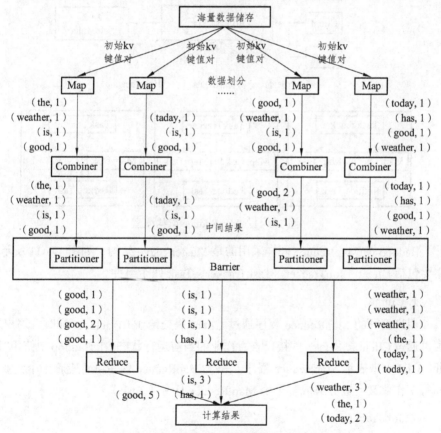

图 1-5-10　MapReduce 程序处理实例

除了 Map 和 Reduce 两个最主要的函数之外,在完整的 MapReduce 程序模型中其实还包括了 Combiner 和 Partitioner 这两个函数,如图 1-5-11 所示。其中

Combiner 函数是在 Map 函数所产生的中间结果数据被传递给 Reducer 函数之前，提前对中间结果数据进行一些合并（Combine）处理，把中间结果数据中具有相同键的键值对的值合并到一起，避免数据的重复传输，可以大幅度减少数据通信的性能开销。

而 Partitioner 函数则是对中间结果数据的键值对需要传输给哪一个 Reduce 函数处理进行划分，也称为对中间结果数据进行分区（Partition），这样做可以保证相关的数据被发送给同一个 Reduce 函数进行处理。

1.5.2.3 MapReduce 的架构

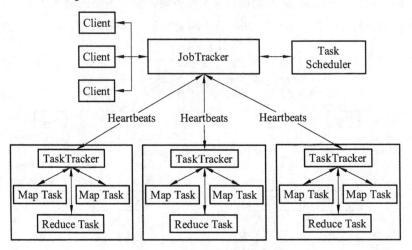

图 1-5-11 MapReduce 架构模型

Hadoop 的 MapReduce 同样采用的是 Master/Slave 架构，如图 1-5-11 所示，主要包括 Client、JobTracker、TaskTracker、Task 四个组件。

1. Client

用户编写的 MapReduce 程序通过 Client 提交给 JobTracker。同时，用户还可以通过 Client 提供的一些接口查看作业的当前运行状态。在 Hadoop 中使用"作业"（Job）来表示 MapReduce 程序。一个 MapReduce 程序可对应若干个作业，而每个作业又会被分解成若干个 Map/Reduce 任务（Task）。

2. JobTracker

JobTracker 主要负责资源监控和作业调度。JobTracker 监控所有 TaskTracker 与作业的健康状况，一旦发现任务出现失败的情况后，会将相应的任务转移到其他节点执行。同时 JobTracker 会跟踪任务的执行进度、资源使用量等，并将这些信息告诉给任务调度器（Task Scheduler）。而任务调度器会在资源出现空闲时，选择合适的任务来使用这些资源。在 Hadoop 中任务调度器是一个可插拔的模块，用户可以根据自己的需要自行设计任务调度器。

3. TaskTracker

TaskTracker 会周期性地通过 Heartbeat 消息将当前节点上资源的使用情况和任务的运行进度汇报给 JobTracker，同时接收 JobTracker 发送过来的命令并执行相应操作，如启动新任务、杀死任务等。TaskTracker 使用 Slot 来等量划分当前节点上的资源量，其中包含了 CPU、内存等计算资源。一个 Task 只有获取到一个 Slot 之后才有机会运行，而 Hadoop 的任务调度器的作用就是将各个 TaskTracker 上的空闲 Slot 分配给 Task 使用。Slot 分为 Map Slot 和 Reduce Slot 两种，分别提供给 Map Task 和 Reduce Task 使用。TaskTracker 通过 Slot 的数量来限制 Task 的并发度，该数量可以通过配置参数来进行配置。

4. Task

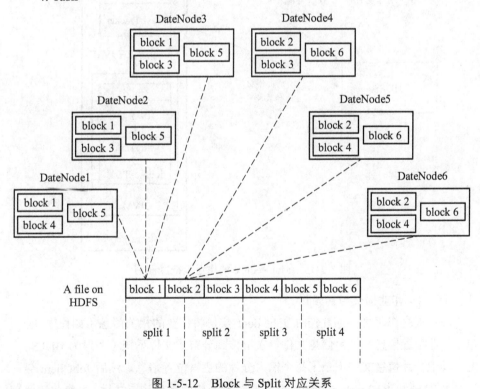

图 1-5-12　Block 与 Split 对应关系

Task 分为 Map Task 和 Reduce Task 两种，均由 TaskTracker 启动。HDFS 中

以固定大小的 Block 为基本单位来存储数据，而对于 MapReduce 而言，其处理数据的最小单位是 Split，两者之间的对应关系如图 1-5-12 所示。Split 是一个逻辑概念，它只包含一些元数据信息，比如数据起始位置、数据长度、数据所在节点等，它的划分方法可以完全由用户自己决定。但需要注意的是，Split 的多少决定了 Map Task 的数量，因为每个 Split 都会交由一个 Map Task 来处理。

1.5.2.4 MapReduce 的作业运行过程（见图 1-5-13）

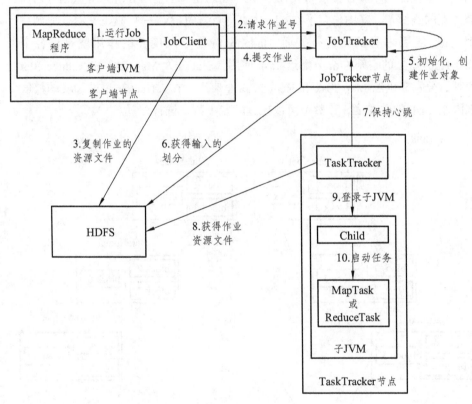

图 1-5-13　MapReduce 的作业运行过程

步骤 1：作业提交与初始化。

用户提交作业之后，首先由 JobClient 实例将作业的相关信息（如程序 Jar 包、作业配置信息、分片信息文件等）上传到分布式文件系统（一般为 HDFS）上，其中分片信息文件记录了每个输入分片的逻辑位置信息。然后 JobClient 会通过 RPC 通知 JobTracker。JobTracker 在收到新作业提交请求之后，由作业调度模块对作业进行初始化。初始化过程会为作业创建一个 JobInProcess 对象以跟踪作业的运行状况，而 JobInProcess 则会为每个 Task 创建一个 TaskInProgress 对象以跟踪每个任务的运行状态。

步骤 2：任务调度与监控。

任务调度与监控均由 JobTracker 来完成。TaskTracker 会周期性地通过 Heartbeat 消息向 JobTracker 汇报当前节点的资源使用情况，一旦出现空闲资源，JobTracker 会按照一定的策略选择一个合适的任务来使用该空闲资源，这个过程由任务调度器来完成。任务调度器是一个可插拔的独立模块，且为双层架构，即首先选择作业，然后从该作业中选择任务，其中选择任务时会重点考虑数据的本地性。此外，JobTracker 跟踪作业的整个运行过程，并为作业的成功运行提供全方位的保障。当 TaskTracker 或者 Task 失败时，JobTracker 会转移计算任务。而当某个 Task 执行进度远落后于同一个作业的其他 Task 时，JobTracker 会为其启动一个相同的 Task，并选择计算快的 Task 结果作为最终结果。

步骤 3：任务运行环境准备。

运行环境准备包括了 JVM 的启动和进行资源隔离，这些均由 TaskTracker 来实现。TaskTracker 为每个 Task 启动一个独立的 JVM 以避免不同 Task 在运行过程中相互影响。同时，TaskTracker 使用操作系统进程来实现资源隔离，以防止 Task 滥用资源。

步骤 4：任务执行。

TaskTracker 为 Task 准备好运行环境之后，便会启动 Task。在运行过程中，每个 Task 的最新进度首先由 Task 通过 RPC 汇报给 TaskTracker，再由 TaskTracker 通过 RPC 汇报给 JobTracker。

步骤 5：作业完成。

待所有 Task 执行完毕之后，整个作业执行完成。

1.5.3 YARN

YARN（Yet Another Resource Negotiator）是 Hadoop 2.0 中新增的通用资源管理系统，可以为上层应用提供统一的资源管理和调度。它的基本设计思想是将资源管理和处理组件分开，将 Hadoop 1.0 的 MapReduce 中的 JobTracker 拆分成了两个独立的服务，一个是代替原有 JobTracker 中集群计算资源管理和分配职责的全局资源管理器 ResourceManager，另一个是代替原有 JobTracker 中协调单个应用程序中所有任务并管理和控制 TaskTracker 职责的每个应用程序所特有的 ApplicationMaster。它的引入为集群在利用率、资源统一管理、数据共享等方面带来了巨大好处。而在 YARN 中，MapReduce 也被降级成为一个分布式应用程序中的一个角色。

1.5.3.1 YARN 的架构

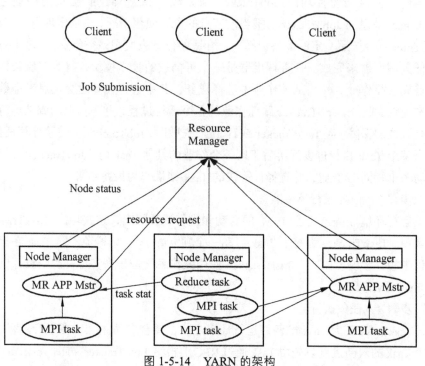

图 1-5-14　YARN 的架构

YARN 主要由 ResourceManager、NodeManager、ApplicationMaster、Container 四个部分构成，整体架构仍然是采用的 Master/Slave 结构，如图 1-5-14 所示，其中 ResourceManager 为 Master，NodeManager 为 Slave。

1. ResourceManager

Resourcemanager 是一个全局的资源管理器，在整个集群中只有一个，通常运行在集群中专用机器上，运行负责整个系统的资源管理和分配。它的工作主要包括处理来自客户端请求、启动并监控 ApplicationMaster、监控 NodeManager、追踪集群中可用的活动节点和资源、分配与调度物理资源等。它主要由两个组件构成，一个是调度器（Scheduler），另一个是应用程序管理器（Applications Manager, ASM）。

1）调度器

调度器会根据容量、队列等限制条件（如每个队列只能获得固定的资源，或是最多执行一定数量的作业等），将系统中的资源分配给各个正在运行的应用程序。需要注意的是，这个调度器是一个纯粹的调度器，它不从事类似于监控或者跟踪应用的执行状态等任何与具体应用程序相关的工作，也不负责重新启动因应用执行失败或硬件故障而产生的失败任务，这些工作均交给了与应用程序相关的 ApplicationMaster 来完成。调度器仅根据各个应用程序的资源需求进行资源分配，而资源分配单位使用一个抽象概念"资源容器"（Resource Container，简称 Container）来表示。Container 是一个动态的资源分配单位，它将内存、CPU、磁盘、网络等资源封装在一起，从而限定每个任务使用的资源量。此外，调度器是一个可插拔的组件，用户可根据自己的需要选择不同的调度器或是设计新的调度器。YARN 自身便提供了多种可直接使用的调度器，比如 Fair Scheduler 和 Capacity Scheduler 等。

2）应用程序管理器

应用程序管理器负责对整个系统中的所有应用程序进行管理，包括应用程序的提交、向调度器申请资源以启动 ApplicationMaster，监控每个 ApplicationMaster 的运行状态，在 ApplicationMaster 出错时进行重新启动等。

2. ApplicationMaster

ApplicationMaster 相当于以前的 MapReduce 中分配给所有作业的单个 JobTracker，与属于它的应用程序的任务，在受 NodeManager 控制的资源容器之中运行，负责管理在 YARN 内运行的应用程序的每个实例。它为应用程序向 ResourceManager 申请资源，然后负责协调获取的资源，并进一步分配给内部任务。同时 ApplicationMaster 还通过 NodeManager 监视任务的执行和资源使用情况，协调应用程序内的所有任务的执行，对出错的任务进行重新启动，推测运行缓慢的任务，并计算应用程序的计数器的值。

ApplicationMaster 不再有老版本 MapReduce 中只能运行 Map 或 Reduce 任务的局限性，而是可以在容器内运行任何类型的任务。例如，MapReduce ApplicationMaster 可以请求一个容器来运行 Map 或 Reduce 任务，而 Giraph ApplicationMaster 可以请求一个容器来运行 Giraph 任务，甚至还可以实现一个自定义的 ApplicationMaster 来运行特定的任务。

3. NodeManager

NodeManager 是可以看作是 TaskTracker 的一种更为高效的版本，在整个集群有多个，负责每个节点的资源和任务的管理。NodeManager 拥有并管理着许多动态创建的资源容器，不再需要设定固定数量的 Map Slots 和 Reduce Slots，而容器的大小取决于节点所包含的资源量。NodeManager 处理来自 ResourceManager 和 ApplicationMaster 的命令，并定期向 ResourceManager 汇报当前节点上的资源使用情况和各个 Container 的运行状态。

4. Container

Container 是 YARN 中对物理资源的抽象表示，它封装了集群中某个节点之上的多种资源，如内存、CPU、磁盘、网络等。当 ApplicationMaster 向 ResourceManager 申请资源时，ResourceManager 为 ApplicationMaster 返回一个使用 Container 表示的资源。YARN 会为每个任务分配一个 Container，且该任务只能使用该 Container 中所描述的资源。需要注意的是，Container 不同于老版本 MapReduce 中的 Slot，它是一个动态的资源划分单位，是根据应用程序实际的需求动态生成的。

1.5.3.2　YARN 的应用程序提交流程（见图 1-5-15）

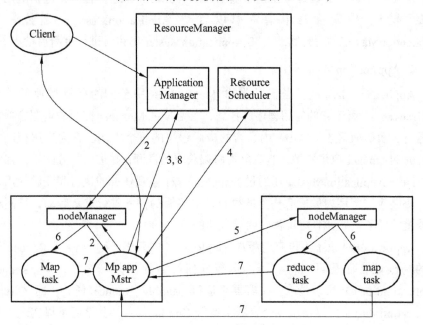

图 1-5-15　YARN 的应用程序提交流程

（1）Client 程序向 YARN 提交应用程序，其中包括用户程序、ApplicationMaster 程序、启动 ApplicationMaster 命令等。

（2）ResourceManager 为该应用程序分配第一个 Container，并与对应的 NodeManager 通信，要求它在这个 Container 中启动应用程序的 ApplicationMaster。

(3) ApplicationMaster 首先向 ResourceManager 进行注册,这样用户便可以直接通过 ResourceManager 查看应用程序的运行状态。然后 ApplicationMaster 将为各个任务申请资源,并监控它们的运行状态,并重复执行步骤(4)~(7),直到运行结束。

(4) ApplicationMaster 采用轮询的方式通过 RPC 协议向 ResourceManager 申请和领取资源。

(5) 一旦 ApplicationMaster 申请到资源之后,便与对应的 Nodemanager 通信,要求它启动任务。

(6) NodeManager 为任务设置运行环境,包括环境变量、JAR 包、二进制程序等,完成后将任务启动命令写到一个脚本之中,并通过运行该脚本启动任务。

(7) 各个任务通过某个 RPC 协议向 ApplicationMaster 汇报自己当前的状态和进度,让 ApplicationMaster 随时掌握各个任务的运行状态,从而可以在任务失败时重新启动任务。在应用程序运行过程中,用户可随时通过 RPC 向 ApplicationMaster 查询应用程序的当前运行状态。

(8) 应用程序运行完成后,ApplicationMaster 向 ResourceManager 注销并关闭自己。

1.5.4 Hadoop 安装规划

1. 环境要求

本地磁盘剩余空间 400 M 以上;
已安装 CentOS 7 1611 64 位操作系统;
已安装 JDK 的 1.8.0_131 版本。

2. 软件版本

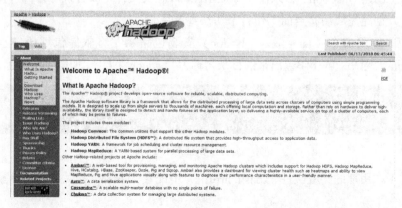

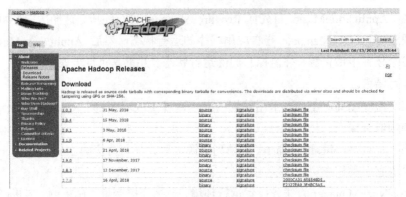

图 1-5-16　获取软件包

选用 Hadoop 的 2.7.3 版本，软件包名为 hadoop-2.7.3.tar.gz，该软件包可以在 Hadoop 项目位于 Apache 的官方网站（http：//hadoop.apache.org）的 Releases 页面（http：//hadoop.apache.org/releases.html）获取。

3. 服务规划

Hadoop 的伪分布模式实际上是在单台主机上使用多个不同的进程来模拟出不同类型的多个服务节点。

Hadoop 的服务包括 HDFS 的 NameNode 服务、SecondaryNameNode 服务和 DataNode 服务，以及 YARN 资源管理器的 ResourceManager 服务和 NodeManager 服务，总共五种类型的服务节点。伪分布模式下只需要满足每个类型的服务节点进程至少有 1 个即可。

1.5.5　Hadoop 基础环境配置

（1）创建"hadoop"目录用于存放 Hadoop 相关文件，该目录可自行选择创建位置，创建完成后将当前工作目录切换到该目录。

（2）使用命令"tar　-xzf　Hadoop 安装包路径"将软件包解压解包到"hadoop"目录下，解压解包出来的目录名称为"hadoop-2.7.3"，如图 1-5-17 所示。

```
[hadoop@localhost hadoop]$ ls -l
total 0
drwxr-xr-x. 9 hadoop hadoop 149 Aug 17  2016 hadoop-2.7.3
```

图 1-5-17　解压软件包

（3）在用户的配置文件".bash_profile"中配置 Hadoop 相关的环境变量，

如图 1-5-18 所示，在文件末尾添加以下内容：

\# hadoop environment

HADOOP_HOME=（Hadoop 软件目录路径）

PATH=$HADOOP_HOME/bin：$HADOOP_HOME/sbin：$PATH

export HADOOP_HOME PATH

Hadoop 软件目录即 Hadoop 软件包解压解包出来的"hadoop-2.7.3"目录，这里需要书写该目录及其所在的完整绝对路径。

```
# hadoop environment
HADOOP_HOME=/home/hadoop/hadoop/hadoop-2.7.3
PATH=$HADOOP_HOME/bin:$HADOOP_HOME/sbin:$PATH
export HADOOP_HOME PATH
```

图 1-5-18　配置环境变量

（4）使用命令"source ~/.bash_profile"使新配置的环境变量立即生效。

（5）使用命令"echo $变量名"查看新添加和修改的环境变量的值是否正确，如图 1-5-19 所示。

```
[hadoop@localhost ~]$ echo $HADOOP_HOME
/home/hadoop/hadoop/hadoop-2.7.3
```

图 1-5-19　查看环境变量

（6）使用命令"hadoop version"验证 Hadoop 的安装配置是否成功。如图 1-5-20 所示，若显示有 Hadoop 的版本信息，并且与所使用的 Hadoop 软件版本相符，则表示 Hadoop 的基本安装已经成功，可以正常使用。

```
[hadoop@localhost ~]$ hadoop version
Hadoop 2.7.3
Subversion https://git-wip-us.apache.org/repos/asf/hadoop.git -r baa91f7rfhc9cb92be5982de4719c1c8af9
1ccff
Compiled by root on 2016-08-18T01:41Z
Compiled with protoc 2.5.0
From source with checksum 2e4ce5f957ea4db193bce3734ff29ff4
This command was run using /home/hadoop/hadoop/hadoop-2.7.3/share/hadoop/common/hadoop-common-2.7.3.
jar
```

图 1-5-20　验证配置是否成功

1.5.6　配置本机免密码登录

（1）使用命令"ssh-keygen -t rsa"生成当前系统登录用户密钥文件，包括公钥文件和私钥文件。如图 1-5-21 所示，密钥文件的生成过程为交互模式，会提示用户输入一些相关设置信息,包括密钥文件的存放位置以及密钥文件的密码。这里可以不用进行输入,全部使用回车键直接跳过即可,在跳过的情况下系统会默认在用户的家目录中创建一个名为".ssh"的隐藏目录,然后将生成的密钥文件存放于其中。

```
[hadoop@localhost ~]$ ssh-keygen -t rsa
Generating public/private rsa key pair.
Enter file in which to save the key (/home/hadoop/.ssh/id_rsa):
Created directory '/home/hadoop/.ssh'.
Enter passphrase (empty for no passphrase):
Enter same passphrase again:
Your identification has been saved in /home/hadoop/.ssh/id_rsa.
Your public key has been saved in /home/hadoop/.ssh/id_rsa.pub.
The key fingerprint is:
fc:ee:49:91:b1:e6:cf:eb:e3:bf:c3:b8:e3:0b:2f:ad hadoop@localhost.localdomain
The key's randomart image is:
+--[ RSA 2048]----+
|                 |
|                 |
|           .     |
|          . +    |
|       S =       |
|      + .        |
|      =. o       |
|     o.*= o      |
|     .E*@Boo     |
+-----------------+
```

图 1-5-21 生成密钥文件

(2)进入密钥文件的存放目录,使用命令"cp id_rsa.pub authorized_keys"利用生成的公钥文件来生成免密码登录文件,如图 1-5-22 所示。

```
[hadoop@localhost .ssh]$ ls -l
total 12
-rw-r--r--. 1 hadoop hadoop  410 Jul 20 04:03 authorized_keys
-rw-------. 1 hadoop hadoop 1679 Jul 20 04:00 id_rsa
-rw-r--r--. 1 hadoop hadoop  410 Jul 20 04:00 id_rsa.pub
```

图 1-5-22 生成免密码登录文件

注意,免密码登录文件的文件名的书写要保证正确,否则系统将因为找不到免密码登录文件而导致免密码登录失败。

(3)使用命令"ssh localhost"进行本地主机的登录操作。如图 1-5-23 所示,第一次登录时可能会因为还没有将"localhost"主机名加入已知主机列表中,出现是否想继续连接的提示信息,输入"yes"后按回车键即可,系统会自动生成已知主机列表文件并把"localhost"添加到其中。

```
[hadoop@localhost ~]$ ssh localhost
The authenticity of host 'localhost (::1)' can't be established.
ECDSA key fingerprint is c4:2b:a8:5a:12:3f:eb:41:a9:3d:2b:68:b3:52:ed:1b.
Are you sure you want to continue connecting (yes/no)? yes
Warning: Permanently added 'localhost' (ECDSA) to the list of known hosts.
Last login: Sun Apr 22 21:51:56 2018
```

图 1-5-23　第一次本地主机登录操作

若"localhost"主机名已经存在于已知主机列表中，则会直接进行登录，如图 1-5-24 所示。

```
[hadoop@localhost ~]$ ssh localhost
Last login: Wed Apr 25 03:53:13 2018
```

图 1-5-24　直接登录

若可以进行正常的免密码登录则表示配置成功，若提示输入密码则表示配置失败，需要删除用户家目录下的".ssh"目录之后再重新执行整个免密码登录的配置过程。

1.5.7　Hadoop 伪分布模式配置

（1）进入 Hadoop 相关文件的目录，如图 1-5-25 所示，分别创建 Hadoop 的临时文件的存放目录"tmp"、HDFS 的元数据文件的存放目录"name"、以及 HDFS 的数据文件的存放目录"data"。

```
[hadoop@localhost hadoop]$ ls -l
total 0
drwxrwxr-x. 2 hadoop hadoop   6 Apr 25 04:08 data
drwxr-xr-x. 9 hadoop hadoop 149 Aug 17  2016 hadoop-2.7.3
drwxrwxr-x. 2 hadoop hadoop   6 Apr 25 04:08 name
drwxrwxr-x. 2 hadoop hadoop   6 Apr 25 04:08 tmp
```

图 1-5-25　创建目录

（2）Hadoop 的配置文件位于其软件目录的"etc/hadoop"目录下，如图 1-5-26 所示。进入该目录，对其中的配置文件进行编辑。

```
[hadoop@localhost hadoop]$ pwd
/home/hadoop/hadoop/hadoop-2.7.3/etc/hadoop
```

图 1-5-26　进入"etc/hadoop"目录

（3）编辑配置文件"hadoop-env.sh"，在该配置文件中可以对 Hadoop 平台运行中所需要使用到的相关环境变量进行设置。如图 1-5-27 所示，找到其中的配置项"JAVA_HOME"的所在行，将其改为以下的内容：

export JAVA_HOME=JDK 软件目录路径

JDK 软件目录即 JDK 软件包解压解包出来的"jdk1.8.0_131"目录，这里需要书写该目录及其所在的完整绝对路径。

```
19 # The only required environment variable is JAVA_HOME. All others are
20 # optional.  When running a distributed configuration it is best to
21 # set JAVA_HOME in this file, so that it is correctly defined on
22 # remote nodes.
23
24 # The java implementation to use.
25 export JAVA_HOME=/home/hadoop/java/jdk1.8.0_131
```

图 1-5-27　修改配置文件"hadoop-env.sh"

（4）编辑配置文件"core-site.xml"，在该配置文件可以对 Hadoop 平台的一些基本配置项进行设置。在标签"<configuration>"和"</configuration>"之间添加如下的内容：

<configuration>
　<property>
　　<name>fs.defaultFS</name>
　　<value>hdfs：//localhost：9000</value>
　</property>
　<property>
　　<name>hadoop.tmp.dir</name>
　　<value>Hadoop 的临时文件的存放目录"tmp"的路径</value>
　</property>
</configuration>

配置项说明：

fs.defaultFS——指定 HDFS 的文件系统访问路径，即 HDFS 的 NameNode 服务器的地址或主机名，以及访问的端口号。

hadoop.tmp.dir——指定 Hadoop 的临时文件在本地文件系统中的存放路径。

（5）编辑配置文件"hdfs-site.xml"，在该配置文件中可以对 HDFS 相关的配置项进行设置。在标签"<configuration>"和"</configuration>"之间添加如下的内容：

<configuration>
　<property>

         <name>dfs.namenode.name.dir</name>
         <value>HDFS 的 NameNode 节点的元数据文件的存放目录"name"的路径</value>
      </property>
      <property>
         <name>dfs.datanode.data.dir</name>
         <value>HDFS 的 DataNode 节点的数据文件的存放目录"data"的路径</value>
      </property>
      <property>
         <name>dfs.replication</name>
         <value>1</value>
      </property>
</configuration>

配置项说明：

dfs.namenode.name.dir——指定 HDFS 的 NameNode 节点中的元数据文件在本地文件系统中的存放路径。

dfs.datanode.data.dir——指定 HDFS 的 DataNode 节点中的数据文件在本地文件系统中的存放路径。

dfs.replication——指定 HDFS 中数据分块的备份数量，在不对该配置项进行设置的情况下，默认的数据分块的备份数量为 3。由于当前配置的伪分布模式只会有一个存放数据分块的 DataNode 节点，所以将之设置为 1。

（6）将 MapReduce 配置文件的模板文件"mapred-site.xml.template"复制一份并将其重命名为"mapred-site.xml"。然后编辑该配置文件，在该配置文件中可以对 MapReduce 相关的配置项进行设置。在标签"<configuration>"和"</configuration>"之间添加如下的内容：

<configuration>
   <property>
      <name>mapreduce.framework.name</name>
      <value>yarn</value>
   </property>
</configuration>

配置项说明：

mapreduce.framework.name——指定 MapReduce 所使用的外部管理框架，这里使用的是 Hadoop 的 2.7.3 版本自带的 YARN 资源管理器。

（7）编辑配置文件"yarn-env.sh"，在该配置文件中可以对 MapReduce 所使用的外部管理框架 YARN 资源管理器在运行中所需要使用到的相关环境变量进行设置。如图 1-5-28 所示，找到其中的配置项"JAVA_HOME"的所在行，将其改为以下的内容：

export JAVA_HOME=JDK 软件目录路径

JDK 软件目录即 JDK 软件包解压解包出来的"jdk1.8.0_131"目录，这里需要书写该目录及其所在的绝对路径。

```
22 # some Java parameters
23 export JAVA_HOME=/home/hadoop/java/jdk1.8.0_131
```

图 1-5-28　修改配置文件"yarn-env.sh"

（8）编辑配置文件"yarn-site.xml"，在该配置文件中可以对 YARN 资源管理器的相关配置项进行设置。在标签"<configuration>"和"</configuration>"之间添加如下的内容：

    <configuration>
      <property>
        <name>yarn.nodemanager.aux-services</name>
        <value>mapreduce_shuffle</value>
      </property>
    </configuration>

配置项说明：

yarn.nodemanager.aux-services——指定 YARN 的 NodeManager 节点上运行的附属服务的类型。

1.5.8　Hadoop 伪分布模式格式化和启动

（1）使用命令"hadoop namenode -format"对 HDFS 的文件系统进行格式化，如图 1-5-29 所示。若格式化过程中没有出现报错提示信息，则表示格式化过程成功完成。

```
[hadoop@localhost ~]$ hadoop namenode -format
DEPRECATED: Use of this script to execute hdfs command is deprecated.
Instead use the hdfs command for it.
```

项目1 搭建单节点 Hadoop 整合平台

```
op/hadoop-2.7.3/share/hadoop/hdfs/lib/htrace-core-3.1.0-incubating.jar:/home/hadoop/hadoop/hadoop-2.
7.3/share/hadoop/hdfs/lib/commons-daemon-1.0.13.jar:/home/hadoop/hadoop/hadoop-2.7.3/share/hadoop/hd
fs/lib/netty-all-4.0.23.Final.jar:/home/hadoop/hadoop/hadoop-2.7.3/share/hadoop/hdfs/lib/xercesImpl
-2.9.1.jar:/home/hadoop/hadoop/hadoop-2.7.3/share/hadoop/hdfs/lib/xml-apis-1.3.04.jar:/home/hadoop/ha
doop/hadoop-2.7.3/share/hadoop/hdfs/lib/leveldbjni-all-1.8.jar:/home/hadoop/hadoop/hadoop-2.7.3/shar
e/hadoop/hdfs/hadoop-hdfs-2.7.3.jar:/home/hadoop/hadoop/hadoop-2.7.3/share/hadoop/hdfs/hadoop-hdfs-2
.7.3-tests.jar:/home/hadoop/hadoop/hadoop-2.7.3/share/hadoop/hdfs/hadoop-hdfs-nfs-2.7.3.jar:/home/ha
doop/hadoop/hadoop-2.7.3/share/hadoop/yarn/lib/zookeeper-3.4.6-tests.jar:/home/hadoop/hadoop/hadoop-
2.7.3/share/hadoop/yarn/lib/commons-lang-2.6.jar:/home/hadoop/hadoop/hadoop-2.7.3/share/hadoop/yarn/
lib/guava-11.0.2.jar:/home/hadoop/hadoop/hadoop-2.7.3/share/hadoop/yarn/lib/jsr305-3.0.0.jar:/home/h
adoop/hadoop/hadoop-2.7.3/share/hadoop/yarn/lib/commons-logging-1.1.3.jar:/home/hadoop/hadoop/hadoop
-2.7.3/share/hadoop/yarn/lib/protobuf-java-2.5.0.jar:/home/hadoop/hadoop/hadoop-2.7.3/share/hadoop/y
arn/lib/commons-cli-1.2.jar:/home/hadoop/hadoop/hadoop-2.7.3/share/hadoop/yarn/lib/log4j-1.2.17.jar:
/home/hadoop/hadoop/hadoop-2.7.3/share/hadoop/yarn/lib/jaxb-api-2.2.2.jar:/home/hadoop/hadoop/hadoop
-2.7.3/share/hadoop/yarn/lib/stax-api-1.0-2.jar:/home/hadoop/hadoop/hadoop-2.7.3/share/hadoop/yarn/l
ib/activation-1.1.jar:/home/hadoop/hadoop/hadoop-2.7.3/share/hadoop/yarn/lib/commons-compress-1.4.1.
jar:/home/hadoop/hadoop/hadoop-2.7.3/share/hadoop/yarn/lib/xz-1.0.jar:/home/hadoop/hadoop/hadoop-2.7
.3/share/hadoop/yarn/lib/servlet-api-2.5.jar:/home/hadoop/hadoop/hadoop-2.7.3/share/hadoop/yarn/lib/
commons-codec-1.4.jar:/home/hadoop/hadoop/hadoop-2.7.3/share/hadoop/yarn/lib/jetty-util-6.1.26.jar:/
home/hadoop/hadoop/hadoop-2.7.3/share/hadoop/yarn/lib/jersey-core-1.9.jar:/home/hadoop/hadoop/hadoop
-2.7.3/share/hadoop/yarn/lib/jersey-client-1.9.jar:/home/hadoop/hadoop/hadoop-2.7.3/share/hadoop/yar
n/lib/jackson-core-asl-1.9.13.jar:/home/hadoop/hadoop/hadoop-2.7.3/share/hadoop/yarn/lib/jackson-map
per-asl-1.9.13.jar:/home/hadoop/hadoop/hadoop-2.7.3/share/hadoop/yarn/lib/jackson-jaxrs-1.9.13.jar:/
home/hadoop/hadoop/hadoop-2.7.3/share/hadoop/yarn/lib/jackson-xc-1.9.13.jar:/home/hadoop/hadoop/hado
op-2.7.3/share/hadoop/yarn/lib/guice-servlet-3.0.jar:/home/hadoop/hadoop/hadoop-2.7.3/share/hadoop/y
arn/lib/guice-3.0.jar:/home/hadoop/hadoop/hadoop-2.7.3/share/hadoop/yarn/lib/javax.inject-1.jar:/hom
e/hadoop/hadoop/hadoop-2.7.3/share/hadoop/yarn/lib/aopalliance-1.0.jar:/home/hadoop/hadoop/hadoop-2.
7.3/share/hadoop/yarn/lib/commons-io-2.4.jar:/home/hadoop/hadoop/hadoop-2.7.3/share/hadoop/yarn/lib/
jersey-server-1.9.jar:/home/hadoop/hadoop/hadoop-2.7.3/share/hadoop/yarn/lib/asm-3.2.jar:/home/hadoo
p/hadoop/hadoop-2.7.3/share/hadoop/yarn/lib/jersey-json-1.9.jar:/home/hadoop/hadoop/hadoop-2.7.3/sha
re/hadoop/yarn/lib/jettison-1.1.jar:/home/hadoop/hadoop/hadoop-2.7.3/share/hadoop/yarn/lib/jaxb-impl
-2.2.3-1.jar:/home/hadoop/hadoop/hadoop-2.7.3/share/hadoop/yarn/lib/jersey-guice-1.9.jar:/home/hadoo
p/hadoop/hadoop-2.7.3/share/hadoop/yarn/lib/zookeeper-3.4.6.jar:/home/hadoop/hadoop/hadoop-2.7.3/sha
re/hadoop/yarn/lib/netty-3.6.2.Final.jar:/home/hadoop/hadoop/hadoop-2.7.3/share/hadoop/yarn/lib/leve
ldbjni-all-1.8.jar:/home/hadoop/hadoop/hadoop-2.7.3/share/hadoop/yarn/lib/commons-collections-3.2.2.
jar:/home/hadoop/hadoop/hadoop-2.7.3/share/hadoop/yarn/lib/jetty-6.1.26.jar:/home/hadoop/hadoop/hado
op-2.7.3/share/hadoop/yarn/hadoop-yarn-api-2.7.3.jar:/home/hadoop/hadoop/hadoop-2.7.3/share/hadoop/y
arn/hadoop-yarn-common-2.7.3.jar:/home/hadoop/hadoop/hadoop-2.7.3/share/hadoop/yarn/hadoop-yarn-serv
er-common-2.7.3.jar:/home/hadoop/hadoop/hadoop-2.7.3/share/hadoop/yarn/hadoop-yarn-server-nodemanage
r-2.7.3.jar:/home/hadoop/hadoop/hadoop-2.7.3/share/hadoop/yarn/hadoop-yarn-server-web-proxy-2.7.3.ja
r:/home/hadoop/hadoop/hadoop-2.7.3/share/hadoop/yarn/hadoop-yarn-server-applicationhistoryservice-2.
7.3.jar:/home/hadoop/hadoop/hadoop-2.7.3/share/hadoop/yarn/hadoop-yarn-server-resourcemanager-2.7.3.
jar:/home/hadoop/hadoop/hadoop-2.7.3/share/hadoop/yarn/hadoop-yarn-server-tests-2.7.3.jar:/home/hado
op/hadoop/hadoop-2.7.3/share/hadoop/yarn/hadoop-yarn-client-2.7.3.jar:/home/hadoop/hadoop/hadoop-2.7
.3/share/hadoop/yarn/hadoop-yarn-server-sharedcachemanager-2.7.3.jar:/home/hadoop/hadoop/hadoop-2.7.
3/share/hadoop/yarn/hadoop-yarn-applications-distributedshell-2.7.3.jar:/home/hadoop/hadoop/hadoop-2
.7.3/share/hadoop/yarn/hadoop-yarn-applications-unmanaged-am-launcher-2.7.3.jar:/home/hadoop/hadoop/
hadoop-2.7.3/share/hadoop/yarn/hadoop-yarn-registry-2.7.3.jar:/home/hadoop/hadoop/hadoop-2.7.3/share
/hadoop/mapreduce/lib/protobuf-java-2.5.0.jar:/home/hadoop/hadoop/hadoop-2.7.3/share/hadoop/mapreduc
e/lib/avro-1.7.4.jar:/home/hadoop/hadoop/hadoop-2.7.3/share/hadoop/mapreduce/lib/jackson-core-asl-1.
9.13.jar:/home/hadoop/hadoop/hadoop-2.7.3/share/hadoop/mapreduce/lib/jackson-mapper-asl-1.9.13.jar:/
home/hadoop/hadoop/hadoop-2.7.3/share/hadoop/mapreduce/lib/paranamer-2.3.jar:/home/hadoop/hadoop/had
oop-2.7.3/share/hadoop/mapreduce/lib/snappy-java-1.0.4.1.jar:/home/hadoop/hadoop/hadoop-2.7.3/share/
hadoop/mapreduce/lib/commons-compress-1.4.1.jar:/home/hadoop/hadoop/hadoop-2.7.3/share/hadoop/mapred
uce/lib/xz-1.0.jar:/home/hadoop/hadoop/hadoop-2.7.3/share/hadoop/mapreduce/lib/hadoop-annotations-2.
7.3.jar:/home/hadoop/hadoop/hadoop-2.7.3/share/hadoop/mapreduce/lib/commons-io-2.4.jar:/home/hadoop/
hadoop/hadoop-2.7.3/share/hadoop/mapreduce/lib/jersey-core-1.9.jar:/home/hadoop/hadoop/hadoop-2.7.3/
share/hadoop/mapreduce/lib/jersey-server-1.9.jar:/home/hadoop/hadoop/hadoop-2.7.3/share/hadoop/mapre
duce/lib/asm-3.2.jar:/home/hadoop/hadoop/hadoop-2.7.3/share/hadoop/mapreduce/lib/log4j-1.2.17.jar:/h
ome/hadoop/hadoop/hadoop-2.7.3/share/hadoop/mapreduce/lib/netty-3.6.2.Final.jar:/home/hadoop/hadoop/
hadoop-2.7.3/share/hadoop/mapreduce/lib/leveldbjni-all-1.8.jar:/home/hadoop/hadoop/hadoop-2.7.3/shar
e/hadoop/mapreduce/lib/guice-3.0.jar:/home/hadoop/hadoop/hadoop-2.7.3/share/hadoop/mapreduce/lib/jav
ax.inject-1.jar:/home/hadoop/hadoop/hadoop-2.7.3/share/hadoop/mapreduce/lib/aopalliance-1.0.jar:/hom
e/hadoop/hadoop/hadoop-2.7.3/share/hadoop/mapreduce/lib/jersey-guice-1.9.jar:/home/hadoop/hadoop/had
oop-2.7.3/share/hadoop/mapreduce/lib/guice-servlet-3.0.jar:/home/hadoop/hadoop/hadoop-2.7.3/share/ha
doop/mapreduce/lib/junit-4.11.jar:/home/hadoop/hadoop/hadoop-2.7.3/share/hadoop/mapreduce/lib/hamcre
st-core-1.3.jar:/home/hadoop/hadoop/hadoop-2.7.3/share/hadoop/mapreduce/hadoop-mapreduce-client-core
-2.7.3.jar:/home/hadoop/hadoop/hadoop-2.7.3/share/hadoop/mapreduce/hadoop-mapreduce-client-common-2.
7.3.jar:/home/hadoop/hadoop/hadoop-2.7.3/share/hadoop/mapreduce/hadoop-mapreduce-client-shuffle-2.7.
3.jar:/home/hadoop/hadoop/hadoop-2.7.3/share/hadoop/mapreduce/hadoop-mapreduce-client-app-2.7.3.jar:
/home/hadoop/hadoop/hadoop-2.7.3/share/hadoop/mapreduce/hadoop-mapreduce-client-hs-2.7.3.jar:/home/h
adoop/hadoop/hadoop-2.7.3/share/hadoop/mapreduce/hadoop-mapreduce-client-jobclient-2.7.3.jar:/home/h
adoop/hadoop/hadoop-2.7.3/share/hadoop/mapreduce/hadoop-mapreduce-client-hs-2.7.3.jar:/home/h
adoop/hadoop/hadoop-2.7.3/share/hadoop/mapreduce/hadoop-mapreduce-client-jobclient-2.7.3.jar:/home/h
adoop/hadoop/hadoop-2.7.3/share/hadoop/mapreduce/hadoop-mapreduce-client-hs-plugins-2.7.3.jar:/home/
```

项目1 搭建单节点 Hadoop 整合平台

```
hadoop/hadoop/hadoop-2.7.3/share/hadoop/mapreduce/hadoop-mapreduce-examples-2.7.3.jar:/home/hadoop/h
adoop/hadoop-2.7.3/share/hadoop/mapreduce/hadoop-mapreduce-client-jobclient-2.7.3-tests.jar:/home/ha
doop/hadoop/hadoop-2.7.3/contrib/capacity-scheduler/*.jar:/home/hadoop/hadoop/hadoop-2.7.3/contrib/c
apacity-scheduler/*.jar
STARTUP_MSG:   build = https://git-wip-us.apache.org/repos/asf/hadoop.git -r baa91f7c6bc9cb92be5982d
e4719c1c8af91ccff; compiled by 'root' on 2016-08-18T01:41Z
STARTUP_MSG:   java = 1.8.0_131
************************************************************/
18/04/26 04:33:15 INFO namenode.NameNode: registered UNIX signal handlers for [TERM, HUP, INT]
18/04/26 04:33:15 INFO namenode.NameNode: createNameNode [-format]
18/04/26 04:33:17 WARN common.Util: Path /home/hadoop/hadoop/name should be specified as a URI in co
nfiguration files. Please update hdfs configuration.
18/04/26 04:33:17 WARN common.Util: Path /home/hadoop/hadoop/name should be specified as a URI in co
nfiguration files. Please update hdfs configuration.
Formatting using clusterid: CID-6f2ec86d-5abe-4dd7-ad4a-afcb00c35a1b
18/04/26 04:33:18 INFO namenode.FSNamesystem: No KeyProvider found.
18/04/26 04:33:18 INFO namenode.FSNamesystem: fsLock is fair:true
18/04/26 04:33:18 INFO blockmanagement.DatanodeManager: dfs.block.invalidate.limit=1000
18/04/26 04:33:18 INFO blockmanagement.DatanodeManager: dfs.namenode.datanode.registration.ip-hostna
me-check=true
18/04/26 04:33:18 INFO blockmanagement.BlockManager: dfs.namenode.startup.delay.block.deletion.sec i
s set to 000:00:00:00.000
18/04/26 04:33:18 INFO blockmanagement.BlockManager: The block deletion will start around 2018 Apr 2
6 04:33:18
18/04/26 04:33:18 INFO util.GSet: Computing capacity for map BlocksMap
18/04/26 04:33:18 INFO util.GSet: VM type       = 64-bit
18/04/26 04:33:18 INFO util.GSet: 2.0% max memory 966.7 MB = 19.3 MB
18/04/26 04:33:18 INFO util.GSet: capacity      = 2^21 = 2097152 entries
18/04/26 04:33:18 INFO blockmanagement.BlockManager: dfs.block.access.token.enable=false
18/04/26 04:33:18 INFO blockmanagement.BlockManager: defaultReplication         = 1
18/04/26 04:33:18 INFO blockmanagement.BlockManager: maxReplication             = 512
18/04/26 04:33:18 INFO blockmanagement.BlockManager: minReplication             = 1
18/04/26 04:33:18 INFO blockmanagement.BlockManager: maxReplicationStreams      = 2
18/04/26 04:33:18 INFO blockmanagement.BlockManager: replicationRecheckInterval = 3000
18/04/26 04:33:18 INFO blockmanagement.BlockManager: encryptDataTransfer        = false
18/04/26 04:33:18 INFO blockmanagement.BlockManager: maxNumBlocksToLog          = 1000
18/04/26 04:33:18 INFO namenode.FSNamesystem: fsOwner             = hadoop (auth:SIMPLE)
18/04/26 04:33:18 INFO namenode.FSNamesystem: supergroup          = supergroup
18/04/26 04:33:18 INFO namenode.FSNamesystem: isPermissionEnabled = true
18/04/26 04:33:18 INFO namenode.FSNamesystem: HA Enabled: false
18/04/26 04:33:18 INFO namenode.FSNamesystem: Append Enabled: true
18/04/26 04:33:19 INFO util.GSet: Computing capacity for map INodeMap
18/04/26 04:33:19 INFO util.GSet: VM type       = 64-bit
18/04/26 04:33:19 INFO util.GSet: 1.0% max memory 966.7 MB = 9.7 MB
18/04/26 04:33:19 INFO util.GSet: capacity      = 2^20 = 1048576 entries
18/04/26 04:33:19 INFO namenode.FSDirectory: ACLs enabled? false
18/04/26 04:33:19 INFO namenode.FSDirectory: XAttrs enabled? true
18/04/26 04:33:19 INFO namenode.FSDirectory: Maximum size of an xattr: 16384
18/04/26 04:33:19 INFO namenode.NameNode: Caching file names occuring more than 10 times
18/04/26 04:33:19 INFO util.GSet: Computing capacity for map cachedBlocks
18/04/26 04:33:19 INFO util.GSet: VM type       = 64-bit
18/04/26 04:33:19 INFO util.GSet: 0.25% max memory 966.7 MB = 2.4 MB
18/04/26 04:33:19 INFO util.GSet: capacity      = 2^18 = 262144 entries
18/04/26 04:33:19 INFO namenode.FSNamesystem: dfs.namenode.safemode.threshold-pct = 0.9990000127460
33
18/04/26 04:33:19 INFO namenode.FSNamesystem: dfs.namenode.safemode.min.datanodes = 0
18/04/26 04:33:19 INFO namenode.FSNamesystem: dfs.namenode.safemode.extension     = 30000
18/04/26 04:33:19 INFO metrics.TopMetrics: NNTop conf: dfs.namenode.top.window.num.buckets = 10
18/04/26 04:33:19 INFO metrics.TopMetrics: NNTop conf: dfs.namenode.top.num.users = 10
18/04/26 04:33:19 INFO metrics.TopMetrics: NNTop conf: dfs.namenode.top.windows.minutes = 1,5,25
18/04/26 04:33:19 INFO namenode.FSNamesystem: Retry cache on namenode is enabled
18/04/26 04:33:19 INFO namenode.FSNamesystem: Retry cache will use 0.03 of total heap and retry cach
e entry expiry time is 600000 millis
18/04/26 04:33:19 INFO util.GSet: Computing capacity for map NameNodeRetryCache
18/04/26 04:33:19 INFO util.GSet: VM type       = 64-bit
18/04/26 04:33:19 INFO util.GSet: 0.029999999329447746% max memory 966.7 MB = 297.0 KB
18/04/26 04:33:19 INFO util.GSet: capacity      = 2^15 = 32768 entries
18/04/26 04:33:19 INFO namenode.FSImage: Allocated new BlockPoolId: BP-904689104-127.0.0.1-152473159
9271
18/04/26 04:33:19 INFO common.Storage: Storage directory /home/hadoop/hadoop/name has been successfu
lly formatted.
18/04/26 04:33:19 INFO namenode.FSImageFormatProtobuf: Saving image file /home/hadoop/hadoop/name/cu
rrent/fsimage.ckpt_0000000000000000000 using no compression
18/04/26 04:33:19 INFO namenode.FSImageFormatProtobuf: Image file /home/hadoop/hadoop/name/current/f
```

```
simage.ckpt_0000000000000000000 of size 353 bytes saved in 0 seconds.
18/04/26 04:33:19 INFO namenode.NNStorageRetentionManager: Going to retain 1 images with txid >= 0
18/04/26 04:33:19 INFO util.ExitUtil: Exiting with status 0
18/04/26 04:33:19 INFO namenode.NameNode: SHUTDOWN_MSG:
/************************************************************
SHUTDOWN_MSG: Shutting down NameNode at localhost/127.0.0.1
************************************************************/
```

图 1-5-29　对 HDFS 文件系统格式化

（2）使用命令"start-dfs.sh"启动 HDFS 文件系统，如图 1-5-30 所示。

```
[hadoop@localhost ~]$ start-dfs.sh
Starting namenodes on [localhost]
localhost: starting namenode, logging to /home/hadoop/hadoop/hadoop-2.7.3/logs/hadoop-hadoop-namenode-localhost.localdomain.out
localhost: starting datanode, logging to /home/hadoop/hadoop/hadoop-2.7.3/logs/hadoop-hadoop-datanode-localhost.localdomain.out
Starting secondary namenodes [0.0.0.0]
0.0.0.0: starting secondarynamenode, logging to /home/hadoop/hadoop/hadoop-2.7.3/logs/hadoop-hadoop-secondarynamenode-localhost.localdomain.out
```

图 1-5-30　启动 HDFS 文件系统

（3）使用命令"jps"查看 Java 进程信息，如图 1-5-31 所示，若有分别名为"NameNode"、"SecondaryNameNode"、"DataNode"的三个进程，则表示 HDFS 文件系统启动成功。

```
[hadoop@localhost ~]$ jps
9122 NameNode
9220 DataNode
9636 Jps
9398 SecondaryNameNode
```

图 1-5-31　查看 Java 进程信息

（4）使用命令"start-yarn.sh"启动 YARN 资源管理器，如图 1-5-32 所示，YARN 资源管理器启动之后便可以使用 MapReduce 的相关功能。

```
[hadoop@localhost ~]$ start-yarn.sh
starting yarn daemons
starting resourcemanager, logging to /home/hadoop/hadoop/hadoop-2.7.3/logs/yarn-hadoop-resourcemanager-localhost.localdomain.out
localhost: starting nodemanager, logging to /home/hadoop/hadoop/hadoop-2.7.3/logs/yarn-hadoop-nodemanager-localhost.localdomain.out
```

图 1-5-32　启动 YARN 资源管理器

（5）使用命令"jps"查看 Java 进程信息，如图 1-5-33 所示，若有名为"ResourceManager""NodeManager"的两个进程，则表示 YARN 资源管理器启动成功。

```
[hadoop@localhost ~]$ jps
9122 NameNode
9220 DataNode
9876 NodeManager
9398 SecondaryNameNode
9783 ResourceManager
10172 Jps
```

图 1-5-33　查看 Java 进程信息

1.5.9 Hadoop 伪分布模式验证

（1）通过在 Hadoop 中创建目录和运行 Hadoop 自带的示例程序，可以验证安装和配置的 Hadoop 伪分布模式平台中的 HDFS 和 MapReduce 是否能够正常使用。

（2）首先验证 HDFS 是否能够正常使用。使用命令"hadoop fs -mkdir /test"在 HDFS 中创建一个测试目录。

（3）使用命令"hadoop fs -ls /"查看 HDFS 中的根目录下的文件和目录，如图 1-5-34 所示，若可以看到刚才创建的"test"目录，则代表 HDFS 文件系统可以正常使用。

```
[hadoop@localhost ~]$ hadoop fs -ls /
Found 1 items
drwxr-xr-x   - hadoop supergroup          0 2018-04-27 04:26 /test
```

图 1-5-34　查看文件和目录

（4）然后验证 MapReduce 是否能够正常使用。Hadoop 的示例程序软件包位于其软件所在目录下的"share/hadoop/mapreduce"目录中，如图 1-5-35 所示，示例程序软件包包名为"hadoop-mapreduce-examples-2.7.3.jar"，使用命令"hadoop jar 示例程序软件包路径 pi 2 1000"来运行蒙地卡罗法计算圆周率 PI 值的示例程序。在"pi"参数之后的第一个参数表示程序运行 map 任务的次数，第二个参数表示每个 map 任务的取样个数。如图 1-5-36 所示，若示例程序运行过程中没有报错，并且正确的得出圆周率 PI 值得结果，则代表 MapReduce 可以正常使用。

```
[hadoop@localhost mapreduce]$ pwd
/home/hadoop/hadoop/hadoop-2.7.3/share/hadoop/mapreduce
[hadoop@localhost mapreduce]$ ls -l
total 4972
-rw-r--r--. 1 hadoop hadoop  537521 Jul 12 08:43 hadoop-mapreduce-client-app-2.7.3.jar
-rw-r--r--. 1 hadoop hadoop  773501 Jul 12 08:43 hadoop-mapreduce-client-common-2.7.3.jar
-rw-r--r--. 1 hadoop hadoop 1554595 Jul 12 08:43 hadoop-mapreduce-client-core-2.7.3.jar
-rw-r--r--. 1 hadoop hadoop  189714 Jul 12 08:43 hadoop-mapreduce-client-hs-2.7.3.jar
-rw-r--r--. 1 hadoop hadoop   27598 Jul 12 08:43 hadoop-mapreduce-client-hs-plugins-2.7.3.jar
-rw-r--r--. 1 hadoop hadoop   61745 Jul 12 08:43 hadoop-mapreduce-client-jobclient-2.7.3.jar
-rw-r--r--. 1 hadoop hadoop 1551594 Jul 12 08:43 hadoop-mapreduce-client-jobclient-2.7.3-tests.jar
-rw-r--r--. 1 hadoop hadoop   71310 Jul 12 08:43 hadoop-mapreduce-client-shuffle-2.7.3.jar
-rw-r--r--. 1 hadoop hadoop  295812 Jul 12 08:43 hadoop-mapreduce-examples-2.7.3.jar
drwxr-xr-x. 2 hadoop hadoop    4096 Jul 12 08:43 lib
drwxr-xr-x. 2 hadoop hadoop      30 Jul 12 08:43 lib-examples
drwxr-xr-x. 2 hadoop hadoop    4096 Jul 12 08:43 sources
```

图 1-5-35　查看示例程序软件包

```
[hadoop@localhost ~]$ hadoop jar hadoop/hadoop-2.7.3/share/hadoop/mapreduce/hadoop-mapreduce-example
s-2.7.3.jar pi 2 1000
Number of Maps  = 2
Samples per Map = 1000
Wrote input for Map #0
Wrote input for Map #1
Starting Job
18/04/28 04:38:04 INFO client.RMProxy: Connecting to ResourceManager at /0.0.0.0:8032
18/04/28 04:38:06 INFO input.FileInputFormat: Total input paths to process : 2
18/04/28 04:38:06 INFO mapreduce.JobSubmitter: number of splits:2
18/04/28 04:38:07 INFO mapreduce.JobSubmitter: Submitting tokens for job: job_1524900953431_0002
18/04/28 04:38:08 INFO impl.YarnClientImpl: Submitted application application_1524900953431_0002
18/04/28 04:38:08 INFO mapreduce.Job: The url to track the job: http://localhost:8088/proxy/applicat
ion_1524900953431_0002/
18/04/28 04:38:08 INFO mapreduce.Job: Running job: job_1524900953431_0002
18/04/28 04:38:35 INFO mapreduce.Job: Job job_1524900953431_0002 running in uber mode : false
18/04/28 04:38:35 INFO mapreduce.Job:  map 0% reduce 0%
18/04/28 04:39:46 INFO mapreduce.Job:  map 100% reduce 0%
18/04/28 04:40:44 INFO mapreduce.Job:  map 100% reduce 100%
18/04/28 04:40:46 INFO mapreduce.Job: Job job_1524900953431_0002 completed successfully
18/04/28 04:40:47 INFO mapreduce.Job: Counters: 49
        File System Counters
                FILE: Number of bytes read=50
                FILE: Number of bytes written=356931
                FILE: Number of read operations=0
                FILE: Number of large read operations=0
                FILE: Number of write operations=0
                HDFS: Number of bytes read=530
                HDFS: Number of bytes written=215
                HDFS: Number of read operations=11
                HDFS: Number of large read operations=0
                HDFS: Number of write operations=3
        Job Counters
                Launched map tasks=2
                Launched reduce tasks=1
                Data-local map tasks=2
                Total time spent by all maps in occupied slots (ms)=158014
                Total time spent by all reduces in occupied slots (ms)=28067
                Total time spent by all map tasks (ms)=158014
                Total time spent by all reduce tasks (ms)=28067
                Total vcore-milliseconds taken by all map tasks=158014
                Total vcore-milliseconds taken by all reduce tasks=28067
                Total megabyte-milliseconds taken by all map tasks=161806336
                Total megabyte-milliseconds taken by all reduce tasks=28740608
        Map-Reduce Framework
                Map input records=2
                Map output records=4
                Map output bytes=36
                Map output materialized bytes=56
                Input split bytes=294
                Combine input records=0
                Combine output records=0
                Reduce input groups=2
                Reduce shuffle bytes=56
                Reduce input records=4
                Reduce output records=0
                Spilled Records=8
                Shuffled Maps =2
                Failed Shuffles=0
                Merged Map outputs=2
                GC time elapsed (ms)=1597
                CPU time spent (ms)=5840
                Physical memory (bytes) snapshot=452407296
                Virtual memory (bytes) snapshot=6234807424
                Total committed heap usage (bytes)=263901184
        Shuffle Errors
                BAD_ID=0
                CONNECTION=0
                IO_ERROR=0
                WRONG_LENGTH=0
                WRONG_MAP=0
                WRONG_REDUCE=0
        File Input Format Counters
                Bytes Read=236
        File Output Format Counters
                Bytes Written=97
Job Finished in 162.945 seconds
Estimated value of Pi is 3.14400000000000000000
```

图 1-5-36　计算你圆周率

1.5.10　安装与配置的常见问题及解决方法

Hadoop 平台的安装和配置过程中的常见问题主要就是节点进程启动失败，而导致这问题的原因，大多时候都是由于配置过程中输入错误或漏掉某个步骤造成的。出现问题或错误时可以先检查配置文件中的配置项的名称和值是否书写正确，比如主机名、指定功能目录、配置文件格式等。然后检查整个安装和配置过程中的步骤是否有所遗漏，比如没有配置本机的免密码登录，没有对 HDFS 文件系统进行格式化等。

另一种查找问题原因的更主要方式是查看日志文件，从日志信息的内容来查看导致服务节点启动失败的错误信息，从而为找到问题的原因，并为找到问题的解决方法提供帮助。Hadoop 的日志文件在没有手动进行配置的情况下，默认位于 Hadoop 自身软件目录下的"logs"目录中。在该日志文件目录下，每个服务节点都有自己的日志文件，如图 1-5-37 所示，其中 HDFS 的三个服务节点的日志文件名称格式为"hadoop-用户名-服务节点名-主机名.log"，而 YARN 的两个服务节点的日志文件的名称格式为"yarn-用户名-服务节点名-主机名.log"。

图 1-5-37　hadoop 日志文件

另一种和查看日志文件类似的查找问题原因的方法是使用对应服务节点的命令来以前台运行的方式单独启动该服务节点。通过该方式启动服务节点，会在控制台中打印出启动过程的详细信息，这些信息其实和日志文件中的日志信息是一样的。通过这些信息可以查看导致服务节点启动失败的原因，从而为找到解决问题的方法提供帮助。

服务节点"NameNode"对应的启动命令为"hadoop　namenode"。

服务节点"SecondaryNameNode"对应的启动命令为"hadoop secondary namenode"。

服务节点"DataNode"对应的启动命令为"hadoop　datanode"。

服务节点"ResourceManager"对应的启动命令为"yarn　resourcemanager"。

服务节点"NodeManager"对应的启动命令为"yarn nodemanager"。

若是因为配置文件书写错误而修改了任意配置文件中的内容，特别是修改了主机名、文件目录路径等方面的内容，建议将创建的三个功能目录"tmp""name""data"中的内容清空，并重新将 HDFS 文件系统进行格式化之后，再启动 Hadoop 平台服务。

任务 1.6　安装 HBase 伪分布模式

1.6.1　HBase

HBase 是 Apache Hadoop 的子项目，是一个分布式的、面向列的开源数据库，其实现原理源自 Google 在 2006 年发表的有关于 BigTable 的一篇论文，可以认为 HBase 是 BigTable 的一个开源实现版本。

HBase 依托于 Hadoop 的 HDFS 作为基本存储单元，并且不同于一般的关系型数据库，是一个适合于非结构化数据存储的非关系型数据库。另一个与关系型数据库不同的是，HBase 是基于列的模式而不是基于行的模式。HBase 非常适合于存储大表数据，表的规模可以达到数十亿行以及数百万列，并且对大表数据的读写访问可以达到实时级别。

1.6.1.1　HBase 的架构

HBase 需要运行在 Hadoop 平台的 HDFS 之上，以 HDFS 作为其基础的存储设施。HBase 上层提供了访问数据的 Java API 层，供应用访问存储在 HBase 的数据。在 HBase 的集群中主要由 Master 和 Region Server 组成，并使用 Zookeeper 作为集群的协调器，具体架构如图 1-6-1 所示。

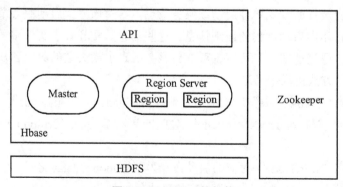

图 1-6-1　HBase 的架构

1. Master

HBase 的 Master 节点用于协调多个 Region Server 节点，侦测各个 Region Server 节点之间的状态，同时平衡各个 Region Server 节点之间的负载。Master 节点还有一个职责就是负责分配 Region 给 Region Server 节点。HBase 允许多个 Master 节点共存，但是这需要 Zookeeper 的帮助。不过当多个 Master 节点共存时，只有一个主 Master 节点是用于提供服务的，其他的备用 Master 节点都是处于待命的状态。当正在工作的主 Master 节点宕机时，其他的备用 Master 节点则会通过选举出新的主 Master 节点接管 HBase 集群来提供服务。

2. Region Server

对于一个 Region Server 节点而言，其中包括了多个 Region。Region Server 节点的作用只是管理表格，以及实现 HBase 的读写操作。Client 会直接连接到 Region Server 节点，并与其通信获取 HBase 中的数据。对于 Region 而言，则是真实存放 HBase 数据的地方，也就说 Region 是 HBase 的可用性和分布式的基本单位。如果当 HBase 中的一个表格很大，并由多个 Column Family 组成时，那么表的数据将存放在多个 Region 之中，并且在每个 Region 中会关联多个存储单元（Store）。

3. Zookeeper

对于 HBase 而言，Zookeeper 的作用是至关重要的。首先，Zookeeper 是作为 HBase 的 Master 节点的 HA 解决方案。也就是说，是 Zookeeper 保证了至少有一个 HBase 的 Master 节点处于运行状态，同时 Zookeeper 还负责 Region Server 节点和 Region 的注册。其实 Zookeeper 发展到目前为止，已经逐渐成为分布式大数据框架中容错性的标准框架。不光是 HBase，几乎所有的分布式大数据相关的开源框架，都依赖于 Zookeeper 实现 HA。

1.6.1.2 HBase 的数据模型

HBase 基于 BigTable 的原理实现，使用的也是和 BigTable 相同的数据模型。用户存储数据行在一个表里。一个数据行里拥有一个可选择的键和任意数量的列，一个或多个列组成一个 Column Family，一个 Column Family 下的列位于一个 HFile 中，易于缓存数据。表是疏松存储的，因此用户可以给行定义各种不同的列。在 HBase 中数据按主键排序，同时表按主键划分为多个 Region。具体的数据模型逻辑视图如图 1-6-2 所示。

Row Key	Timestamp	Column Family 1		Column Family 1		
		URI	Content	Column 1	Column 2	
row 1	t2	www.huawei.com	"<html>"	...	...	Region
	t1	www.huawei.com	"<html>"	...	...	
...	...	...	...	...	...	
row M						
row M+1	t1	...	...	...	...	Region
row M+2	t3	...	...	...	...	
	t2	...	...	...	...	
	t1	...	...	...	...	
...		...	...	...	...	
row N	t1	...	...	...	...	Region
...						

图 1-6-2　HBase 的数据模型

1. Table（表格）

一个 HBase 表格由多个行（Row）组成。

2. Row（行）

HBase 中的行里面包含一个行键（Row Key）、一个或者多个列（Column），行按照行键的字母顺序存储在表格中。

3. Row Key

表中每条记录的主键，是一个二进制类型的数据。在数据存储中，相似的数据一般要存储在一起，这样才能更为高效地进行数据检索，所以行键的设计非常重要。一个常见的 HBase 行键设计的例子是网站信息中的网络域名，一般是将域名进行反转之后再进行存储，如将"www.apache.org"存储为"org.apache.www"、将"mail.apache.org"存储为"org.apache.mail"。这样的话，所有 Apache 相关的网络域名便会存储在一起，否则按照原域名的形式进行存储，数据将分散在表中各处。

4. Column（列）

HBase 中的列使用列族名和列限定符来表示，两者之间用":"符号分隔开。

5. Column Family（列族）

列族拥有一个名称，在物理上包含了一组列和它们的值，也可以认为是一级列。每一个列族还拥有一系列的存储属性,例如值是否缓存在内存中、数据是否要压缩、行键是否要加密等。表格中的每一行拥有相同的列族,但一个给定的行可以不存储任何数据在一个给定的列族中。

6. Column Qualifier(列限定符)

列限定符是列族中数据的索引,也可以认为是二级列。数据通过列限定符与列族名组成的"列族名:列限定符"格式的二进制数据来进行映射,也可以将列限定符理解为存储数据的键值对中的键。相对于列族在创建表格时便是确定的,列限定符及其对应的值是可以动态的增加和删除的。所以表中的每一行具有相同的列族,但每一行的列族里面的列限定符可能会有很大差异。

7. Timestamp(时间戳)

时间戳是写在值旁边的一个用于区分值的版本的数据,是一个 Long 型的数值类型数据。默认情况下,时间戳的值是数据写入时 Region Sever 节点的当前时间,但也可以在写入数据时手动指定一个不同的时间戳。

8. Cell(单元)

单元是由行、列族、列限定符、值、时间戳组成,其中的值为二进制数据。

将逻辑视图中的内容对应到现实中的物理节点和文件,则每个列族会被存储在 HDFS 上的一个单独文件之中,其中空值不会被保存,而行键和时间戳在每个列族中均会保存一份。HBase 会为每个值维护多级的索引,即表在行的方向上会被分割为多个 Region。Region 是 HBase 中分布式存储和负载均衡的最小单元,其按照大小进行分割,并且在因数据增多而增大到一个阈值的时候,被分成两个新的 Region。

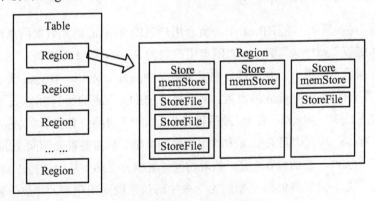

图 1-6-3　表的逻辑存储结构

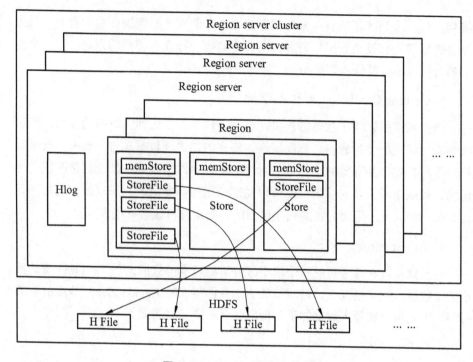

图 1-6-4 Region 的存储结构

Region 虽然是分布式存储的最小单元,但并不是存储的最小单元。如图 1-6-3 和图 1-6-4 所示,每个 Region 中包含了多个 Store 对象,每个 Store 又包含了一个 MemStore 以及若干个 StoreFile,而 StoreFile 又包含了一个或多个 HFile。其中 MemStore 存放在内存中,StoreFile 存储在 HDFS 上。

1.6.1.3 HBase 的元数据

如图 1-6-5 所示,在 HBase 中,所有用户表的 Region 的元数据信息都会被存储在系统表".META."表中。而随着用户表的 Region 数量的增多,".META."表中的数据量也会增大,并逐渐分裂成为多个 Region。所以,HBase 为了定位".META."表的各个 Region 的位置,又会把".META."表的所有 Region 的元数据保存在系统表"-ROOT-"表中。最后再将"-ROOT-"表的位置记录在 Zookeeper 之中。所有客户端在访问 HBase 中的用户数据之前,都需要首先访问 Zookeeper 来获得"-ROOT-"表的位置信息。然后访问"-ROOT-"表,从中获得".META."表的位置信息。最后再根据".META."表中的信息确定用户表的存放位置。

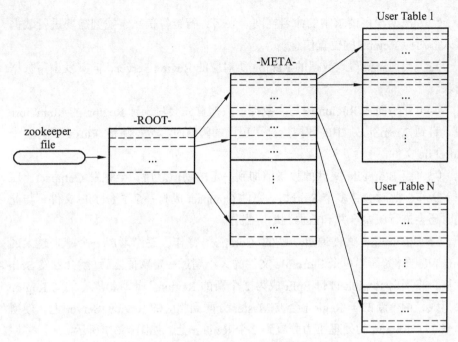

图 1-6-5 HBase 的元数据

1.6.1.4 HBase 的写操作

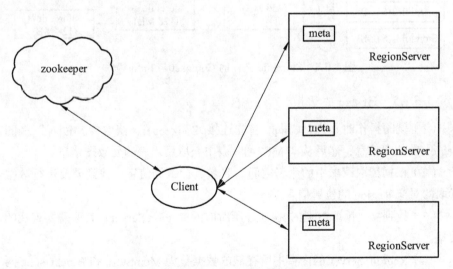

图 1-6-6 HBase 的写操作

（1）Client 首先连接 Zookeeper 获取".META."表的 Region 的位置信息，然后从".META."表中获取用户表的元数据信息。

（2）根据写入请求中的命名空间、表名、行键等信息，找到需要进行数据写入的对应 Region 的位置信息。

（3）找到需要写入数据的 Region 所对应的 Region Server，向其发出写数据请求。

（4）开始向在 Region 中写入数据。数据首先会写入到 Region 的 MemStore 中，直到 MemStore 中的数据达到了预设的阈值时，数据将被 Flush 成为一个 StoreFile 文件。

（5）当 StoreFile 文件的数量增加到一定的阈值之后，会触发 Compact（压缩）操作。此时会将多个 StoreFile 文件 Compact 成为一个 StoreFile 文件，与此同时还会进行版本合并和数据删除等操作。

（6）StoreFile 文件通过不断的 Compact 操作，逐渐形成一个越来越大的 StoreFile 文件。而当单个 StoreFile 文件的大小超过一定阈值之后，就会触发 Split 操作，将当前 Region 进行 Split 成为 2 个新的 Region。此时，原有的父 Region 会下线，两个新的子 Region 会被 Master 分配到相应的 Region Server 上，使得原来 1 个 Region 的处理压力分流到 2 个 Region 上，如图 1-6-7 所示。

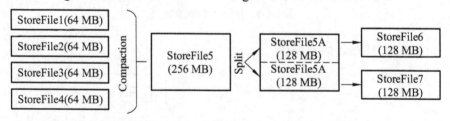

图 1-6-7　StoreFile 文件的 Compact 与 Split 操作

1.6.1.5　HBase 的读操作

（1）同写操作时一样，Client 首先连接 Zookeeper，获取 ".META." 表的 Region 的位置信息，然后从 ".META." 表中获取用户表的元数据信息。

（2）根据读取请求中的命名空间、表名、行键等信息，找到需要进行数据读取的对应 Region 的位置信息。

（3）找到需要读取数据的 Region 所对应的 Region Server，获取需要查找的数据。

（4）Region Server 的内存中所存储的数据分为 MemStore 和 BlockCache 两部分，其中 MemStore 主要用于写数据，而 BlockCache 主要用于读数据。读操作会先在 MemStore 中查找数据，若找不到就到 BlockCache 中查找，再找不到才会从 StoreFile 中读取，同时会并把读取的结果放入到 BlockCache 之中。

项目 1　搭建单节点 Hadoop 整合平台

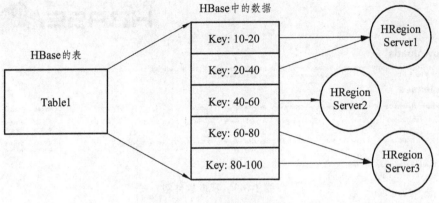

图 1-6-8　HBase 的读操作

1.6.2　HBase 安装规划

1. 环境要求

本地磁盘剩余空间 400M 以上；
已安装 CentOS 7 1611 64 位操作系统；
已安装 JDK 的 1.8.0_131 版本；
已安装 Zookeeper 的 3.4.9 版本；
已安装 Hadoop 的 2.7.3 版本。

2. 软件版本

图 1-6-9 获取软件包

选用 HBase 的 1.2.3 版本，软件包名为 hbase-1.2.3-bin.tar.gz，该软件包可以在 HBase 项目位于 Apache 的官方网站（http：//hbase.apache.org）的 Downloads 页面（http：//hbase.apache.org/downloads.html）获取，如图 1-6-9 所示。

3. 服务规划

（1）Hadoop 的伪分布模式实际上是在单台主机上使用多个不同的进程来模拟出不同类型的多个服务节点。

（2）HBase 的服务包括 Master 节点和 Region 节点共两个类型的服务节点，伪分布模式下只需要满足每个类型的服务节点进程至少有 1 个即可。

1.6.3 HBase 基础环境配置

（1）创建"hbase"目录用于存放 HBase 相关文件，该目录可自行选择创建位置，创建完成后将当前工作目录切换到该目录。

（2）使用命令"tar -xzf HBase 安装包路径"将软件包解压解包到"hbase"目录下，如图 1-6-10 所示，解压解包出来的目录名称为"hbase-1.2.3"。

```
[hadoop@localhost hbase]$ ls -l
total 0
drwxrwxr-x. 7 hadoop hadoop 160 May  3 22:58 hbase-1.2.3
```

图 1-6-10 解压软件包

（3）在用户的配置文件".bash_profile"中配置 HBase 相关的环境变量，如图 1-6-11 所示，在文件末尾添加以下内容：

# hbase environment
HBASE_HOME=（HBase 软件目录路径）
PATH=$HBASE_HOME/bin：$PATH
export HBASE_HOME PATH

HBase 软件目录即 HBase 软件包解压解包出来的"hbase-1.2.3"目录，这里需要书写该目录及其所在的完整绝对路径。

```
# hbase environment
HBASE_HOME=/home/hadoop/hbase/hbase-1.2.3
PATH=$HBASE_HOME/bin:$PATH
export HBASE_HOME PATH
```

图 1-6-11　修改".bash_profile"配置文件

（4）使用命令"source ~/.bash_profile"使新配置的环境变量立即生效。

（5）使用命令"echo $变量名"查看新添加和修改的环境变量的值是否正确，如图 1-6-12 所示。

```
[hadoop@localhost ~]$ echo $HBASE_HOME
/home/hadoop/hbase/hbase-1.2.3
```

图 1-6-12　查看环境变量

（6）使用命令"hbase version"验证 Hadoop 的安装配置是否成功。如图 1-6-13 所示，若显示有 HBase 的版本信息，并且与所使用的 HBase 软件版本相符，则表示 HBase 的基本安装已经成功，可以正常使用。

```
[hadoop@localhost ~]$ hbase version
HBase 1.2.3
Source code repository git://kalashnikov.att.net/Users/stack/checkouts/hbase.git.commit revision=bd6
3744624a26dc3350137b564fe746df7a721a4
Compiled by stack on Mon Aug 29 15:13:42 PDT 2016
From source with checksum 0ca49367ef6c3a680888bbc4f1485d18
```

图 1-6-13　验证是否安装成功

1.6.4　HBase 伪分布模式配置

（1）进入 HBase 相关文件的目录，如图 1-6-14 所示，分别创建 HBase 的临时文件的存放目录"tmp"和 HBase 的日志文件的存放目录"logs"。

```
[hadoop@localhost hbase]$ ls -l
total 0
drwxrwxr-x. 7 hadoop hadoop 160 May  3 22:58 hbase-1.2.3
drwxrwxr-x. 2 hadoop hadoop   6 May  3 23:19 logs
drwxrwxr-x. 2 hadoop hadoop   6 May  3 23:19 tmp
```

图 1-6-14　创建目录

（2）HBase 的配置文件位于其软件目录的"conf"目录下，如图 1-6-15 所示。进入该目录，对其中的配置文件进行编辑。

```
[hadoop@localhost conf]$ pwd
/home/hadoop/hbase/hbase-1.2.3/conf
```

图 1-6-15　编辑 HBase 配置文件

（3）编辑配置文件"hbase-env.sh"，在该配置文件中可以对 HBase 数据库运行中所需要使用到的相关环境变量进行设置，如图 1-6-16 所示，找到下列配置项的所在行并修改其对应内容：

export　JAVA_HOME=JDK 软件目录路径

export　HBASE_CLASSPATH=Hadoop 软件目录路径/etc/hadoop

export　HBASE_LOG_DIR=HBase 的日志文件的存放目录"logs"的路径

export　HBASE_MANAGES_ZK=false

```
26 # The java implementation to use.  Java 1.7+ required.
27 export JAVA_HOME=/home/hadoop/java/jdk1.8.0_131

29 # Extra Java CLASSPATH elements.  Optional.
30 export HBASE_CLASSPATH=/home/hadoop/hadoop/hadoop-2.7.3/etc/hadoop

104 # Where log files are stored.  $HBASE_HOME/logs by default.
105 export HBASE_LOG_DIR=/home/hadoop/hbase/logs

127 # Tell HBase whether it should manage it's own instance of Zookeeper or not.
128 export HBASE_MANAGES_ZK=false
```

图 1-6-16　编辑配置文件

配置项说明：

JAVA_HOME——指定本地系统中 JDK 软件的所在路径。

HBASE_CLASSPATH——指定本地系统中 Hadoop 软件的配置文件的所在路径。

HBASE_LOG_DIR——指定 HBase 的日志文件在本地系统中的存放路径。

HBASE_MANAGES_ZK——指定 HBase 是否使用自带的 Zookeeper 组件，在需要使用独立安装的 Zookeeper 软件的情况下，该项必须设置为关闭状态。

（4）编辑配置文件"hbase-site.xml"，在该配置文件中可以对 HDFS 相关的配置项进行设置。在标签"<configuration>"和"</configuration>"之间添加如下的内容：

<configuration>

　　<property>

```xml
    <name>hbase.rootdir</name>
    <value>hdfs://localhost:9000/user/hadoop/hbase</value>
</property>
<property>
    <name>hbase.tmp.dir</name>
    <value>HBase 的临时文件的存放目录"tmp"的路径</value>
</property>
<property>
    <name>hbase.cluster.distributed</name>
    <value>true</value>
</property>
<property>
    <name>hbase.zookeeper.quorum</name>
    <value>localhost:2181, localhost:2182, localhost:2183</value>
</property>
</configuration>
```

配置项说明：

hbase.rootdir——指定 HBase 数据的存放路径，一般指定为 HDFS 文件系统中的路径，也可以使用本地文件系统中的路径。

hbase.tmp.dir——指定 HBase 的临时文件在本地系统中的存放路径。

hbase.cluster.distributed——指定 HBase 的分布式集群模式的开启状态，该项关闭的情况下是 HBase 的单机模式而不是伪分布模式。

hbase.zookeeper.quorum——指定 Zookeeper 的连接地址及端口号，多个用"，"分隔开，这里的连接地址及端口号的值要与前面 Zookeeper 的安装中配置的。

1.6.5 HBase 伪分布模式启动和验证

（1）在启动 HBase 数据库之前需要保证 Zookeeper 和 Hadoop 的服务都已经启动运行并能够正常使用。然后如图 1-6-17 所示，使用命令"start-hbase.sh"启动 HBase 数据库。

```
[hadoop@localhost ~]$ start-hbase.sh
starting master, logging to /home/hadoop/hbase/logs/hbase-hadoop-master-localhost.localdomain.out
starting regionserver, logging to /home/hadoop/hbase/logs/hbase-hadoop-1-regionserver-localhost.loca
ldomain.out
```

图 1-6-17　启动 HBase 数据库

（2）使用命令“jps”查看 Java 进程信息，如图 1-6-18 所示，若有名为“HMaster”“HRegionServer”的两个进程，则表示 HBase 数据库启动成功。Java 进程信息中还会包含有 Zookeeper 的三个“QuorumPeerMain”服务进程，以及 Hadoop 的“NameNode”“SecondaryNameNode”“DataNode”“ResourceManager”“NodeManager”五个服务进程。

```
[hadoop@localhost ~]$ jps
4256 HMaster
3346 DataNode
3506 SecondaryNameNode
4402 Jps
3811 NodeManager
2901 QuorumPeerMain
4343 HRegionServer
3705 ResourceManager
2926 QuorumPeerMain
3246 NameNode
2879 QuorumPeerMain
```

图 1-6-18　查看 Java 进程信息

（3）在第一次启动 HBase，并且 HBase 的所有服务进程启动成功之后，使用 Zookeeper 的客户端工具连接到 Zookeeper 服务。然后查看 Zookeeper 中的目录节点信息，如图 1-6-19 所示，可以看到其中新增加了一个名为“hbase”的目录节点，同时该目录节点下还有很多子目录节点。

```
[zk: localhost:2181(CONNECTED) 0] ls /
[zookeeper, hbase]
[zk: localhost:2181(CONNECTED) 1] get /hbase

cZxid = 0x100000002
ctime = Tue Jul 24 06:25:14 EDT 2018
mZxid = 0x100000002
mtime = Tue Jul 24 06:25:14 EDT 2018
pZxid = 0x10000007b
cversion = 24
dataVersion = 0
aclVersion = 0
ephemeralOwner = 0x0
dataLength = 0
numChildren = 16
[zk: localhost:2181(CONNECTED) 2] ls /hbase
[replication, meta-region-server, rs, splitWAL, backup-masters, table-lock, flush-table-proc, region
-in-transition, online-snapshot, master, running, recovering-regions, draining, namespace, hbaseid,
table]
[zk: localhost:2181(CONNECTED) 3] _
```

图 1-6-19　查看 Zookeeper 目录节点信息

（4）使用命令“hbase　shell”利用 HBase 的客户端工具连接 HBase 服务，如图 1-6-20 所示，若进入 HBase 客户端程序的控制台界面且没有出现报错信息，则表示 HBase 服务连接正常。

```
[hadoop@localhost ~]$ hbase shell
SLF4J: Class path contains multiple SLF4J bindings.
SLF4J: Found binding in [jar:file:/home/hadoop/hbase/hbase-1.2.3/lib/slf4j-log4j12-1.7.5.jar!/org/sl
f4j/impl/StaticLoggerBinder.class]
SLF4J: Found binding in [jar:file:/home/hadoop/hadoop/hadoop-2.7.3/share/hadoop/common/lib/slf4j-log
4j12-1.7.10.jar!/org/slf4j/impl/StaticLoggerBinder.class]
SLF4J: See http://www.slf4j.org/codes.html#multiple_bindings for an explanation.
SLF4J: Actual binding is of type [org.slf4j.impl.Log4jLoggerFactory]
HBase Shell; enter 'help<RETURN>' for list of supported commands.
Type "exit<RETURN>" to leave the HBase Shell
Version 1.2.3, rbd63744624a26dc3350137b564fe746df7a721a4, Mon Aug 29 15:13:42 PDT 2016

hbase(main):001:0> _
```

图 1-6-20　连接 HBase 服务

（5）在 HBase 控制台界面中使用命令“create　’表名’,’列名 1’,’列名 2’，…”在 HBase 数据库中创建数据表。如图 1-6-21 所示，若没有出现报错信息，并且显示变更行数和执行时间的信息，则表示数据表创建成功，HBase 数据库可以正常使用。

```
hbase(main):001:0> create 'test','col1','col2'
0 row(s) in 25.1970 seconds

=> Hbase::Table - test
```

图 1-6-21　创建数据表

（6）在 HBase 控制台界面中使用命令“exit”可以退出 HBase 客户端程序，返回操作系统的控制台界面，如图 1-6-22 所示。

```
hbase(main):001:0> exit
[hadoop@Cluster-01 ~]$ _
```

图 1-6-22　退出 HBase 客户端程序

1.6.6　安装与配置的常见问题及解决方法

常见问题如下：

HBase 的服务节点进程“HMaster”或“HRegionServer”启动失败。

HBase 的服务节点进程“HMaster”或“HRegionServer”在成功启动之后经过一段时间自动关闭。

不能正常进入 HBase 客户端的控制台界面，或进入之后有报错信息。

不能正常创建数据库表。

在这些常见问题中，前两种问题一般在操作系统的控制台界面中不会打印出详细的问题描述和报错信息，而后两种问题在 HBase 的控制台界面会打印出一些错误信息和问题描述。但不管是哪一类问题，大多都是由于配置过程中输入错误或漏掉某个步骤造成的。可以先检查配置文件中的配置项的名称和值是否书写正确，比如 JDK 软件路径、Hadoop 配置文件路径、Zookeeper 连接地址和端口号、配置文件格式等。

另一种查找问题原因的更主要方式是查看日志文件，从日志信息的内容来查看导致各种问题的错误信息，从而为找到问题的原因，并为找到问题的解决方法提供帮助。HBase 的日志文件位于之前在配置过程中指定的日志文件的存放目录之中。在该日志文件目录下，每个服务节点都有自己的日志文件，如图 1-6-22 所示，HBase 的服务节点进程“HMaster”对应的日志文件的名称格式为“hbase-用户名-master-主机名.log”，而服务节点进程“HRegionServer”对应的日志文件的名称格式为“hbase-用户名-regionserver-主机名.log”。

```
[hadoop@localhost logs]$ ls -l
total 64
-rw-rw-r--. 1 hadoop hadoop 40163 Jul 24 06:28 hbase-hadoop-master-localhost.localdomain.log
-rw-rw-r--. 1 hadoop hadoop   700 Jul 24 06:28 hbase-hadoop-master-localhost.localdomain.out
-rw-rw-r--. 1 hadoop hadoop 13838 Jul 24 06:28 hbase-hadoop-regionserver-localhost.localdomain.log
-rw-rw-r--. 1 hadoop hadoop   205 Jul 24 06:28 hbase-hadoop-regionserver-localhost.localdomain.out
-rw-rw-r--. 1 hadoop hadoop     0 Jul 24 06:28 SecurityAuth.audit
```

图 1-6-22　查看日志文件

若是因为配置文件书写错误而修改了任意配置文件中的内容，特别是修改了主机名、文件目录路径等方面的内容，建议将创建的功能目录“tmp”中的内容清空之后再重新启动 HBase 服务。

项目 2　搭建高可用 Hadoop 整合平台

前面介绍了单节点模式的 Hadoop 整合平台的搭建，该平台的搭建只需要使用一台计算机即可完成，整个搭建的过程也相对比较简单，主要是用于学习和了解整个 Hadoop 平台。同时单节点模式的 Hadoop 整合平台对硬件要求低、搭建过程简单的特点，也经常被用于搭建 Hadoop 相关程序的开发和测试环境。

而在真实的应用环境下，Hadoop 整合平台肯定是构建在由多台计算机构成的集群之上的，因为只有集群才能提供真正的分布式环境，也只有将 Hadoop 整合平台搭建在真正的分布式环境之上，才能充分发挥 HDFS 分布式文件系统、MapReduce 分布式计算框架、HBase 分布式数据库的高性能、高可用性、高可靠性、高可扩展性等方面的特点。这里将使用由五台计算所构成的集群来搭建 Hadoop 整合平台。

Hadoop 整合平台中不管是 Hadoop 软件自身还是 HBase 数据库都有两种集群部署模式。一种是最原始的只有单个 Master 节点的普通集群模式，对应 Hadoop 软件自身就是集群中只有一个 NameNode 服务节点，同时有一个 SecondaryNameNode 服务节点来辅助 NameNode 服务节点，而对应 HBase 数据库就是只有一个 HMaster 服务节点；另一种是具备多个 Master 节点的高可用集群模式，这种模式下虽然对外提供服务的 Master 节点仍然只有一个，但是一个或多个备用的 Master 节点可以在提供服务的 Master 节点失效时进行动态切换，有效避免了普通集群模式中 Master 节点的单点故障问题。Hadoop 软件自身是在 Hadoop 2.0 版本之后，才实现了高可用集群模式，也就是集群中有多个 NameNode 服务节点，同时需要多个 JournalNode 服务节点来对高可用的实现进行辅助，而对应 HBase 数据库就是有多个 HMaster 服务节点。同时不管是 Hadoop 软件自身还是 HBase 数据库，其高可用集群模式都需要使用到 Zookeeper 的协调服务来实现 Master 节点的自动切换。接下来将对 Hadoop 整合平台的两种集群模式的搭建分别进行介绍。

任务 2.1 集群网络属性配置

2.1.1 静态网络地址配置

Hadoop 的集群模式是由多台计算机构建而成，而每台计算机之间通过网络进行连接和通信，所以需要对集群中每台主机的网络属性进行配置。

CentOS 操作系统默认的网络服务使用的是 DHCP 方式获取 IP 地址，这种方式获取的 IP 地址可能会因为主机或路由器的重启而发生变化，从而使得原本配置的主机名和 IP 地址的映射关系失效，导致集群中的主机之间无法进行连接和通信。管理员需要重新编辑地址映射关系文件或 DNS 服务器中的主机名与 IP 地址的映射关系，才能使集群恢复正常运行。这无疑增加了集群管理的难度和复杂度，所以需要将集群中所有主机的网络配置为静态 IP 地址来避免非计划性的 IP 地址变更。

在给集群中主机配置静态 IP 地址之前，应该对集群中所有主机、路由器、交换机可能使用到的网络、IP 地址、端口进行一个合理的规划，尽可能在最大程度上减少后期的使用和维护过程中可能出现的变更情况。

（1）使用"root"用户登录到主机的操作系统。

（2）CentOS 7 的网络配置文件位于系统中的"/etc/sysconfig/network- scripts/"目录下，如图 2-1-1 所示，配置文件的文件名为"ifcfg-*"，其中"*"部分为网络的名称，默认有一个名为"lo"的环回网络，而名为"enp*"的网络为网卡所对应的网络，每个物理或虚拟网卡默认都有一个对应的网络配置文件，根据所使用的网卡选择对应的配置文件进行编辑。

```
[root@localhost network-scripts]# pwd
/etc/sysconfig/network-scripts
[root@localhost network-scripts]# ls -l ifcfg-*
-rw-r--r--. 1 root root 310 Apr  9 12:47 ifcfg-ens33
-rw-r--r--. 1 root root 254 Sep 12  2016 ifcfg-lo
```

图 2-1-1 网络配置文件

（3）根据事先制定的集群网络规划，如图 2-1-2 所示，在配置文件中编辑或添加以下配置项，"#"号之后的内容为对应配置项或配置内容的说明，不需要写入到配置文件之中：

设置采用静态模式获取 IP 地址

项目 2 搭建高可用 Hadoop 整合平台

BOOTPROTO=static
# 设置该网络为开机自动启动
ONBOOT=yes
# 设置该网络通过配置文件管理，而不通过网络管理器管理
NM_CONTROLLED=no
# 设置网络的静态 IP 地址，请根据实际规划的 IP 地址更改配置项的值
IPADDR=192.168.60.130
# 设置网络的子网掩码，请根据实际规划的子网属性更改配置项的值
NETMASK=255.255.255.0
# 设置网络的网关地址，请根据实际规划的网络属性更改配置项的值
GATEWAY=192.168.60.2

```
TYPE=Ethernet
BOOTPROTO=static
DEFROUTE=yes
PEERDNS=yes
PEERROUTES=yes
IPV4_FAILURE_FATAL=no
IPV6INIT=yes
IPV6_AUTOCONF=yes
IPV6_DEFROUTE=yes
IPV6_PEERDNS=yes
IPV6_PEERROUTES=yes
IPV6_FAILURE_FATAL=no
IPV6_ADDR_GEN_MODE=stable-privacy
NAME=ens33
UUID=8a58772f-46bb-4d51-8978-973764a7c333
DEVICE=ens33
ONBOOT=yes
NM_CONTROLLED=no
IPADDR=192.168.60.130
NETMASK=255.255.255.0
GATEWAY=192.168.60.2
```

图 2-1-2　添加配置项

（4）使用命令 "service network restart" 或 "systemctl restart network" 重新启动网络服务，如图 2-1-3 所示，使新修改的网络配置生效。

```
[root@localhost network-scripts]# service network restart
Restarting network (via systemctl):                        [  OK  ]
```

图 2-1-3　重新启动网络服务

（5）可以在处于同一网络中的其他计算机上使用“ping”命令来测试网络配置是否正确，如图 2-1-4 所示，若能正确收到该主机的返回信息，则表示新配置的网络可用。

网络配置中的常见问题和解决方式：

配置静态 IP 地址模式的网络需要手动填写 IP 地址、子网掩码、网关，所以需要事先通过网络规划获取这些相关信息。若事先不清楚当前网络的这些相关信息，可以让系统通过 DHCP 方式自动从当前网络中获取 IP 地址，然后查看当前的网络信息。需要注意的是通过这种方式获取 IP 地址要求当前网络中的路由器或交换机开启了相应功能。

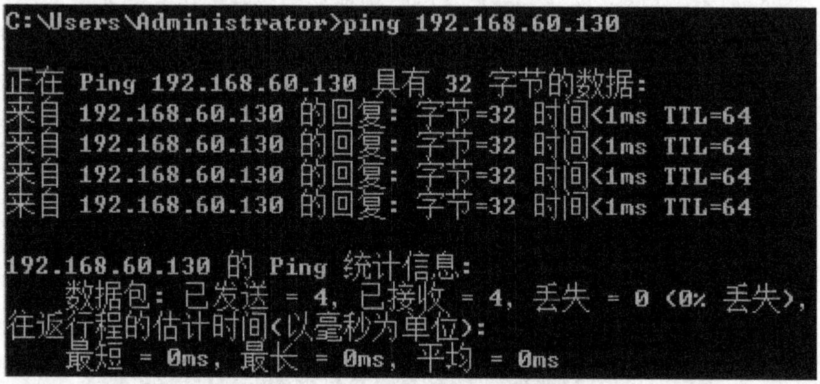

```
C:\Users\Administrator>ping 192.168.60.130

正在 Ping 192.168.60.130 具有 32 字节的数据:
来自 192.168.60.130 的回复: 字节=32 时间<1ms TTL=64
来自 192.168.60.130 的回复: 字节=32 时间<1ms TTL=64
来自 192.168.60.130 的回复: 字节=32 时间<1ms TTL=64
来自 192.168.60.130 的回复: 字节=32 时间<1ms TTL=64

192.168.60.130 的 Ping 统计信息:
    数据包: 已发送 = 4, 已接收 = 4, 丢失 = 0 (0% 丢失),
往返行程的估计时间(以毫秒为单位):
    最短 = 0ms, 最长 = 0ms, 平均 = 0ms
```

图 2-1-4　测试网络配置是否可用

使用命令“ip　addr”可以查看系统当前网络的相关信息，如图 2-1-5 所示，网卡“ens33”对应的 IP 地址为“192.168.60.130”，IP 地址后面的“24”则表示当前网络的子网掩码为“255.255.255.0”。

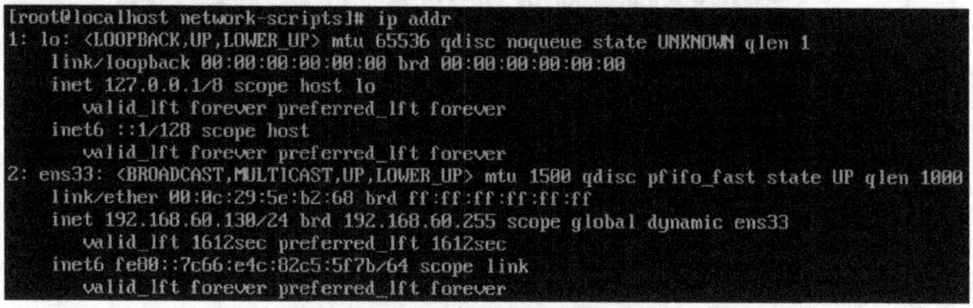

```
[root@localhost network-scripts]# ip addr
1: lo: <LOOPBACK,UP,LOWER_UP> mtu 65536 qdisc noqueue state UNKNOWN qlen 1
    link/loopback 00:00:00:00:00:00 brd 00:00:00:00:00:00
    inet 127.0.0.1/8 scope host lo
       valid_lft forever preferred_lft forever
    inet6 ::1/128 scope host
       valid_lft forever preferred_lft forever
2: ens33: <BROADCAST,MULTICAST,UP,LOWER_UP> mtu 1500 qdisc pfifo_fast state UP qlen 1000
    link/ether 00:0c:29:5e:b2:68 brd ff:ff:ff:ff:ff:ff
    inet 192.168.60.130/24 brd 192.168.60.255 scope global dynamic ens33
       valid_lft 1612sec preferred_lft 1612sec
    inet6 fe80::7c66:e4c:82c5:5f7b/64 scope link
       valid_lft forever preferred_lft forever
```

图 2-1-5　查看当前网络信息

使用命令“ip　route”可以查看当前网络的路由信息，如图 2-1-6 所示，路由信息中“default”一行显示的 IP 地址便是当前网络的网关地址“192.168.60.2”。

```
[root@localhost network-scripts]# ip route
default via 192.168.60.2 dev ens33 proto static metric 100
192.168.60.0/24 dev ens33 proto kernel scope link src 192.168.60.130 metric 100
```

图 2-1-6　查看当前网络的路由信息

2.1.2　主机名配置

在 Hadoop 的安装和配置过程中，通常不直接使用 IP 地址来标识集群中的各个主机，而是使用主机名。然后通过在地址映射关系文件或 DNS 服务器中配置主机名与 IP 地址的映射关系，来使得 Hadoop 能够通过主机名找到对应主机的 IP 地址，从而进行连接访问。这样做的好处在于不仅可以通过合理规划的主机名更方便地识别出集群中主机的分组和主要功能，还可以当集群中某台主机的 IP 地址发生变化时，只更改地址映射关系文件或 DNS 服务器中的主机名与 IP 地址的映射关系即可，而不需要去更改烦琐的 Hadoop 配置文件。

在给集群中主机配置主机名之前也应该进行合理的规划，在主机名中尽量表现出当前主机的所属分组和主要功能职责，使得在后期的使用和维护过程中能够更容易地找到所需要的对应主机。

（1）使用"root"用户登录到主机的操作系统。

（2）方法一：编辑配置文件。

CentOS 7 的主机名配置文件位于系统的"/etc"目录下，如图 2-1-7 所示，文件名为"hostname"。编辑该配置文件，删除配置文件中的原有内容，然后添加需要设定的主机名即可。

```
[root@localhost etc]# ls -l host*
-rw-r--r--. 1 root root   9 Jun  7  2013 host.conf
-rw-r--r--. 1 root root  22 Apr  9 09:16 hostname
-rw-r--r--. 1 root root 158 Jun  7  2013 hosts
-rw-r--r--. 1 root root 370 Jun  7  2013 hosts.allow
-rw-r--r--. 1 root root 460 Jun  7  2013 hosts.deny
```

图 2-1-7　编辑该配置文件

（3）方法二：使用 hostnamectl 命令。

CentOS 7 中的新命令"hostnamectl"可以用来对主机名进行设置，其中使用命令"hostnamectl　set-hostname　主机名"可以永久性的修改主机名。

（4）这里将集群中主机的主机名设定为"Cluster-**"，其中"**"表示主机的编号。修改完成之后使用命令"reboot"重新启动操作系统，使新修改的主机名生效。

（5）重启操作系统完成之后，如图 2-1-8 所示，在命令行提示符中可以看到新设置的主机名，也可以使用命令"hostname"来查看新设置的主机名。

```
[root@Cluster-01 ~]# hostname
Cluster-01
```

图 2-1-8　查看新设置的主机名

2.1.3　防火墙配置

Linux 操作系统一般在所有安装模式中都自带了防火墙功能，并且在安装完成之后默认为跟随操作系统启动而自动启动。在系统自带防火墙的默认安全策略中，关闭了所有端口的外部访问，需要手动进行端口策略的设置之后，网络中的其他主机才能访问到对应的端口，否则即便网络属性配置正确，也无法访问到相应端口的服务。所以在 Hadoop 的安装和配置过程中，需要对集群中所有主机的防火墙进行配置，以保证集群中主机之间能够相互连接并访问 Hadoop 相关服务。

CentOS 操作系统在 6 及之前的版本中默认使用的自带防火墙是 iptables，而从版本 7 开始自带防火墙更换为了 firewall。由于当前使用的是 CentOS 7 1611 版本，所以只介绍 firewall 防火墙的设置方法。

对于集群中主机的防火墙设置有两种方法，其中一种是在防火墙中对每一个将要使用到的端口单独配置端口策略，具体方法如下。

（1）使用"root"用户登录到主机的操作系统。

（2）使用命令"firewall-cmd　--zone=public　--add-port=端口号/tcp --permanent"添加防火墙的端口策略，对外开启指定端口，如图 2-1-9 所示，若显示"success"则表示策略添加成功。

```
[root@Cluster-01 ~]# firewall-cmd --zone=public --add-port=9000/tcp --permanent
success
```

图 2-1-9　对外开启指定端口

（3）使用命令"firewall-cmd　--reload"重启防火墙服务，使防火墙新添加的端口策略生效，如图 2-1-10 所示，若显示"success"则表示防火墙重启成功。

```
[root@Cluster-01 ~]# firewall-cmd --reload
success
```

图 2-1-10　重启防火墙

（4）Hadoop 平台相关组件的常用端口如下：

① Zookeeper 的常用端口：2181、2888、3888。

② Hadoop 的常用端口：8019、8020、8031、8032、8033、8040、8041、8042、8088、8480、8485、9000、9001、0020、19588、50010、50020、50070、50075、50470、50475

③ HBase 的常用端口：2181、2888、3888、60000、60010、60020、60030。

HBase 的常用端口和 Zookeeper 的常用端口有重复是因为 HBase 自带 Zookeeper 组件，使用独立 Zookeeper 时这些端口不会被启用，也就不会造成端口冲突。

④ Hive 的常用端口：9083、10000。

⑤ MySQL 的常用端口：1186、2202、3306。

（5）使用命令 "firewall-cmd --query-port=端口号/tcp" 可以查看指定的端口是否已经开启，如图 2-1-11 所示，若显示 "yes" 则表示指定端口已经开启，可以允许外部的访问。

```
[root@Cluster-01 ~]# firewall-cmd --query-port=9000/tcp
yes
```

图 2-1-11　查看指定的端口是否已经开启

而防火墙设置的另一种方法是直接关闭防火墙的运行，并且同时禁用掉防火墙，使之不会随着操作系统的启动而自动启动，具体方法如下。

（1）使用 "root" 用户登录到主机的操作系统。

（2）使用命令 "systemctl stop firewalld" 关闭防火墙服务。

（3）使用命令 "systemctl disable firewalld" 禁用防火墙服务，如图 2-1-12 所示，使之不会随着操作系统的启动而自动启动。

```
[root@Cluster-01 ~]# systemctl disable firewalld.service
Removed symlink /etc/systemd/system/dbus-org.fedoraproject.FirewallD1.service.
Removed symlink /etc/systemd/system/basic.target.wants/firewalld.service.
```

图 2-1-12　禁用防火墙服务

（4）使用命令 "systemctl status firewalld" 可以查看防火墙服务的状态，如图 2-1-13 所示，其中包括了当前运行状态以及自启动状态。

```
[root@Cluster-01 ~]# systemctl status firewalld.service
● firewalld.service - firewalld - dynamic firewall daemon
   Loaded: loaded (/usr/lib/systemd/system/firewalld.service; disabled; vendor preset: enabled)
   Active: inactive (dead)
     Docs: man:firewalld(1)
```

图 2-1-13　查看防火墙服务状态

关于两种方法的说明：

第一种方法中由于保持了防火墙的运行及开机自动启动状态，所以主机的安全性较高，能够有效屏蔽其他无用的网络访问。但其缺点也很明显，由于需要对每个将要使用到的端口进行单独的端口策略设置，整个过程会比较烦琐。Hadoop 的平台相关组件的常用端口加起来有几十个之多，——单独设置端口策略会花费较长的时间。

而第二种方法设置起来相对比较简单快捷，能够节约大量的时间。但是由于完全关闭了防火墙的运行，会降低主机自身在网络中的安全性。

对于两种方法请根据实际情况需要进行选择。正式的商用 Hadoop 平台建议选择前者，而学习、开发、测试用的 Hadoop 平台可以选择后者。

2.1.4 主机名与 IP 地址映射关系配置

前面在进行主机名配置时已经说明了配置主机名的好处，但在实际进行网络连接时依然使用的是 IP 地址。那么要使得 Hadoop 平台能够通过主机名找到并连接访问集群中的所有主机，便需要配置相应的主机名与 IP 地址之间的映射关系。该映射关系可以通过系统的相应配置文件进行配置，也可以在集群中搭建专用的 DNS 服务器。

通过配置文件进行配置相对比较简单，但因为需要在每台主机上都进行配置，并且每次集群中主机名与 IP 地址映射关系发生变化时，也需要在每台主机上都进行配置的更改，所以该方法仅适用于集群中主机数量较少时。

而在集群中搭建专用的 DNS 服务器相对比较复杂，但集群中其他主机上只需要在配置网络属性时添加对应的 DNS 服务器即可，不需要进行任何额外的配置。在集群中的主机名和 IP 地址映射关系发生变化时，也仅需要在 DNS 服务器上进行配置的更改即可。所以该方法适用于集群中主机数量较多时。

由于当前所搭建的平台仅需要满足 Hadoop 高可用平台最低要求的 5 台主机的集群即可，所以选择相对简单的配置系统中的地址映射关系文件。

（1）使用"root"用户登录到主机的操作系统。

（2）CentOS 7 的地址映射关系配置文件位于系统的"/etc"目录下，如图 2-1-14 所示，文件名为"hosts"。

（3）编辑该配置文件，添加集群中所有主机的 IP 地址与主机名的映射关系，文件中每一行代表一台主机的 IP 地址及其对应的主机名，具体书写格式如下：

192.168.60.***　　　　　Cluster-**

192.168.60.***　　　　　　Cluster-**
......

```
[root@localhost etc]# ls -l host*
-rw-r--r--. 1 root root    9 Jun  7  2013 host.conf
-rw-r--r--. 1 root root   22 Apr  9 09:16 hostname
-rw-r--r--. 1 root root  158 Jun  7  2013 hosts
-rw-r--r--. 1 root root  370 Jun  7  2013 hosts.allow
-rw-r--r--. 1 root root  460 Jun  7  2013 hosts.deny
```

图 2-1-14　查看地址映射关系配置文件

（4）使用命令"scp　/etc/hosts　root@目标主机名或 IP 地址：/etc/"将地址映射关系配置文件"hosts"发送给集群中所有其他主机，如图 2-1-15 所示，发送文件时由于目标主机的 root 用户没有配置免密码登录，所以会提示输入 root 用户的密码。在正确输入密码之后便会开始进行文件的传输。当需要传输的所有文件全部显示 100%之后，便代表文件传输过程全部完成。

```
[root@Cluster-01 ~]# scp /etc/hosts root@Cluster-02:/etc/hosts
root@cluster-02's password:
hosts                                           100%  212     0.2KB/s   00:00
```

图 2-1-15　发送"hosts"文件给其他主机

使用该方式进行远程文件传输会自动覆盖目标主机上目标路径中的同名文件，所以在使用此方法向目标主机传输"hosts"文件之前，需要确保目标主机使用的该文件与待传输的文件相同。若目标主机因为用于其他用途而在"hosts"文件中添加有其他主机名与 IP 地址映射关系的条目，则需要手动将 Hadoop 集群中主机的主机名和 IP 地址映射关系的条目添加到文件中。

2.1.5　免密码登录配置

在前面的伪分布模式平台的搭建过程中，由于 Hadoop 的所有服务只需要在一台主机上运行，也就是只需要进行本机的免密码登录，所以采用了一种相对方便快捷的方式来配置免密码登录。而在完全分布模式下，需要真正远程登录到集群中的其他主机，所以需要使用另一种方式来配置免密码登录。

（1）使用"hadoop"用户登录到主机的操作系统，在之后的各个软件的安装和配置过程中，若没有特殊说明都默认使用"hadoop"用户登录到主机的操作系统进行操作。

（2）使用命令"ssh-keygen　-t　rsa"生成当前系统登录用户密钥文件，包括公钥文件和私钥文件。密钥文件的生成过程为交互模式，如图 2-1-16 所示，会提示用户输入一些相关设置信息，包括密钥文件的存放位置以及密钥文件的密码。这里可以不用进行输入，全部使用回车键直接跳过即可，在跳过的情况下系统会默认在用户的家目录中创建一个名为“.ssh”的隐藏目录，然后将生成的密钥文件存放于其中。

```
[hadoop@Cluster-01 ~]$ ssh-keygen -t rsa
Generating public/private rsa key pair.
Enter file in which to save the key (/home/hadoop/.ssh/id_rsa):
Created directory '/home/hadoop/.ssh'.
Enter passphrase (empty for no passphrase):
Enter same passphrase again:
Your identification has been saved in /home/hadoop/.ssh/id_rsa.
Your public key has been saved in /home/hadoop/.ssh/id_rsa.pub.
The key fingerprint is:
69:fb:dd:7a:84:3e:4a:1a:36:2b:ac:16:c1:95:94:55 hadoop@Cluster-01
The key's randomart image is:
+--[ RSA 2048]----+
|      ..+..E     |
|       +         |
|   . .           |
|    o     .      |
|     .  S    .   |
|     .  .  .  .  |
|     o  = . .    |
|    . o. B .o..  |
|   ... .o o.o+.  |
+-----------------+
```

图 2-1-16　密钥文件的生成过程

（3）使用命令“ssh-copy-id　-i　~/.ssh/id_rsa.pub　目标用户@目标主机名或 IP 地址”将公钥文件拷贝到需要进行免密码登录的目标主机和目标用户。第一次连接时由于目标主机还未被添加到已知主机列表中，会提示目标主机未知是否继续连接，如图 2-1-17 所示，输入“yes”后按回车键确认进行连接。接着会提示输入目标主机对应目标用户的登录密码，正确输入密码之后按回车键便能够进行公钥文件的拷贝，并且在拷贝完成之后自动退出命令的执行。

```
[hadoop@Cluster-01 ~]$ ssh-copy-id -i ~/.ssh/id_rsa.pub hadoop@Cluster-02
The authenticity of host 'cluster-02 (192.168.60.131)' can't be established.
ECDSA key fingerprint is 3a:40:45:93:a4:ba:d8:9e:14:f4:43:92:45:be:ea:e2.
Are you sure you want to continue connecting (yes/no)? yes
/bin/ssh-copy-id: INFO: attempting to log in with the new key(s), to filter out any that are already
 installed
/bin/ssh-copy-id: INFO: 1 key(s) remain to be installed -- if you are prompted now it is to install
the new keys
hadoop@cluster-02's password:

Number of key(s) added: 1

Now try logging into the machine, with:   "ssh 'hadoop@Cluster-02'"
and check to make sure that only the key(s) you wanted were added.
```

图 2-1-17　拷贝公钥文件

由于 Hadoop 会默认使用当前登录用户的用户名来进行远程登录访问，所以集群中所有主机都应该有一个用户名相同的专门用于Hadoop集群的用户，并且在配置免密码登录时目标用户全部采用该用户的用户名。

（4）使用命令"ssh 目标主机名"远程登录到目标主机，验证免密码登录配置是否成功。若可以进行正常的免密码登录则表示配置成功，若提示输入密码则表示配置失败，需要删除用户家目录下的".ssh"目录之后再重新执行整个免密码登录的配置过程。

在Hadoop完全分布模式中，主节点需要能够免密码登录到所有其他节点以及自身。若是拥有多个主节点的高可用完全分布模式，则多个主节点之间还需要能够相互进行免密码登录。

任务2.2 集群下JDK的便捷安装

2.2.1 JDK安装规划

1. 环境要求

集群中所有主机的本地磁盘剩余空间400 MB以上；

集群中所有主机已安装CentOS 7 1611 64位操作系统；

集群中所有主机已完成基础网络环境配置。

2. 软件版本

与单节点模式Hadoop整合平台搭建中相同，选用JDK的1.8.0_131版本，并对应64位的操作系统选用该版本JDK的64位版本，其软件包名为jdk-8u131-linux-x64.tar.gz。

2.2.2 在集群主机上安装JDK

（1）选择集群中的一台主机，这里尽量选用后面将会用作主节点的主机，这样可以使用配置的免密码登录方便的远程操作集群中其他主机。然后在其上的Hadoop专用用户下完成JDK的安装和配置。安装配置方法与单节点Hadoop平台的搭建过程中JDK的安装配置方法相同，先将操作系统原有的JDK及相关软件包卸载，然后再安装和配置新的JDK。

（2）使用命令"ssh root@主机名或IP地址"远程登录到集群中所有其他主机，分别在这些主机上执行原有JDK的卸载操作。由于之前并没有配置root

用户的免密码登录，所以在远程登录目标主机的 root 用户时需要输入用户密码。

所有涉及远程登录到另一台主机的步骤中，在完成所有操作之后，都需要使用命令"logout"或"exit"退出当前登录之后，再登录到其他主机进行操作。

（3）使用命令"scp ~/.bash_profile 用户名@目标主机名或 IP 地址:/home/用户名/"和"scp -r ~/java 用户名@目标主机名或 IP 地址:/home/用户名/"依次将用户的环境变量设置文件和 Java 的专用目录发送到集群中所有其他主机的相应用户家目录下。

任务 2.3　安装 Zookeeper 完全分布模式

2.3.1　Zookeeper 完全分布模式安装规划

1. 环境要求

集群中所有主机的本地磁盘剩余空间 100 MB 以上；
集群中所有主机已安装 CentOS 7 1611 64 位操作系统；
集群中所有主机已完成基础网络环境配置；
集群中所有主机已安装 JDK 的 1.8.0_131 版本。

2. 软件版本

与单节点模式 Hadoop 整合平台搭建中相同，选用 Zookeeper 的 3.4.9 版本，其软件包名为 zookeeper-3.4.9.tar.gz。

3. 服务规划

Zookeeper 采用了 Paxos 投票算法，要求至少 3 个及以上的服务节点数量，并且服务节点总数为奇数时最佳。这里将使用集群中的全部五台主机作为 Zookeeper 的服务节点，服务进程名称为 QuorumPeerMain。具体 IP 地址和服务节点规划如表 2-3-1 所示。

表 2-3-1　IP 地址和服务节点规划

主机名	IP 地址	服务描述
Cluster-01	192.168.60.130	Zookeeper 服务 （QuorumPeerMain）

续表

主机名	IP 地址	服务描述
Cluster-02	192.168.60.131	Zookeeper 服务（QuorumPeerMain）
Cluster-03	192.168.60.132	Zookeeper 服务（QuorumPeerMain）
Cluster-04	192.168.60.133	Zookeeper 服务（QuorumPeerMain）
Cluster-05	192.168.60.134	Zookeeper 服务（QuorumPeerMain）

2.3.2　Zookeeper 完全分布模式配置

（1）选择集群中的一台后面将会用作主节点的主机，在其上的 Hadoop 专用用户下完成 Zookeeper 的基础安装和环境配置。安装配置方法与单节点 Hadoop 平台的搭建过程中 Zookeeper 的基础环境配置方法相同，解压 Zookeeper 软件包并配置环境变量文件。

（2）进入 Zookeeper 相关文件的目录，如图 2-3-1 所示，分别创建数据文件的存放目录"data"和日志文件的存放目录"logs"。

```
[hadoop@Cluster-01 zookeeper]$ ls -l
total 4
drwxrwxr-x. 3 hadoop hadoop   63 Jun  8 04:21 data
drwxrwxr-x. 3 hadoop hadoop   23 Jun  8 04:21 logs
drwxr-xr-x. 10 hadoop hadoop 4096 Aug 23  2016 zookeeper-3.4.9
```

图 2-3-1　创建目录

（3）进入 Zookeeper 软件目录下的配置文件所在目录"conf"，如图 2-3-2 所示，复制并重命名配置文件的模板文件"zoo_sample.cfg"，生成 Zookeeper 的配置文件"zoo.cfg"。

```
[hadoop@Cluster-01 conf]$ ls -l
total 16
-rw-rw-r--. 1 hadoop hadoop  535 Aug 23  2016 configuration.xsl
-rw-rw-r--. 1 hadoop hadoop 2161 Aug 23  2016 log4j.properties
-rw-rw-r--. 1 hadoop hadoop  922 Jun  8 03:11 zoo.cfg
-rw-rw-r--. 1 hadoop hadoop  922 Aug 23  2016 zoo_sample.cfg
```

图 2-3-2　生成配置文件

（4）编辑生成的 Zookeeper 配置文件"zoo.cfg"，如图 2-3-3 所示，首先找到配置项"dataDir"，将其值改为数据文件的存放目录"data"的路径。

```
 9 # the directory where the snapshot is stored.
10 # do not use /tmp for storage, /tmp here is just
11 # example sakes.
12 dataDir=/home/hadoop/zookeeper/data
```

图 2-3-3　修改配置文件

然后如图 2-3-4 所示，在配置文件末尾添加如下内容：
dataLogDir=（日志文件的存放目录"logs"的路径）

server.1=Cluster-01：2888：3888

server.2=Cluster-02：2888：3888

server.3=Cluster-03：2888：3888

server.4=Cluster-04：2888：3888

server.5=Cluster-05：2888：3888

```
dataLogDir=/home/hadoop/zookeeper/logs
server.1=Cluster-01:2888:3888
server.2=Cluster-02:2888:3888
server.3=Cluster-03:2888:3888
server.4=Cluster-04:2888:3888
server.5=Cluster-05:2888:3888
```

图 2-3-4　在配置文件中添加内容

（5）进入 Zookeeper 软件目录下的脚本文件所在目录"bin"，对 Zookeeper 的脚本文件"zkServer.sh"进行编辑，如图 2-3-5 所示，找到"ZOO_LOG_DIR"关键字所在行，在其上面一行添加如下脚本代码：

ZOO_LOG_DIR=

"$（ $GREP　"^[[：space：]]*dataLogDir"　"$ZOOCFG"　|　sed　–e 's/.*=//'）"

```
125 ZOO_LOG_DIR="$($GREP "^[[:space:]]*dataLogDir" "$ZOOCFG" | sed -e 's/.*=//')"
126 if [ ! -w "$ZOO_LOG_DIR" ] ; then
127 mkdir -p "$ZOO_LOG_DIR"
128 fi
```

图 2-3-5　编辑脚本文件

（6）使用命令"scp　~/.bash_profile　用户名@目标主机名或 IP 地址:/home/

用户名/"和"scp -r ~/zookeeper 用户名@目标主机名或 IP 地址：/home/用户名/"依次将新修改的用户环境变量设置文件和 Zookeeper 的专用目录发送到集群中所有其他主机的相应用户家目录下。

（7）进入 Zookeeper 的数据文件的存放目录"data"，在其中创建 Zookeeper 的节点标识文件"myid"，并在文件中写入当前主机的 Zookeeper 的节点编号，如图 2-3-6 所示，该编号与 Zookeeper 配置文件中"server.*"的"*"部分对应。

```
[hadoop@Cluster-01 zookeeper]$ cat data/myid
1
```

图 2-3-6 创建节点标识文件"myid"

（8）分别远程登录到集群中所有其他主机，创建 Zookeeper 的节点标识文件，并在文件中添加各个主机的节点编号，如图 2-3-7 所示。

```
[hadoop@Cluster-01 ~]$ ssh Cluster-02
Last login: Thu Jun 7 23:13:27 2018 from cluster-01
[hadoop@Cluster-02 ~]$ cd zookeeper/data/
[hadoop@Cluster-02 data]$ echo 2 > myid
[hadoop@Cluster-02 data]$ cat myid
2

[hadoop@Cluster-01 ~]$ ssh Cluster-03
Last login: Fri Jun 8 08:14:21 2018 from cluster-01
[hadoop@Cluster-03 ~]$ cd zookeeper/data/
[hadoop@Cluster-03 data]$ echo 3 > myid
[hadoop@Cluster-03 data]$ cat myid
3

[hadoop@Cluster-01 ~]$ ssh Cluster-04
Last login: Fri Jun 8 09:56:15 2018 from 192.168.60.130
[hadoop@Cluster-04 ~]$ cd zookeeper/data/
[hadoop@Cluster-04 data]$ echo 4 > myid
[hadoop@Cluster-04 data]$ cat myid
4

[hadoop@Cluster-01 ~]$ ssh Cluster-05
Last login: Fri Jun 8 11:36:05 2018 from cluster-01
[hadoop@Cluster-05 ~]$ cd zookeeper/data/
[hadoop@Cluster-05 data]$ echo 5 > myid
[hadoop@Cluster-05 data]$ cat myid
5
```

图 2-3-7 添加各个主机节点编号

2.3.3 Zookeeper 完全分布模式启动和验证

（1）在集群中所有主机上使用命令"zkServer.sh start"启动 Zookeeper 服务，如图 2-3-8 所示，若启动过程中没有报错并显示"STARTED"则表示启动成功。然后查看 Java 进程信息，如图 2-3-9 所示，若存在一个名为"QuorumPeerMain"的进程，则表示 Zookeeper 服务正常运行。

```
[hadoop@Cluster-01 ~]$ zkServer.sh start
ZooKeeper JMX enabled by default
Using config: /home/hadoop/zookeeper/zookeeper-3.4.9/bin/../conf/zoo.cfg
Starting zookeeper ... STARTED
```

图 2-3-8　启动 Zookeeper 服务

```
[hadoop@Cluster-01 ~]$ jps
5139 QuorumPeerMain
5158 Jps
```

图 2-3-9　查看 Java 进程信息

（2）在集群中所有主机上使用命令"zkServer.sh status"查看该节点当前 Zookeeper 服务的状态，如图 2-3-10 所示，可以看到集群中有一个"leader"节点，其余的均为"follower"节点。

```
[hadoop@Cluster-01 ~]$ zkServer.sh status
ZooKeeper JMX enabled by default
Using config: /home/hadoop/zookeeper/zookeeper-3.4.9/bin/../conf/zoo.cfg
Mode: follower

[hadoop@Cluster-02 ~]$ zkServer.sh status
ZooKeeper JMX enabled by default
Using config: /home/hadoop/zookeeper/zookeeper-3.4.9/bin/../conf/zoo.cfg
Mode: follower

[hadoop@Cluster-03 ~]$ zkServer.sh status
ZooKeeper JMX enabled by default
Using config: /home/hadoop/zookeeper/zookeeper-3.4.9/bin/../conf/zoo.cfg
Mode: leader

[hadoop@Cluster-04 ~]$ zkServer.sh status
ZooKeeper JMX enabled by default
Using config: /home/hadoop/zookeeper/zookeeper-3.4.9/bin/../conf/zoo.cfg
Mode: follower

[hadoop@Cluster-05 ~]$ zkServer.sh status
ZooKeeper JMX enabled by default
Using config: /home/hadoop/zookeeper/zookeeper-3.4.9/bin/../conf/zoo.cfg
Mode: follower
```

图 2-3-10　查看所有主机的 Zookeeper 服务状态

（3）使用命令"zkCli.sh -server 主机名：2181"并利用 Zookeeper 的客

户端工具连接到 Zookeeper 集群，可以选择集群中的任意一台主机进行连接，如图 2-3-11 所示，若没有出现错误信息并正常进入 Zookeeper 客户端程序的控制台界面，同时显示包含“CONNECTED”内容的命令行提示符，则表示 Zookeeper 服务连接正常。在 Zookeeper 控制台界面中使用命令“quit”可以退出客户端程序。

```
[hadoop@Cluster-01 ~]$ zkCli.sh -server Cluster-02:2181
Connecting to Cluster-02:2181
2018-06-10 21:08:12,424 [myid:] - INFO  [main:Environment@100] - Client environment:zookeeper.versio
n=3.4.9-1757313, built on 08/23/2016 06:50 GMT
2018-06-10 21:08:12,436 [myid:] - INFO  [main:Environment@100] - Client environment:host.name=Cluste
r-01
2018-06-10 21:08:12,437 [myid:] - INFO  [main:Environment@100] - Client environment:java.version=1.8
.0_131
2018-06-10 21:08:12,442 [myid:] - INFO  [main:Environment@100] - Client environment:java.vendor=Orac
le Corporation
2018-06-10 21:08:12,444 [myid:] - INFO  [main:Environment@100] - Client environment:java.home=/home/
hadoop/java/jdk1.8.0_131/jre
2018-06-10 21:08:12,448 [myid:] - INFO  [main:Environment@100] - Client environment:java.class.path=
/home/hadoop/zookeeper/zookeeper-3.4.9/bin/../build/classes:/home/hadoop/zookeeper/zookeeper-3.4.9/b
in/../build/lib/*.jar:/home/hadoop/zookeeper/zookeeper-3.4.9/bin/../lib/slf4j-log4j12-1.6.1.jar:/hom
e/hadoop/zookeeper/zookeeper-3.4.9/bin/../lib/slf4j-api-1.6.1.jar:/home/hadoop/zookeeper/zookeeper-3
.4.9/bin/../lib/netty-3.10.5.Final.jar:/home/hadoop/zookeeper/zookeeper-3.4.9/bin/../lib/log4j-1.2.1
6.jar:/home/hadoop/zookeeper/zookeeper-3.4.9/bin/../lib/jline-0.9.94.jar:/home/hadoop/zookeeper/zook
eeper-3.4.9/bin/../zookeeper-3.4.9.jar:/home/hadoop/zookeeper/zookeeper-3.4.9/bin/../src/java/lib/*.
jar:/home/hadoop/zookeeper/zookeeper-3.4.9/bin/../conf:.:/home/hadoop/java/jdk1.8.0_131/lib/tools.ja
r:/home/hadoop/java/jdk1.8.0_131/lib/dt.jar
```

```
2018-06-10 21:08:12,456 [myid:] - INFO  [main:Environment@100] - Client environment:java.library.pat
h=/usr/java/packages/lib/amd64:/usr/lib64:/lib64:/lib:/usr/lib
2018-06-10 21:08:12,458 [myid:] - INFO  [main:Environment@100] - Client environment:java.io.tmpdir=/
tmp
2018-06-10 21:08:12,460 [myid:] - INFO  [main:Environment@100] - Client environment:java.compiler=<N
A>
2018-06-10 21:08:12,463 [myid:] - INFO  [main:Environment@100] - Client environment:os.name=Linux
2018-06-10 21:08:12,464 [myid:] - INFO  [main:Environment@100] - Client environment:os.arch=amd64
2018-06-10 21:08:12,465 [myid:] - INFO  [main:Environment@100] - Client environment:os.version=3.10.
0-514.el7.x86_64
2018-06-10 21:08:12,468 [myid:] - INFO  [main:Environment@100] - Client environment:user.name=hadoop
2018-06-10 21:08:12,470 [myid:] - INFO  [main:Environment@100] - Client environment:user.home=/home/
hadoop
2018-06-10 21:08:12,473 [myid:] - INFO  [main:Environment@100] - Client environment:user.dir=/home/h
adoop
2018-06-10 21:08:12,478 [myid:] - INFO  [main:ZooKeeper@438] - Initiating client connection, connect
String=Cluster-02:2181 sessionTimeout=30000 watcher=org.apache.zookeeper.ZooKeeperMain$MyWatcher@277
050dc
Welcome to ZooKeeper!
2018-06-10 21:08:12,548 [myid:] - INFO  [main-SendThread(Cluster-02:2181):ClientCnxn$SendThread@1032
] - Opening socket connection to server Cluster-02/192.168.60.131:2181. Will not attempt to authenti
cate using SASL (unknown error)
JLine support is enabled
2018-06-10 21:08:12,770 [myid:] - INFO  [main-SendThread(Cluster-02:2181):ClientCnxn$SendThread@876]
 - Socket connection established to Cluster-02/192.168.60.131:2181, initiating session
2018-06-10 21:08:12,803 [myid:] - INFO  [main-SendThread(Cluster-02:2181):ClientCnxn$SendThread@1299
] - Session establishment complete on server Cluster-02/192.168.60.131:2181, sessionid = 0x263ec4350
930001, negotiated timeout = 30000

WATCHER::

WatchedEvent state:SyncConnected type:None path:null
[zk: Cluster-02:2181(CONNECTED) 0]
```

图 2-3-11　连接主机

2.3.4　安装与配置的常见问题及解决方法

集群模式的 Zookeeper 的安装与配置和单节点模式的 Zookeeper 的安装和配置的常见问题基本相同，主要就是服务节点的服务进程启动失败，使用客户端工具连接 Zookeeper 服务失败等问题。查找问题原因与解决问题方式也与单节点模式的 Zookeeper 的安装和配置中查找问题原因与解决问题方式基本相同，主要就是检查配置文件中各个配置项的名称和值是否书写正确，或通过查看 Zookeeper 的日志文件来确定问题和错误的详细信息。不过需要注意的是，在集群模式下每个 Zookeeper 服务节点的主机都会各自有一个日志文件存放于本地文件系统的指定日志文件的存放目录中，所以当某个节点的安装与配置出现问题时，只需要查看该问题节点主机上的日志文件即可。

任务 2.4　安装 Hadoop 完全分布模式

2.4.1　Hadoop 完全分布模式安装规划

1. 环境要求

集群中所有主机的本地磁盘剩余空间 400 MB 以上；
集群中所有主机已安装 CentOS 7 1611 64 位操作系统；
集群中所有主机已完成基础网络环境配置；
集群中所有主机已安装 JDK 的 1.8.0_131 版本。

2. 软件版本

与单节点模式 Hadoop 整合平台搭建中相同，选用 Hadoop 的 2.7.3 版本，软件包名为 hadoop-2.7.3.tar.gz。

3. 服务规划

（1）Hadoop 完全分布模式包含的服务和伪分布模式包含的服务一样，包括 HDFS 的 NameNode、SecondaryNameNode、DataNode 和 YARN 的 ResourceManager、NodeManager，共五种类型的服务节点。

（2）HDFS 的 NameNode 和 SecondaryNameNode 以及 YARN 的 ResourceManager 三类服务对主机的性能要求较高，相互之间会发生争抢资源的情况。在完全分布模式下集群中若有足够的主机数量，一般推荐每个服务各自使用一台独立主机来运行。

（3）剩下的 HDFS 的 DataNode 和 YARN 的 NodeManager 两类服务作为 HDFS 的数据存储服务和 MapReduce 的任务管理服务，默认同时运行在同一台主机上，以达到 MapReduce 中数据本地化处理的目的。同时在完全分布模式下若要满足分布式并行处理的需求,则需要集群中具备至少两个或以上的对应服务节点。

(4)对于当前的总共五台主机的集群,具备了将 HDFS 的 NameNode 和 SecondaryNameNode 以及 YARN 的 ResourceManager 分别独立分配一台主机,同时 HDFS 的 DataNode 和 YARN 的 NodeManager 拥有两个主机来运行的条件,所以具体 IP 地址和服务节点规划如表 2-4-1 所示。

表 2-4-1　IP 地址与服务节点规划

主机名	IP 地址	服务描述
Cluster-01	192.168.60.130	HDFS 的 NameNode
Cluster-02	192.168.60.131	HDFS 的 SecondaryNameNode
Cluster-03	192.168.60.132	YARN 的 ResourceManager
Cluster-04	192.168.60.133	HDFS 的 DataNode YARN 的 NodeManager
Cluster-05	192.168.60.134	HDFS 的 DataNode YARN 的 NodeManager

2.4.2　Hadoop 完全分布模式配置

(1)配置集群主机之间的免密码登录,要求 HDFS 的 NameNode 服务节点能够免密码登录到集群中包括自身在内的所有主机,同时 YARN 的 ResourceManager 服务节点能够免密码登录除 HDFS 的 SecondaryNameNode 服务节点之外的集群中(包括自身在内)的所有主机。

(2)选择集群中用作 HDFS 的 NameNode 服务节点的主机,在其上的 Hadoop 专用用户下完成 Hadoop 的基础安装和环境配置。安装配置方法与单节点 Hadoop 平台的搭建过程中 Hadoop 的基础环境配置方法相同,解压 Hadoop 软件包并配置环境变量文件。

```
[hadoop@Cluster-01 hadoop]$ ls -l
total 0
drwx------.  3 hadoop hadoop  21 May  4 04:54 data
drwxr-xr-x. 10 hadoop hadoop 161 Apr 26 04:52 hadoop-2.7.3
drwxrwxr-x.  3 hadoop hadoop  21 May  4 04:54 name
drwxrwxr-x.  4 hadoop hadoop  37 Apr 26 22:04 tmp
```

图 2-4-1　创建目录

(3)进入 Hadoop 相关文件的目录,如图 2-4-1 所示,分别创建 Hadoop 的临时文件的存放目录"tmp"、HDFS 的元数据文件的存放目录"name"、HDFS 的数据文件的存放目录"data"。

（4）进入 Hadoop 软件目录下的配置文件所在目录"etc/had oop"。

（5）编辑配置文件"hadoop-env.sh"，编辑的内容与伪分布模式相同，如图 2-4-2 所示，找到配置项"JAVA_HOME"并将其对应的值改为 JDK 软件目录所在路径。

```
19 # The only required environment variable is JAVA_HOME. All others are
20 # optional.  When running a distributed configuration it is best to
21 # set JAVA_HOME in this file, so that it is correctly defined on
22 # remote nodes.
23
24 # The java implementation to use.
25 export JAVA_HOME=/home/hadoop/java/jdk1.8.0_131
```

图 2-4-2　编辑配置文件"hadoop-env.sh"

（6）编辑配置文件"core-site.xml"，在其中配置以下内容：

<configuration>
  <property>
    <name>fs.defaultFS</name>
    <value>hdfs：//Cluster-01：9000</value>
  </property>
  <property>
    <name>hadoop.tmp.dir</name>
    <value>Hadoop 的临时文件的存放目录"tmp"的路径</value>
  </property>
  <property>
    <name>fs.checkpoint.period</name>
    <value>3600</value>
  </property>
  <property>
    <name>fs.checkpoint.size</name>
    <value>67108864</value>
  </property>
</configuration>

配置项说明：

fs.checkpoint.period——指定 HDFS 的 SecondaryNameNode 节点触发检查点

操作的时间间隔条件，单位为秒，超过设定时间间隔便会执行检查点操作。检查点操作即合并事务日志文件 edits 和映像文件 fsimage 的操作。

fs.checkpoint.size——指定 HDFS 的 SecondaryNameNode 节点触发检查点操作的文件大小条件，单位为字节。这里的文件大小指的是事务文件 edits 的大小，超过设定的文件大小便会执行检查点操作。与之前的时间间隔条件一起，只要满足两者的其中一个，便会执行检查点操作。

（7）编辑配置文件"hdfs-site.xml"，在其中配置以下内容：

```
<configuration>
    <property>
        <name>dfs.namenode.name.dir</name>
        <value>HDFS 的 NameNode 节点的元数据文件的存放目录"name"的路径</value>
    </property>
    <property>
        <name>dfs.datanode.data.dir</name>
        <value>HDFS 的 DataNode 节点的数据文件的存放目录"data"的路径</value>
    </property>
    <property>
        <name>dfs.replication</name>
        <value>2</value>
    </property>
    <property>
        <name>dfs.http.address</name>
        <value>Cluster-01：50070</value>
    </property>
    <property>
        <name>dfs.namenode.secondary.http-address</name>
        <value>Cluster-02：50090</value>
    </property>
</configuration>
```

配置项说明：

dfs.http.address——指定 HDFS 的 NameNode 节点的 Web 管理平台的访问地

址和端口。

dfs.namenode.secondary.http-address——指定 HDFS 的 Secondary Name Node 节点的 Tracker 页面的访问地址和端口。

（8）从 MapReduce 配置文件的模板文件生成配置文件"mapred-site.xml"，然后编辑该配置文件，在其中配置以下内容：

```
<configuration>
  <property>
    <name>mapreduce.framework.name</name>
    <value>yarn</value>
  </property>
</configuration>
```

（9）编辑配置文件"yarn-env.sh"，编辑的内容与伪分布模式相同，如图 2-4-3 所示，找到配置项"JAVA_HOME"并将其对应的值改为 JDK 软件目录所在路径。

```
22 # some Java parameters
23 export JAVA_HOME=/home/hadoop/java/jdk1.8.0_131
```

图 2-4-3　编辑配置文件"yarn-env.sh"

（10）编辑配置文件"yarn-site.xml"，在其中配置以下内容：

```
<configuration>
  <property>
    <name>yarn.nodemanager.aux-services</name>
    <value>mapreduce_shuffle</value>
  </property>
  <property>
    <name>yarn.resourcemanager.address</name>
    <value>Cluster-03：8032</value>
  </property>
  <property>
    <name>yarn.resourcemanager.scheduler.address</name>
    <value>Cluster-03：8030</value>
  </property>
  <property>
    <name>yarn.resourcemanager.resource-tracker.address</name>
```

　　　　<value>Cluster-03：8031</value>
　　</property>
</configuration>
配置项说明：

　　yarn.resourcemanager.address——指定 YARN 的 ResourceManager 服务节点的地址和端口。

　　yarn.resourcemanager.scheduler.address——指定 YARN 的 Resource Manager 服务节点的计划服务的地址和端口。

　　yarn.resourcemanager.resource-tracker.address——指定 YARN 的 Resource Manager 服务节点的资源服务的地址和端口。

　　（11）创建并编辑配置文件"masters"，然后在其中添加 HDFS 的 SecondaryNameNode 节点的主机名"Cloud-02"。

　　（12）编辑配置文件"slaves"，删除文件中原有的所有内容，然后在其中按照每行一个主机名的方式，添加 HDFS 的 DataNode 节点与 YARN 的 NodeManager 节点的主机名"Cluster-04"和"Cluster-05"。

　　（13）使用命令"scp　~/.bash_profile　用户名@目标主机名或 IP 地址：/home/用户名/"和"scp　-r　~/hadoop　用户名@目标主机名或 IP 地址：/home/用户名/"依次将新修改的用户环境变量设置文件和 Hadoop 的专用目录发送到集群中所有其他主机的相应用户家目录下。

2.4.3　Hadoop 完全分布模式格式化和启动

　　（1）使用命令"hadoop　namenode　-format"对 HDFS 的文件系统进行格式化，如图 2-4-4 所示，若格式化过程中没有出现报错提示信息，则表示格式化过程成功完成。

项目2 搭建高可用 Hadoop 整合平台

```
.3/share/hadoop/yarn/lib/servlet-api-2.5.jar:/home/hadoop/hadoop-2.7.3/share/hadoop/yarn/lib/
commons-codec-1.4.jar:/home/hadoop/hadoop-2.7.3/share/hadoop/yarn/lib/jetty-util-6.1.26.jar:/
home/hadoop/hadoop-2.7.3/share/hadoop/yarn/lib/jersey-core-1.9.jar:/home/hadoop/hadoop
-2.7.3/share/hadoop/yarn/lib/jersey-client-1.9.jar:/home/hadoop/hadoop-2.7.3/share/hadoop/yar
n/lib/jackson-core-asl-1.9.13.jar:/home/hadoop/hadoop-2.7.3/share/hadoop/yarn/lib/jackson-map
per-asl-1.9.13.jar:/home/hadoop/hadoop-2.7.3/share/hadoop/yarn/lib/jackson-jaxrs-1.9.13.jar:/
home/hadoop/hadoop-2.7.3/share/hadoop/yarn/lib/jackson-xc-1.9.13.jar:/home/hadoop/hado
op-2.7.3/share/hadoop/yarn/lib/guice-servlet-3.0.jar:/home/hadoop/hadoop-2.7.3/share/hadoop/y
arn/lib/guice-3.0.jar:/home/hadoop/hadoop-2.7.3/share/hadoop/yarn/lib/javax.inject-1.jar:/hom
e/hadoop/hadoop-2.7.3/share/hadoop/yarn/lib/aopalliance-1.0.jar:/home/hadoop/hadoop-2.
7.3/share/hadoop/yarn/lib/commons-io-2.4.jar:/home/hadoop/hadoop-2.7.3/share/hadoop/yarn/lib/
jersey-server-1.9.jar:/home/hadoop/hadoop-2.7.3/share/hadoop/yarn/lib/asm-3.2.jar:/home/hadoo
p/hadoop/hadoop-2.7.3/share/hadoop/yarn/lib/jersey-json-1.9.jar:/home/hadoop/hadoop-2.7.3/sha
re/hadoop/yarn/lib/jettison-1.1.jar:/home/hadoop/hadoop-2.7.3/share/hadoop/yarn/lib/jaxb-impl
-2.2.3-1.jar:/home/hadoop/hadoop-2.7.3/share/hadoop/yarn/lib/jersey-guice-1.9.jar:/home/hadoo
p/hadoop/hadoop-2.7.3/share/hadoop/yarn/lib/zookeeper-3.4.6.jar:/home/hadoop/hadoop-2.7.3/sha
re/hadoop/yarn/lib/netty-3.6.2.Final.jar:/home/hadoop/hadoop-2.7.3/share/hadoop/yarn/lib/leve
ldbjni-all-1.8.jar:/home/hadoop/hadoop-2.7.3/share/hadoop/yarn/lib/commons-collections-3.2.2.
jar:/home/hadoop/hadoop-2.7.3/share/hadoop/yarn/lib/jetty-6.1.26.jar:/home/hadoop/hado
op-2.7.3/share/hadoop/yarn/hadoop-yarn-api-2.7.3.jar:/home/hadoop/hadoop-2.7.3/share/hadoop/y
arn/hadoop-yarn-common-2.7.3.jar:/home/hadoop/hadoop-2.7.3/share/hadoop/yarn/hadoop-yarn-serv
er-common-2.7.3.jar:/home/hadoop/hadoop-2.7.3/share/hadoop/yarn/hadoop-yarn-server-nodemanage
r-2.7.3.jar:/home/hadoop/hadoop-2.7.3/share/hadoop/yarn/hadoop-yarn-server-web-proxy-2.7.3.ja
r:/home/hadoop/hadoop-2.7.3/share/hadoop/yarn/hadoop-yarn-server-applicationhistoryservice-2.
7.3.jar:/home/hadoop/hadoop-2.7.3/share/hadoop/yarn/hadoop-yarn-server-resourcemanager-2.7.3.
jar:/home/hadoop/hadoop-2.7.3/share/hadoop/yarn/hadoop-yarn-server-tests-2.7.3.jar:/home/hado
op/hadoop/hadoop-2.7.3/share/hadoop/yarn/hadoop-yarn-client-2.7.3.jar:/home/hadoop/hadoop-2.7
.3/share/hadoop/yarn/hadoop-yarn-server-sharedcachemanager-2.7.3.jar:/home/hadoop/hadoop-2.7.
3/share/hadoop/yarn/hadoop-yarn-applications-distributedshell-2.7.3.jar:/home/hadoop/hadoop-2
.7.3/share/hadoop/yarn/hadoop-yarn-applications-unmanaged-am-launcher-2.7.3.jar:/home/hadoop/
hadoop-2.7.3/share/hadoop/yarn/hadoop-yarn-registry-2.7.3.jar:/home/hadoop/hadoop-2.7.3/share
/hadoop/mapreduce/lib/protobuf-java-2.5.0.jar:/home/hadoop/hadoop-2.7.3/share/hadoop/mapreduc
e/lib/avro-1.7.4.jar:/home/hadoop/hadoop-2.7.3/share/hadoop/mapreduce/lib/jackson-core-asl-1.
9.13.jar:/home/hadoop/hadoop-2.7.3/share/hadoop/mapreduce/lib/jackson-mapper-asl-1.9.13.jar:/
home/hadoop/hadoop-2.7.3/share/hadoop/mapreduce/lib/paranamer-2.3.jar:/home/hadoop/had
oop-2.7.3/share/hadoop/mapreduce/lib/snappy-java-1.0.4.1.jar:/home/hadoop/hadoop-2.7.3/share/
```

```
oop-2.7.3/share/hadoop/mapreduce/lib/snappy-java-1.0.4.1.jar:/home/hadoop/hadoop-2.7.3/share/
hadoop/mapreduce/lib/commons-compress-1.4.1.jar:/home/hadoop/hadoop-2.7.3/share/hadoop/mapred
uce/lib/xz-1.0.jar:/home/hadoop/hadoop-2.7.3/share/hadoop/mapreduce/lib/hadoop-annotations-2.
7.3.jar:/home/hadoop/hadoop-2.7.3/share/hadoop/mapreduce/lib/commons-io-2.4.jar:/home/hadoop/
hadoop-2.7.3/share/hadoop/mapreduce/lib/jersey-core-1.9.jar:/home/hadoop/hadoop-2.7.3/
share/hadoop/mapreduce/lib/jersey-server-1.9.jar:/home/hadoop/hadoop-2.7.3/share/hadoop/mapre
duce/lib/asm-3.2.jar:/home/hadoop/hadoop-2.7.3/share/hadoop/mapreduce/lib/log4j-1.2.17.jar:/h
ome/hadoop/hadoop/hadoop-2.7.3/share/hadoop/mapreduce/lib/netty-3.6.2.Final.jar:/home/hadoop/
hadoop-2.7.3/share/hadoop/mapreduce/lib/leveldbjni-all-1.8.jar:/home/hadoop/hadoop-2.7.3/shar
e/hadoop/mapreduce/lib/guice-3.0.jar:/home/hadoop/hadoop-2.7.3/share/hadoop/mapreduce/lib/jav
ax.inject-1.jar:/home/hadoop/hadoop-2.7.3/share/hadoop/mapreduce/lib/aopalliance-1.0.jar:/hom
e/hadoop/hadoop-2.7.3/share/hadoop/mapreduce/lib/jersey-guice-1.9.jar:/home/hadoop/had
oop-2.7.3/share/hadoop/mapreduce/lib/guice-servlet-3.0.jar:/home/hadoop/hadoop-2.7.3/share/ha
doop/mapreduce/lib/junit-4.11.jar:/home/hadoop/hadoop-2.7.3/share/hadoop/mapreduce/lib/hamcre
st-core-1.3.jar:/home/hadoop/hadoop-2.7.3/share/hadoop/mapreduce/hadoop-mapreduce-client-core
-2.7.3.jar:/home/hadoop/hadoop-2.7.3/share/hadoop/mapreduce/hadoop-mapreduce-client-common-2.
7.3.jar:/home/hadoop/hadoop-2.7.3/share/hadoop/mapreduce/hadoop-mapreduce-client-shuffle-2.7.
3.jar:/home/hadoop/hadoop-2.7.3/share/hadoop/mapreduce/hadoop-mapreduce-client-app-2.7.3.jar:
/home/hadoop/hadoop-2.7.3/share/hadoop/mapreduce/hadoop-mapreduce-client-hs-2.7.3.jar:/home/h
adoop/hadoop/hadoop-2.7.3/share/hadoop/mapreduce/hadoop-mapreduce-client-jobclient-2.7.3.jar:/home/h
adoop/hadoop/hadoop-2.7.3/share/hadoop/mapreduce/hadoop-mapreduce-client-hs-plugins-2.7.3.jar:/home/
hadoop/hadoop/hadoop-2.7.3/share/hadoop/mapreduce/hadoop-mapreduce-examples-2.7.3.jar:/home/h
adoop/hadoop/hadoop-2.7.3/share/hadoop/mapreduce/hadoop-mapreduce-client-jobclient-2.7.3-tests.jar:/home/ha
doop/hadoop/hadoop-2.7.3/contrib/capacity-scheduler/*.jar:/home/hadoop/hadoop-2.7.3/contrib/c
apacity-scheduler/*.jar
```

```
STARTUP_MSG:   build = https://git-wip-us.apache.org/repos/asf/hadoop.git -r baa91f7c6bc9cb92be5982d
e4719c1c8af91ccff; compiled by 'root' on 2016-08-18T01:41Z
STARTUP_MSG:   java = 1.8.0_131
************************************************************/
18/06/11 03:50:49 INFO namenode.NameNode: registered UNIX signal handlers for [TERM, HUP, INT]
18/06/11 03:50:49 INFO namenode.NameNode: createNameNode [-format]
18/06/11 03:50:50 WARN common.Util: Path /home/hadoop/hadoop/name should be specified as a URI in co
nfiguration files. Please update hdfs configuration.
18/06/11 03:50:50 WARN common.Util: Path /home/hadoop/hadoop/name should be specified as a URI in co
nfiguration files. Please update hdfs configuration.
Formatting using clusterid: CID-e4e6134b-907c-4551-b329-b520e1cf949a
18/06/11 03:50:50 INFO namenode.FSNamesystem: No KeyProvider found.
18/06/11 03:50:50 INFO namenode.FSNamesystem: fsLock is fair:true
18/06/11 03:50:50 INFO blockmanagement.DatanodeManager: dfs.block.invalidate.limit=1000
18/06/11 03:50:50 INFO blockmanagement.DatanodeManager: dfs.namenode.datanode.registration.ip-hostna
```

```
me-check=true
18/06/11 03:50:50 INFO blockmanagement.BlockManager: dfs.namenode.startup.delay.block.deletion.sec i
s set to 000:00:00:00.000
18/06/11 03:50:50 INFO blockmanagement.BlockManager: The block deletion will start around 2018 Jun 1
1 03:50:50
18/06/11 03:50:50 INFO util.GSet: Computing capacity for map BlocksMap
18/06/11 03:50:50 INFO util.GSet: VM type       = 64-bit
18/06/11 03:50:50 INFO util.GSet: 2.0% max memory 966.7 MB = 19.3 MB
18/06/11 03:50:50 INFO util.GSet: capacity      = 2^21 = 2097152 entries
18/06/11 03:50:51 INFO blockmanagement.BlockManager: dfs.block.access.token.enable=false
18/06/11 03:50:51 INFO blockmanagement.BlockManager: defaultReplication         = 2
18/06/11 03:50:51 INFO blockmanagement.BlockManager: maxReplication             = 512
18/06/11 03:50:51 INFO blockmanagement.BlockManager: minReplication             = 1
18/06/11 03:50:51 INFO blockmanagement.BlockManager: maxReplicationStreams      = 2
18/06/11 03:50:51 INFO blockmanagement.BlockManager: replicationRecheckInterval = 3000
18/06/11 03:50:51 INFO blockmanagement.BlockManager: encryptDataTransfer        = false
18/06/11 03:50:51 INFO blockmanagement.BlockManager: maxNumBlocksToLog          = 1000
18/06/11 03:50:51 INFO namenode.FSNamesystem: fsOwner             = hadoop (auth:SIMPLE)
18/06/11 03:50:51 INFO namenode.FSNamesystem: supergroup          = supergroup
18/06/11 03:50:51 INFO namenode.FSNamesystem: isPermissionEnabled = true
18/06/11 03:50:51 INFO namenode.FSNamesystem: HA Enabled: false
18/06/11 03:50:51 INFO namenode.FSNamesystem: Append Enabled: true
18/06/11 03:50:51 INFO util.GSet: Computing capacity for map INodeMap
18/06/11 03:50:51 INFO util.GSet: VM type       = 64-bit
18/06/11 03:50:51 INFO util.GSet: 1.0% max memory 966.7 MB = 9.7 MB
18/06/11 03:50:51 INFO util.GSet: capacity      = 2^20 = 1048576 entries
18/06/11 03:50:51 INFO namenode.FSDirectory: ACLs enabled? false
18/06/11 03:50:51 INFO namenode.FSDirectory: XAttrs enabled? true
18/06/11 03:50:51 INFO namenode.FSDirectory: Maximum size of an xattr: 16384
18/06/11 03:50:51 INFO namenode.NameNode: Caching file names occuring more than 10 times
18/06/11 03:50:51 INFO util.GSet: Computing capacity for map cachedBlocks
18/06/11 03:50:51 INFO util.GSet: VM type       = 64-bit
18/06/11 03:50:51 INFO util.GSet: 0.25% max memory 966.7 MB = 2.4 MB
18/06/11 03:50:51 INFO util.GSet: capacity      = 2^18 = 262144 entries
18/06/11 03:50:51 INFO namenode.FSNamesystem: dfs.namenode.safemode.threshold-pct = 0.9990000128746
033
18/06/11 03:50:51 INFO namenode.FSNamesystem: dfs.namenode.safemode.min.datanodes = 0
18/06/11 03:50:51 INFO namenode.FSNamesystem: dfs.namenode.safemode.extension     = 30000
18/06/11 03:50:51 INFO metrics.TopMetrics: NNTop conf: dfs.namenode.top.window.num.buckets = 10
18/06/11 03:50:51 INFO metrics.TopMetrics: NNTop conf: dfs.namenode.top.num.users = 10
18/06/11 03:50:51 INFO metrics.TopMetrics: NNTop conf: dfs.namenode.top.windows.minutes = 1,5,25
18/06/11 03:50:51 INFO namenode.FSNamesystem: Retry cache on namenode is enabled
18/06/11 03:50:51 INFO namenode.FSNamesystem: Retry cache will use 0.03 of total heap and retry cach
e entry expiry time is 600000 millis
18/06/11 03:50:51 INFO util.GSet: Computing capacity for map NameNodeRetryCache
18/06/11 03:50:51 INFO util.GSet: VM type       = 64-bit
18/06/11 03:50:51 INFO util.GSet: 0.029999999329447746% max memory 966.7 MB = 297.0 KB
18/06/11 03:50:51 INFO util.GSet: capacity      = 2^15 = 32768 entries
18/06/11 03:50:52 INFO namenode.FSImage: Allocated new BlockPoolId: BP-1012000605-192.168.60.130-152
8703451995
18/06/11 03:50:52 INFO common.Storage: Storage directory /home/hadoop/hadoop/name has been successfu
lly formatted.
18/06/11 03:50:52 INFO namenode.FSImageFormatProtobuf: Saving image file /home/hadoop/hadoop/name/cu
rrent/fsimage.ckpt_0000000000000000000 using no compression
18/06/11 03:50:52 INFO namenode.FSImageFormatProtobuf: Image file /home/hadoop/hadoop/name/current/f
simage.ckpt_0000000000000000000 of size 353 bytes saved in 0 seconds.
18/06/11 03:50:52 INFO namenode.NNStorageRetentionManager: Going to retain 1 images with txid >= 0
18/06/11 03:50:52 INFO util.ExitUtil: Exiting with status 0
18/06/11 03:50:52 INFO namenode.NameNode: SHUTDOWN_MSG:
/************************************************************
SHUTDOWN_MSG: Shutting down NameNode at Cluster-01/192.168.60.130
************************************************************/
```

图 2-4-4　对 HDFS 文件系统进行格式化

（2）如图 2-4-5 所示，在提供 HDFS 的 NameNode 服务的主机 Cluster-01 上使用命令“start-dfs.sh”启动 Hadoop 平台的 HDFS 文件系统。

（3）如图 2-4-6 所示，在提供 YARN 的 ResourceManager 服务的主机 Cluster-03 上使用命令“start-yarn.sh”启动 Hadoop 平台的 YARN 资源管理器，可以远程登录到该主机执行启动命令。

```
[hadoop@Cluster-01 mapreduce]$ start-dfs.sh
Starting namenodes on [Cluster-01]
Cluster-01: starting namenode, logging to /home/hadoop/hadoop/hadoop-2.7.3/logs/hadoop-hadoop-nameno
de-Cluster-01.out
Cluster-04: starting datanode, logging to /home/hadoop/hadoop/hadoop-2.7.3/logs/hadoop-hadoop-datano
de-Cluster-04.out
Cluster-05: starting datanode, logging to /home/hadoop/hadoop/hadoop-2.7.3/logs/hadoop-hadoop-datano
de-Cluster-05.out
Starting secondary namenodes [Cluster-02]
Cluster-02: starting secondarynamenode, logging to /home/hadoop/hadoop/hadoop-2.7.3/logs/hadoop-hado
op-secondarynamenode-Cluster-02.out
```

图 2-4-5　启动 HDFS 文件

```
[hadoop@Cluster-03 ~]$ start-yarn.sh
starting yarn daemons
starting resourcemanager, logging to /home/hadoop/hadoop/hadoop-2.7.3/logs/yarn-hadoop-resourcemanag
er-Cluster-03.out
Cluster-04: starting nodemanager, logging to /home/hadoop/hadoop/hadoop-2.7.3/logs/yarn-hadoop-nodem
anager-Cluster-04.out
Cluster-05: starting nodemanager, logging to /home/hadoop/hadoop/hadoop-2.7.3/logs/yarn-hadoop-nodem
anager-Cluster-05.out
```

图 2-4-6　远程登录主机

（4）查看集群中各个主机的 Java 进程信息，如图 2-4-7 所示，若满足以下的主机与进程名的对应关系，则表示 Hadoop 完全分布模式的集群启动成功。

```
[hadoop@Cluster-01 ~]$ jps
2827 QuorumPeerMain
7228 NameNode
10766 Jps

[hadoop@Cluster-02 ~]$ jps
4406 QuorumPeerMain
8554 Jps
5195 SecondaryNameNode

[hadoop@Cluster-03 logs]$ jps
10803 QuorumPeerMain
12827 Jps
11948 ResourceManager

[hadoop@Cluster-04 logs]$ jps
12211 NodeManager
10084 QuorumPeerMain
12101 DataNode
13373 Jps

[hadoop@Cluster-05 logs]$ jps
2418 QuorumPeerMain
5330 Jps
4169 NodeManager
4060 DataNode
```

图 2-4-7　查看各主机的 Java 进程信息

Cluster-01　　　　NameNode
Cluster-02　　　　SecondaryNameNode

Cluster-03	ResourceManager
Cluster-04	DataNode、NodeManager
Cluster-05	DataNode、NodeManager

（5）验证 Hadoop 的完全分布模式平台是否能够正常使用，验证方法与单节点 Hadoop 平台的验证方法相同，在通过创建并查看目录来验证 HDFS 文件系统，然后使用 Hadoop 自带的示例程序包来验证通过 YARN 资源管理器运行 MapReduce。

2.4.4 安装和配置的常见问题及解决方法

完全分布模式的 Hadoop 的安装与配置和单节点模式的 Hadoop 的安装和配置的常见问题基本相同，主要就是服务节点的服务进程启动失败。

查找问题原因与解决问题方式也与单节点模式的 Hadoop 的安装和配置中查找问题原因与解决问题方式基本相同，主要就是检查配置文件中各个配置项的名称和值是否书写正确，或通过查看 Hadoop 的日志文件来确定问题和错误的详细信息。不过需要注意的是，在集群模式下，每个 Hadoop 服务节点的主机上只有自身所提供的 Hadoop 服务所对应的日志文件，如 NameNode 服务节点上只有 NameNode 的日志文件"hadoop-用户名-namenode-主机名.log"，ResourceManager 服务节点上只有 ResourceManager 的日志文件"yarn-用户名-resourcemanager-主机名.log"，而对于同时作为 HDFS 的 DataNode 服务节点和 YARN 的 NodeManager 服务节点的主机，则会有"hadoop-用户名-datanode-主机名.log"和"yarn-用户名-nodemanager-主机名.log"两个日志文件。所以当某个节点的安装与配置出现问题时，只需要查看该问题节点的主机上的对应日志文件即可。

同样，若是修改了主机名、文件目录路径等方面的内容，建议将创建的三个功能目录"tmp""name""data"中的内容清空，并重新将 HDFS 文件系统进行格式化之后，再启动 Hadoop 平台服务。

任务 2.5 安装 Hadoop 高可用模式

2.5.1 HDFS 高可用架构

在 Hadoop 2.x 版本中，HDFS 架构解决了 NameNode 的单点故障问题，引入了双 NameNode 架构,如图 2-5-1 所示。同时借助共享存储系统来进行元数据的同步,共享存储系统类型一般有几类,如 Shared NAS+NFS、BookKeeper、BackupNode、Quorum Journal Manager(QJM),Hadoop 2.x 中使用的是 QJM 作为共享存储组件,通过搭建奇数个 JournalNode 服务节点来实现主备 NameNode 服务节点之间的元数据操作信息同步。

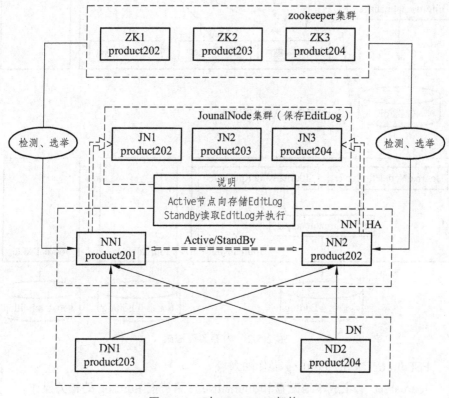

图 2-5-1 对 NameNode 架构

2.5.1.1 QJM 原理

QJM 全称为 Quorum Journal Manager,由多个 JournalNode(JN)服务节点组成,遵循的是 Paxos 协议,所以节点数量一般为奇数。每个 JournalNode 服务节点对外提供一个简易的 RPC 接口,让 NameNode 服务读写 EditLog 到 JournalNode 服务节点的本地磁盘。当需要写入 EditLog 时,NameNode 服务会同时向所有 JournalNode 服务节点并行写入文件,只要有 $N/2+1$ 个节点的写入操作成功,则认为此次写入操作成功。其内部实现框架如图 2-5-2 所示。

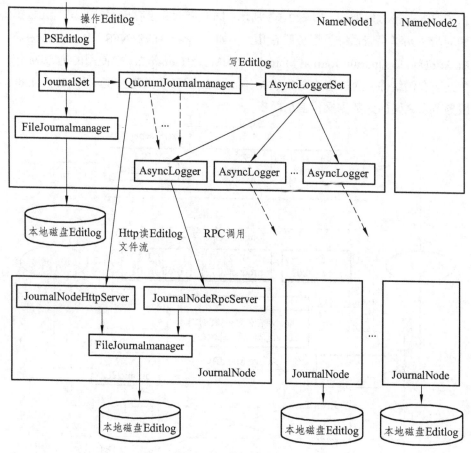

图 2-5-2　内部实现框架

FSEditLog：所有 EditLog 操作的入口。

JournalSet：集成本地磁盘和 JournalNode 集群上 EditLog 的相关操作。

FileJournalManager：实现本地磁盘上的 EditLog 操作。

QuorumJournalManager：实现 JournalNode 集群上的 EditLog 操作。

AsyncLoggerSet：实现 JournalNode 集群上的 EditLog 的写操作集合。

AsyncLogger：发起 RPC 请求到 JournalNode，执行具体的日志同步功能。

JournalNodeRpcServer：运行在 JournalNode 节点进程中的 RPC 服务，接收 NameNode 的 AsyncLogger 的 RPC 请求。

JournalNodeHttpServer：运行在 JournalNode 节点进程中的 Http 服务，用于接收处于 Standby 状态的 NameNode 和其他 JournalNode 的同步 EditLog 文件流的请求。

2.5.1.2 QJM 的写操作

NameNode 服务会把 EditLog 同时写入到本地磁盘和 JournalNode 服务节点之中。写入本地磁盘的路径由配置项"dfs.namenode.name.dir"来控制，写 JournalNode 服务节点的路径由配置项"dfs.namenode.shared.edits.dir"来控制。NameNode 在写 EditLog 时，会由两个不同的输出流来控制日志的写操作过程，分别为本地磁盘的输出流"EditLogFileOutputStream"和 JournalNode 服务节点的输出流"QuorumOutputStream"。写 EditLog 操作也不是直接写入到磁盘之中，为保证整个系统的高吞吐量，NameNode 会为每个输出流分别定义两个同等大小的 Buffer，一个是写 Buffer（buffCurrent），一个是同步 Buffer（buffReady），这样可以一边写一边同步，所以 EditLog 是一个异步写操作过程，但同时也是一个批量同步的过程，这避免了每写一次数据就同步一次日志。

一边写入一边同步的实现依赖于一个缓冲区之间的交换过程，即 bufferCurrent 和 buffReady 在达到一定条件时会触发交换。如 bufferCurrent 内存放的数据达到设定的阈值时，同一时刻 bufferReady 内存放的数据又完成了同步，此时 bufferReady 内存放的数据会被清空，然后会将 bufferCurrent 指针指向之前的 bufferReady 以满足继续写，另外还会将 bufferReady 指针指向之前 bufferCurrent 以提供继续同步 EditLog。上述过程如图 2-5-3 所示。

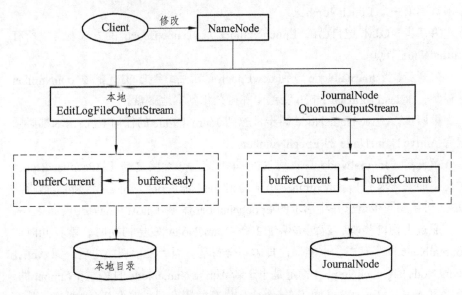

图 2-5-3　QJM 的写操作

EditLog 的写操作是异步进行的，为了保证缓存中的数据不丢失，所有日志都需要通过 logSync 同步成功后才会给 Client 返回成功码。假设某一时刻 NameNode 服务节点失效，其内存中的数据其实是未同步成功的，所以 Client 会认为这部分数据未写入成功。

2.5.1.3 QJM 的数据一致性

在 Active NameNode 每次同步 EditLog 到 JournalNode 时，首先要保证不会有两个 NameNode 同时向 JournalNode 同步日志。这里会使用到一个叫 "Epoch Numbers" 的重要概念，在很多分布式系统都有用到。Epoch Numbers 具有以下几个特性：

（1）当 NameNode 成为活动节点时，其会被赋予一个 EpochNumber。
（2）每个 EpochNumber 是唯一的，不会有相同的 EpochNumber 出现。
（3）EpochNumber 有严格的顺序保证，每次 NameNode 进行切换之后，其 EpochNumber 都会自增 1，所以后面生成的 EpochNumber 始终都会大于前面的 EpochNumber。

QJM 也是通过 Epoch Numbers 来保证不会有两个 NameNode 同时向 JournalNode 同步日志。具体步骤如下：

第一步，在对 EditLog 做任何修改前，NameNode 上的 QuorumJournalManager 必须被赋予一个 EpochNumber。

第二步，QJM 把自己的 EpochNumber 通过 newEpoch 的方式发送给所有 JournalNode 节点。

第三步，当 JournalNode 收到 newEpoch 请求后，会把 QJM 的 EpochNumber 保存到一个 lastPromisedEpoch 变量中并持久化到本地磁盘。

第四步，Active NameNode 同步日志到 JournalNode 的任何 RPC 请求都必须包含 Active NameNode 的 EpochNumber。

第五步，JournalNode 在收到 RPC 请求之后，会将之与 lastPromisedEpoch 对比，如果请求的 EpochNumber 小于 lastPromisedEpoch，将会拒绝同步请求。反之，会接受同步请求并将请求的 EpochNumber 保存在 lastPromisedEpoch 之中。

通过上面的步骤，就能够保证主备 NameNode 发生切换时，就算同时向 JournalNode 同步日志，也能保证日志不会写乱。因为发生切换之后，原 Active NameNode 的 EpochNumber 肯定是小于新 Active NameNode 的 EpochNumber 的，所以原 Active NameNode 向 JournalNode 发起的所有同步请求都会被拒绝，从而实现了隔离功能，防止了脑裂问题。

2.5.1.4 QJM 的读操作

这个读操作是面向 Standby NameNode 的。Standby NameNode 会定期检查 JournalNode 上 EditLog 的变化，然后将 EditLog 获取到本地。同时 Standby NameNode 还会定期将其上的 FSImage 和 EditLog 进行合并，并将合并完成的 FSImage 文件传送给 Active NameNode，这类似于 Hadoop 1.x 版本中 SecondaryNameNode 所做的工作，也就是 Checkpointing 过程。下面我们来看下 Checkpointing 是怎么进行的。

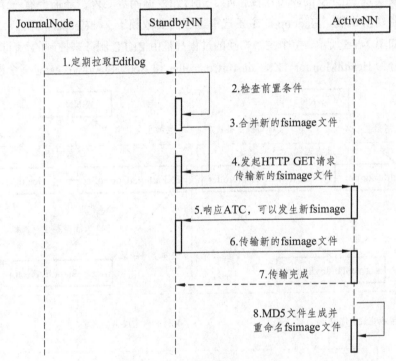

图 2-5-4 Checkpointing 过程

Checkpointing 的整个过程如图 2-5-4 所示。首先是在 Standby NameNode 上检查前置条件，包括了两个方面，一个是距离上次 Checkpointing 的时间间隔，另一个是 EditLog 中事务条数的限制。前置条件中有任何一个满足时都会触发 Checkpointing 操作。然后 Standby NameNode 会将最新的 NameSpace 数据即 Standby NameNode 内存中当前状态的元数据保存到一个临时的 fsimage 文件 (fsimage.ckpt) 中，然后比对从 JournalNode 上获取到的最新 EditLog 的事务 ID。接着将 fsimage.ckpt 文件中没有而 EditLog 中有的所有元数据修改记录合并到一起并重命名为新的 fsimage 文件，同时生成一个 md5 文件。最后将最新的 fsimage

再通过 HTTP 请求传送给 Active NameNode。Standby NameNode 定期的合并 fsimage 文件，可以避免 EditLog 越来越大，因为合并完成之后会将老的 EditLog 删除。同时由于合并工作不是在 Active NameNode 上进行，可以有效减轻 Active NameNode 的负担。而在 Standby NameNode 上保存一份最新的元数据也可以用于在故障恢复时避免数据丢失。

2.5.1.5 NameNode 主备切换机制

要实现 HDFS 的高可用性，除了进行元数据同步之外，还需要有一个完备的主备切换机制，Hadoop 的主备选举过程需要依赖于 ZooKeeper。

如图 2-5-5 所示，整个主备切换的过程都是由 ZKFC 服务来控制的，而其中又具体分为 HealthMonitor、ZKFailoverController 和 ActiveStandbyElector 三个组件。

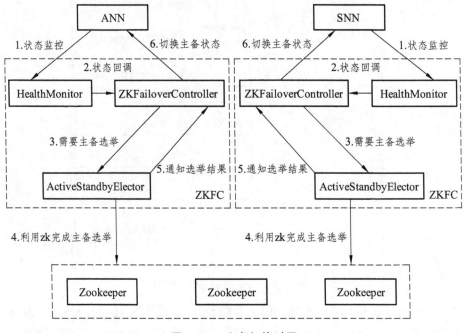

图 2-5-5 主备切换过程

ZKFailoverController 是 HealthMontior 和 ActiveStandbyElector 的母体，执行具体的切换操作。HealthMonitor 负责监控 NameNode 的健康状态，若出现异常状态，就会触发回调 ZKFailoverController 进行自动主备切换的操作。ActiveStandbyElector 则是负责通知 Zookeeper 执行主备选举，若 Zookeeper 完成变更，会回调 ZKFailoverController 相应方法来进行主备之间的状态切换。

在故障切换期间，Zookeeper 的主要作用有以下几点。

第一个是失败保护。集群中每一个 NameNode 都会在 Zookeeper 中维护一个持久的 session，而主机一旦失效，该 session 就会过期，同时故障迁移就会触发。

第二个是 Active NameNode 选择。ZooKeeper 有一个选择 Active NameNode 的机制，一旦现有的 Active NameNode 失效，其他的 Standby NameNode 便可以向 Zookeeper 申请成为下一个 Active NameNode 节点。

第三个是防止脑裂问题。Zookeeper 本身就具有强一致性和高可用性的特点，所以可以用它来保证同一时刻只有一个 Active NameNode。

除了计算机宕机、网络中断等物理上的 Active NameNode 失效问题之外，像 Active NameNode 的 JVM 崩溃或冻结、健康状态异常、ZKFC 服务崩溃等场景也会触发 Active NameNode 的自动切换。另外如果 Hadoop 高可用集群所使用的 Zookeeper 崩溃，所有 NameNode 会进入一个 NeutralMode 模式，同时不改变 Active 和 Standby 的状态，继续发挥作用。只不过此时如果 Active NameNode 也出现故障，那么集群也将不可用。

2.5.2 YARN 高可用实现

YARN 的高可用实现相对比较简单，而且是非常轻量级的。它将共享存储系统抽象成为 RMStateStore，以保存恢复 ResouceManager 所必需的信息。这些信息包括了 Application 的状态信息 ApplicationState、Application 对应的每个 ApplicationAttempt 的状态信息 ApplicationAttemptState、安全令牌的相关状态信息 RMDTSecretManagerState。其中 ApplicationState 内部又包含应用程序的提交描述信息 context、提交时间 submitTime、拥有者 user 三个字段。ApplicationAttemptState 内部又包含 attemptId、所在 Container 的信息 masterContainer、安全 Token 三个字段。RMDTSecretManagerState 内部又包含 delegateionTokenState、masterKeyState、dtSequenceNumber 三个字段。

ResouceManager 本身并不会保存已经分配给每个 ApplicationMaster 的资源信息和每个 NodeManager 的资源使用信息，这些均是可以通过相应的心跳信息获取，从而进行动态重构出来的，这也正是 YARN 的高可用事项非常轻量级的原因。

YARN 提供了四种 RMStateStore 的实现，分别是 NullRMStateStore（不存储任何状态信息，在不启动恢复机制时的默认实现）、MemoryRMStateStore（将状态信息存储到内存中，在启用恢复机制时的默认实现）、FileSystemRMStateStore

（将状态信息存储到 HDFS 中）、ZKRMStateStore（将状态信息存储到 Zookeeper 中）。而实现方式的选择可以通过配置项"yarn.resourcemanager.store.class"来进行指定。就前来说，YARN 的高可用性的最佳实践是采用基于 ZKRMStateStore 的共享存储方案，这也是之后的 Hadoop 高可用模式平台搭建所选择使用的方案。

YARN 的高可用实现的大致流程如下：

（1）ResouceManager 在收到来自客户端发送的提交应用程序的请求之后，在进行响应之前会将应用程序的"application submission context"同步保存下来。这样可以确保一旦客户端成功提交应用程序之后，ResouceManager 无论是重启或是切换时都有足够的信息来对应用程序进行恢复。而且用户提交应用程序的频率一般很低，且 ClientRMService 使用一个独立的 RPC 通道，因此同步记录信息不会产生太大的性能开销。

（2）当一个新的应用程序运行实例被创建时，它的基本信息以及之前失败的运行实例信息都将被保存下来。当 ResouceManager 重启或切换之后，需要获取之前失败的应用程序实例 ID 以产生一个新的运行实例 ID。而当一个应用程序运行结束之后，这些保存的信息都将被移除。

（3）ResouceManager 在重启或切换之后，会首先将之前保存的数据加载到内存中，这些信息将用于构建 RMApp 对象和创建新的运行实例 RMAppAttempt。之后内部服务将被启动，所有操作照旧进行。

（4）ResouceManager 在重启或切换之后，若发现收到的 NodeManager 汇报的心跳信息在所维护的 NodeManager 服务节点信息中并不存在，则会要求对应的 NodeManager 服务节点重新启动并向 ResouceManager 注册，从而将它上面所运行的 container 关闭。

（5）ResouceManager 在重启或切换之后，若发现收到的 ApplicationMaster 汇报的心跳信息中所涉及的运行实例 ID 在所维护的运行实例信息中并不存在，则会向对应的 ApplicationMaster 发送一个 RemoteException 异常，使其异常退出。然后 ResouceManager 会启动新的运行实例。

2.5.3　Hadoop 高可用模式安装规划

1. 环境要求

集群中所有主机的本地磁盘剩余空间 400 MB 以上；

集群中所有主机已安装 CentOS 7 1611 64 位操作系统；

集群中所有主机已完成基础网络环境配置；

集群中所有主机已安装 JDK 的 1.8.0_131 版本；

已完成 Zookeeper 的 3.4.9 版本的完全分布模式的安装和部署。

2. 软件版本

与单节点模式 Hadoop 整合平台搭建中相同，选用 Hadoop 的 2.7.3 版本，软件包名为 hadoop-2.7.3.tar.gz。

3. 服务规划

Hadoop 的高可用模式中除了 HDFS 文件系统和 YARN 资源管理器的相关服务之外，还增加了利用 Zookeeper 实现控制多个 NameNode 服务节点的 Active 与 Standby 状态的切换服务 FailoverController，以及保证多个 NameNode 服务节点之间数据同步的同步通信服务 JournalNode。

在高可用模式中需要满足 HDFS 的 NameNode 服务节点以及 YARN 的 ResourceManager 服务节点有备用的基本要求，所以需要分别有两台或以上的主机来运行这两个服务。并且每台运行 HDFS 的 NameNode 服务的主机都需要运行对应的 FailoverController 服务，已达到对 NameNode 服务进行 Actice 和 Standy 状态的控制切换的目的。

NameNode 的数据同步通信服务 JournalNode 是非常轻量级的服务，可以使用集群中的任意主机，并且不会对原有的服务产生太大影响。但是由于在该服务中存放了 HDFS 的元数据的备份，所以一般不与 HDFS 的 NameNode 服务节点使用相同主机。同时数据同步通信服务 JournalNode 的工作原理同 Zeekeeper 类似，也需要至少 3 个及以上的服务节点数量，并且服务节点总数为奇数时最佳。

余下的 HDFS 的 DataNode 和 YARN 的 NodeManager 两类服务作为 HDFS 的数据存储服务以及 MapReduce 的任务管理服务，默认同时运行在同一台主机上，以达到 MapReduce 中数据本地化处理的目的。同时在高可用模式下若要满足分布式并行处理的需求，则需要集群中具备至少两个或以上的对应服务节点。

对于当前的总共五台主机的集群，使用其中两台主机来运行 HDFS 的 NameNode 服务以及 DFSZKFailoverController 服务以达到 DFS 的高可用的目的。然后使用余下的三台主机运行节点运行数据同步通信服务 JournalNode，使元数据的备份与 NameNode 服务节点分离开。而 YARN 的 ResourceManager 服务则共用 HDFS 的 NameNode 服务的两台以达到高可用的目的，同时 HDFS 的 DataNode 以及 YARN 的 NodeManager 两个服务则使用运行 JournalNode 服务三台主机来运行以实现分布式并行处理。具体的 IP 地址和服务节点规划如表 2-5-1 所示。

表 2-5-1 IP 地址和服务节点规划

主机名	IP 地址	服务描述
Cluster-01	192.168.60.130	HDFS 的 NameNode YARN 的 ResourceManager NameNode 切换服务 DFSZKFailoverController
Cluster-02	192.168.60.131	HDFS 的 NameNode YARN 的 ResourceManager NameNode 切换服务 DFSZKFailoverController
Cluster-03	192.168.60.132	HDFS 的 DataNode YARN 的 NodeManager NameNode 数据同步通信服务 JournalNode
Cluster-04	192.168.60.133	HDFS 的 DataNode YARN 的 NodeManager NameNode 数据同步通信服务 JournalNode
Cluster-05	192.168.60.134	HDFS 的 DataNode YARN 的 NodeManager NameNode 数据同步通信服务 JournalNode

2.5.4 Hadoop 高可用模式配置

（1）进行集群主机之间的免密码登录配置，要求用于运行 HDFS 的 NameNode 服务以及 YARN 的 ResourceManager 服务的所有主机能够免密码登录到集群中包括自身在内的所有主机。

（2）选择集群中一台用作 HDFS 的 NameNode 服务以及 YARN 的 ResourceManager 服务的主机，在其上的 Hadoop 专用用户下完成 Hadoop 的基础安装和环境配置。

（3）在 Hadoop 相关文件的目录中创建 Hadoop 的临时文件的存放目录"tmp"、HDFS 的元数据文件的存放目录"name"、HDFS 的数据文件的存放目录"data"、数据同步通信服务的相关数据的存放目录"journal"，如图 2-5-6 所示。

（4）进入 Hadoop 软件目录下的配置文件所在目录"etc/hadoop"。

（5）编辑配置文件"hadoop-env.sh"和"yarn-env.sh"，将其中配置项"JAVA_HOME"的值改为 JDK 软件目录所在路径。

```
[hadoop@Cluster-01 hadoop]$ ls -l
total 0
drwxrwxr-x. 2 hadoop hadoop   6 Jul 11 05:32 data
drwxr-xr-x. 10 hadoop hadoop 161 Jul 11 06:10 hadoop-2.7.3
drwxrwxr-x. 2 hadoop hadoop   6 Jul 11 05:32 journal
drwxrwxr-x. 3 hadoop hadoop  40 Jul 23 12:45 name
drwxrwxr-x. 2 hadoop hadoop   6 Jul 11 05:32 tmp
```

图 2-5-6 创建目录

（6）编辑配置文件"core-site.xml"，在其中配置以下内容：

<configuration>

  <property>

    <name>fs.defaultFS</name>

    <value>hdfs://hadoop-ha</value>

  </property>

  <property>

    <name>hadoop.tmp.dir</name>

    <value>Hadoop 的临时文件的存放目录"tmp"的路径</value>

  </property>

  <property>

    <name>ha.zookeeper.quorum</name>

    <value>Cluster-01:2181,Cluster-02:2181,Cluster-03:2181,Cluster-04:2181,Cluster-05:2181</value>

  </property>

</configuration>

配置项说明：

fs.defaultFS——指定 Hadoop 的文件系统访问路径，由于高可用模式下有多个 NameNode 服务节点并且会无计划性的切换活动状态，所以不能使用 NameNode 服务节点的主机名或 IP 地址来作为 Hadoop 的文件系统访问路径，而是需要使用后面的 HDFS 高可用配置中定义的命名空间服务的逻辑名称之一来作为文件系统访问路径。

ha.zookeeper.quorum——指定所使用的 Zookeeper 集群中服务节点的连接地址和端口号，多个服务节点之间用","分隔。一般来说 Zookeeper 集群中的每台主机都可以提供服务，这里只需按照所要使用到的服务节点数量来指定连接地址即可，同时该数量也需要满足 Zookeeper 的算法所要求的 3 个及以上的奇数个服务节点数量的要求。

（7）编辑配置文件"hdfs-site.xml"，在其中配置以下内容：
```xml
<configuration>
  <property>
    <name>dfs.nameservices</name>
    <value>hadoop-ha</value>
  </property>
  <property>
    <name>dfs.ha.namenodes.hadoop-ha</name>
    <value>namenode-01,namenode-02</value>
  </property>
  <property>
    <name>dfs.namenode.rpc-address.hadoop-ha.namenode-01</name>
    <value>Cluster-01:9000</value>
  </property>
  <property>
    <name>dfs.namenode.http-address.hadoop-ha.namenode-01</name>
    <value>Cluster-01:50070</value>
  </property>
  <property>
    <name>dfs.namenode.rpc-address.hadoop-ha.namenode-02</name>
    <value>Cluster-02:9000</value>
  </property>
  <property>
    <name>dfs.namenode.http-address.hadoop-ha.namenode-02</name>
    <value>Cluster-02:50070</value>
  </property>
  <property>
    <name>dfs.namenode.shared.edits.dir</name>
    <value>qjournal://Cluster-03:8485;Cluster-04:8485;Cluster-05:8485/hadoop-ha</value>
  </property>
  <property>
    <name>dfs.journalnode.edits.dir</name>
```

        <value>数据同步通信服务的相关数据的存放目录"journal"的路径</value>
    </property>
    <property>
        <name>dfs.ha.automatic-failover.enabled</name>
        <value>true</value>
    </property>
    <property>
        <name>dfs.client.failover.proxy.provider.hadoop-ha</name>
        <value>org.apache.hadoop.hdfs.server.namenode.ha.ConfiguredFailoverProxyProvider</value>
    </property>
    <property>
        <name>dfs.ha.fencing.methods</name>
        <value>
            sshfence
            shell（/bin/true）
        </value>
    </property>
    <property>
        <name>dfs.ha.fencing.ssh.private-key-files</name>
        <value>用户SHH目录下的私钥文件的路径</value>
    </property>
    <property>
        <name>dfs.ha.fencing.ssh.connect-timeout</name>
        <value>30000</value>
    </property>
    <property>
        <name>dfs.namenode.name.dir</name>
        <value>HDFS的NameNode节点的元数据文件的存放目录"name"的路径</value>
    </property>
    <property>

  <name>dfs.datanode.data.dir</name>
  <value>HDFS 的 DataNode 节点的数据文件的存放目录 "data" 的路径</value>
 </property>
 <property>
  <name>dfs.replication</name>
  <value>3</value>
 </property>
</configuration>

配置项说明：

dfs.nameservices——指定 HDFS 的命名空间服务的逻辑名称，该逻辑名称为自定义名称，在需要启用 HDFS 的 Federation 服务时也可以使用该参数指定多个逻辑名称，多个逻辑名称之间用 "," 分隔。

dfs.ha.namenodes.命名空间服务的逻辑名称——指定集群中某个命名空间服务下的所有 NameNode 服务节点的名称的列表，NameNode 服务节点的名称为自定义名称，多个名称之间用 "," 分隔。

dfs.namenode.rpc-address.命名空间服务的逻辑名称.NameNode 服务节点名称——指定某个 NameNode 服务节点的 RPC 通信的地址，可以是主机名也可以是 IP 地址，同时指定所使用的网络端口。集群中的所有 NameNode 服务节点都需要指定 RPC 通信的地址，也就是该配置项需要在配置文件中配置多条。

dfs.namenode.http-address.命名空间服务的逻辑名称.NameNode 服务节点名称——指定某个 NameNode 服务节点的对外 HTTP 通信的地址，可以是主机名也可以是 IP 地址，同时指定所使用的网络端口号。集群中的所有 NameNode 服务节点都需要指定对外 HTTP 通信的地址，也就是该配置项也需要在配置文件中配置多条。

dfs.namenode.shared.edits.di——指定 JournalNode 服务节点的地址列表以及 HDFS 的元数据备份在 JournalNode 服务中的存放目录的名称，该目录为 Linux 系统中的本地目录，一般使用 HDFS 的命名空间服务的逻辑名称即可。地址部分包括主机名或 IP 地址，以及所使用的网络端口号，多个 JournalNode 服务节点的地址之间用 ";" 分隔，具体书写格式为 "qjournal: //地址 1; 地址 2; ……; 地址 n/服务逻辑名称"。

dfs.journalnode.edits.dir——指定 JournalNode 服务的相关数据的本地存放路径。这些数据包括 HDFS 的事务日志 edits，以及数据同步通信服务的逻辑状态数据。

dfs.ha.automatic-failover.enabled——指定 NameNode 服务节点故障时的自动切换功能的开启状态。

dfs.client.failover.proxy.provider.命名空间服务的逻辑名称——指定客户端与处于 Active 状态中的 NameNode 服务节点进行交互的 Java 实现类。HDFS 的客户端程序通过该类来寻找当前处于 Active 状态的 NameNode 服务节点，可以由用户自行编写实现，默认的类为 ConfiguredFailoverProxyProvider。

dfs.ha.fencing.methods——指定主备切换时使用的隔离机制，多个隔离机制之间需要换行书写，即每行一个隔离机制。主备架构在解决单点故障问题的同时面临着出现两个 master 同时对外提供服务，导致系统处于不一致状态，可能导致数据丢失等潜在问题。HDFS 集群的高可用模式在解决单点故障问题的同时也会面临脑裂问题，即出现两个 NameNode 服务节点同时对外提供服务的情况，这将导致系统处于不一致的状态，造成数据丢失等潜在问题。JournalNode 服务虽然只允许一个 NameNode 服务节点对其写入数据，但当主备节点进行切换时，之前处于 Active 状态的 NameNode 服务节点可能仍在处理客户端的 RPC 请求，需要使用隔离机制将该 NameNode 服务节点杀死。HDFS 允许配置多种不同的隔离机制，当发生主备切换时，会依照配置的顺序依次执行这些隔离机制，直到其中一个隔离机制的操作结构返回成功。Hadoop 的 2.0 系列版本中自带了 sshfence 和 shell 两种类型的隔离机制。

① sshfence。

隔离机制 sshfence 通过 SSH 登录到处于 Active 状态的 NameNode 服务节点并将其杀死。该机制的执行需要免密码登录到目标主机，这可以通过在相关配置项中配置对应用户的 SSH 私钥文件路径来实现。同时配置时可以指定登陆目标主机所使用的用户名以及 SSH 的端口，还可以通过相关配置项设定用于判定执行失败的超时时间。

② shell。

隔离机制 shell 通过执行任意一个 shell 命令来达到隔离之前的处于 Active 状态的 NameNode 服务节点的目的。

dfs.ha.fencing.ssh.private-key-files——指定 sshfence 隔离机制在执行免密码登录时所需要使用的当前用户的 SSH 私钥文件路径。

dfs.ha.fencing.ssh.connect-timeout——指定使用 sshfence 隔离机制时用于判定隔离操作失败的超时时间，单位为毫秒。

（8）从 MapReduce 配置文件的模板文件生成配置文件"mapred-site.xml"，然后编辑该配置文件，在其中配置以下内容：

```
<configuration>
  <property>
    <name>mapreduce.framework.name</name>
    <value>yarn</value>
  </property>
</configuration>
```

（9）编辑配置文件"yarn-site.xml"，在其中配置以下内容：

```
<configuration>
  <property>
    <name>yarn.nodemanager.aux-services</name>
    <value>mapreduce_shuffle</value>
  </property>
  <property>
    <name>yarn.resourcemanager.ha.enabled</name>
    <value>true</value>
  </property>
  <property>
    <name>yarn.resourcemanager.cluster-id</name>
    <value>yarn-ha</value>
  </property>
  <property>
    <name>yarn.resourcemanager.ha.rm-ids</name>
    <value>resource-01,resource-02</value>
  </property>
  <property>
    <name>yarn.resourcemanager.hostname.resource-01</name>
    <value>Cluster-01</value>
  </property>
  <property>
    <name>yarn.resourcemanager.hostname.resource-02</name>
    <value>Cluster-02</value>
```

```xml
        </property>
        <property>
            <name>yarn.resourcemanager.address.resource-01</name>
            <value>Cluster-01:8032</value>
        </property>
        <property>
            <name>yarn.resourcemanager.address.resource-02</name>
            <value>Cluster-02:8032</value>
        </property>
        <property>
            <name>yarn.resourcemanager.scheduler.address.resource-01</name>
            <value>Cluster-01:8030</value>
        </property>
        <property>
            <name>yarn.resourcemanager.scheduler.address.resource-02</name>
            <value>Cluster-02:8030</value>
        </property>
        <property>
            <name>yarn.resourcemanager.resource-tracker.address.resource-01</name>
            <value>Cluster-01:8031</value>
        </property>
        <property>
            <name>yarn.resourcemanager.resource-tracker.address.resource-02</name>
<value>Cluster-02:8031</value>
        </property>
        <property>
            <name>yarn.resourcemanager.ha.automatic-failover.recover.enabled</name>
<value>true</value>
        </property>
        <property>
            <name>yarn.resourcemanager.store.class</name>
            <value>org.apache.hadoop.yarn.server.resourcemanager.recovery.ZKRMStateStore</value>
```

</property>
<property>
  <name>yarn.resourcemanager.zk-address</name>
  <value>Cluster-01:2181,Cluster-02:2181,Cluster-03:2181,Cluster-04:2181,Cluster-05:2181</value>
</property>
</configuration>

配置项说明：

yarn.resourcemanager.ha.enabled——指定 YARN 的高可用模式的启用状态。

yarn.resourcemanager.cluster-id——指定 YARN 的高可用模式的集群的逻辑名称，该逻辑名称为自定义名称。

yarn.resourcemanager.ha.rm-ids——指定集群中所有 ResourceManager 服务节点的名称的列表，ResourceManager 服务节点的名称为自定义名称，多个名称之间用","分隔。

yarn.resourcemanager.hostname.ResourceManager 服务节点名称——指定集群中某个 ResourceManager 服务节点对应的主机名或 IP 地址。

yarn.resourcemanager.address.ResourceManager 服务节点名称——指定集群中某个 ResourceManager 服务节点的地址和端口。

yarn.resourcemanager.scheduler.address.ResourceManager 服务节点名称——指定集群中某个 ResourceManager 服务节点的计划服务的地址和端口。

yarn.resourcemanager.resource-tracker.address.ResourceManager 服务节点名称——指定集群中某个 ResourceManager 节点的资源服务的地址和端口。

yarn.resourcemanager.ha.automatic-failover.recover.enabled——指定 YARN 的服务节点 ResourceManager 故障时的自动切换功能的开启状态。

yarn.resourcemanager.store.class——指定 ResourceManager 服务节点中的状态信息的存储方式么，有基于内存（MemStore）和基于 Zookeeper（ZKStore）两种。在有可以使用的 Zookeeper 平台的情况下推荐使用基于 Zookeeper 的存储方式。

yarn.resourcemanager.zk-address——指定所使用的 Zookeeper 集群中服务节点的连接地址和端口号，多个服务节点之间用","分隔，需要满足 Zookeeper 的算法所要求的 3 个及以上的奇数个服务节点数量的要求。

（10）编辑配置文件"slaves"，删除文件中的所有原有内容，然后在其中按照每行一个主机名的方式，添加 HDFS 的 DataNode 服务节点与 YARN 的

NodeManager 服务节点的主机名"Cluster-03""Cluster-04""Cluster-05"。

（11）使用命令"scp ~/.bash_profile 用户名@目标主机名或 IP 地址：/home/用户名/"和"scp -r ~/hadoop 用户名@目标主机名或 IP 地址：/home/用户名/"依次将新修改的用户环境变量设置文件和 Hadoop 的专用目录发送到集群中所有其他主机的相应用户家目录下。

2.5.5 Hadoop 高可用模式格式化和启动

（1）Hadoop 高可用模式下的 HDFS 文件系统格式化操作需要使用到 JournalNode 服务，所以需要先在所有提供数据同步通信服务的主机（Cluster-03、Cluster-04、Cluster-05）上使用命令"hadoop-daemon.sh start journalnode"启动数据同步通信服务。启动完成后查看 Java 进程信息，如图 2-5-7 所示，若有名为"JournalNode"的进程，则表示服务启动成功。

```
[hadoop@Cluster-03 hadoop]$ jps
3554 Jps
3402 QuorumPeerMain
3518 JournalNode
```

图 2-5-7 查看 Java 进程信息

（2）从集群中提供 HDFS 的 NameNode 服务和 YARN 的 ResourceManager 服务的多个主机中选择一个作为主节点，在其上使用命令"hadoop namenode -format"对 HDFS 的文件系统进行格式化，如图 2-5-8 所示，若格式化过程中没有出现报错提示信息，则表示格式化过程成功完成。

```
[hadoop@Cluster-01 hadoop]$ hadoop namenode -format
DEPRECATED: Use of this script to execute hdfs command is deprecated.
Instead use the hdfs command for it.

18/06/26 22:21:14 INFO namenode.NameNode: STARTUP_MSG:
/************************************************************
STARTUP_MSG: Starting NameNode
STARTUP_MSG:   host = Cluster-01/192.168.60.130
STARTUP_MSG:   args = [-format]
STARTUP_MSG:   version = 2.7.3
STARTUP_MSG:   classpath = /home/hadoop/hadoop/hadoop-2.7.3/etc/hadoop:/home/hadoop/hadoop/hadoop-2.7.3/share/hadoop/common/lib/jaxb-impl-2.2.3-1.jar:/home/hadoop/hadoop/hadoop-2.7.3/share/hadoop/common/lib/jaxb-api-2.2.2.jar:/home/hadoop/hadoop/hadoop-2.7.3/share/hadoop/common/lib/stax-api-1.0-2.jar:/home/hadoop/hadoop/hadoop-2.7.3/share/hadoop/common/lib/activation-1.1.jar:/home/hadoop/hadoop/hadoop-2.7.3/share/hadoop/common/lib/jackson-core-asl-1.9.13.jar:/home/hadoop/hadoop/hadoop-2.7.3/share/hadoop/common/lib/jackson-mapper-asl-1.9.13.jar:/home/hadoop/hadoop/hadoop-2.7.3/share/hadoop/common/lib/jackson-jaxrs-1.9.13.jar:/home/hadoop/hadoop/hadoop-2.7.3/share/hadoop/common/lib/jackson-xc-1.9.13.jar:/home/hadoop/hadoop/hadoop-2.7.3/share/hadoop/common/lib/jersey-server-1.9.jar:/home/hadoop/hadoop/hadoop-2.7.3/share/hadoop/common/lib/asm-3.2.jar:/home/hadoop/hadoop/hadoop-2.7.3/share/hadoop/common/lib/log4j-1.2.17.jar:/home/hadoop/hadoop/hadoop-2.7.3/share/hadoop/common/lib/jets3t-0.9.0.jar:/home/hadoop/hadoop/hadoop-2.7.3/share/hadoop/common/lib/httpclient-4.2.5.jar:/home/hadoop/hadoop/hadoop-2.7.3/share/hadoop/common/lib/httpcore-4.2.5.jar:/home/hadoop/hadoop/hadoop-2.7.3/share/hadoop/common/lib/java-xmlbuilder-0.4.jar:/home/hadoop/hadoop/hadoop-2.7.3/share/hadoop/common/lib/c
```

commons-lang-2.6.jar:/home/hadoop/hadoop/hadoop-2.7.3/share/hadoop/common/lib/commons-configuration-1.6.jar:/home/hadoop/hadoop/hadoop-2.7.3/share/hadoop/common/lib/commons-digester-1.8.jar:/home/hadoop/hadoop/hadoop-2.7.3/share/hadoop/common/lib/commons-beanutils-1.7.0.jar:/home/hadoop/hadoop/hadoop-2.7.3/share/hadoop/common/lib/commons-beanutils-core-1.8.0.jar:/home/hadoop/hadoop/hadoop-2.7.3/share/hadoop/common/lib/slf4j-api-1.7.10.jar:/home/hadoop/hadoop/hadoop-2.7.3/share/hadoop/common/lib/slf4j-log4j12-1.7.10.jar:/home/hadoop/hadoop/hadoop-2.7.3/share/hadoop/common/lib/avro-1.7.4.jar:/home/hadoop/hadoop/hadoop-2.7.3/share/hadoop/common/lib/paranamer-2.3.jar:/home/hadoop/hadoop/hadoop-2.7.3/share/hadoop/common/lib/snappy-java-1.0.4.1.jar:/home/hadoop/hadoop/hadoop-2.7.3/share/hadoop/common/lib/commons-compress-1.4.1.jar:/home/hadoop/hadoop/hadoop-2.7.3/share/hadoop/common/lib/xz-1.0.jar:/home/hadoop/hadoop/hadoop-2.7.3/share/hadoop/common/lib/protobuf-java-2.5.0.jar:/home/hadoop/hadoop/hadoop-2.7.3/share/hadoop/common/lib/gson-2.2.4.jar:/home/hadoop/hadoop/hadoop-2.7.3/share/hadoop/common/lib/hadoop-auth-2.7.3.jar:/home/hadoop/hadoop/hadoop-2.7.3/share/hadoop/common/lib/apacheds-kerberos-codec-2.0.0-M15.jar:/home/hadoop/hadoop/hadoop-2.7.3/share/hadoop/common/lib/apacheds-i18n-2.0.0-M15.jar:/home/hadoop/hadoop/hadoop-2.7.3/share/hadoop/common/lib/api-asn1-api-1.0.0-M20.jar:/home/hadoop/hadoop/hadoop-2.7.3/share/hadoop/common/lib/api-util-1.0.0-M20.jar:/home/hadoop/hadoop/hadoop-2.7.3/share/hadoop/common/lib/zookeeper-3.4.6.jar:/home/hadoop/hadoop/hadoop-2.7.3/share/hadoop/common/lib/netty-3.6.2.Final.jar:/home/hadoop/hadoop/hadoop-2.7.3/share/hadoop/common/lib/curator-framework-2.7.1.jar:/home/hadoop/hadoop/hadoop-2.7.3/share/hadoop/common/lib/curator-client-2.7.1.jar:/home/hadoop/hadoop/hadoop-2.7.3/share/hadoop/common/lib/jsch-0.1.42.jar:/home/hadoop/hadoop/hadoop-2.7.3/share/hadoop/common/lib/curator-recipes-2.7.1.jar:/home/hadoop/hadoop/hadoop-2.7.3/share/hadoop/common/lib/htrace-core-3.1.0-incubating.jar:/home/hadoop/hadoop/hadoop-2.7.3/share/hadoop/common/lib/junit-4.11.jar:/home/hadoop/hadoop/hadoop-2.7.3/share/hadoop/common/lib/hamcrest-core-1.3.jar:/home/hadoop/hadoop/hadoop-2.7.3/share/hadoop/common/lib/mockito-all-1.8.5.jar:/home/hadoop/hadoop/hadoop-2.7.3/share/hadoop/common/lib/hadoop-annotations-2.7.3.jar:/home/hadoop/hadoop/hadoop-2.7.3/share/hadoop/common/lib/guava-11.0.2.jar:/home/hadoop/hadoop/hadoop-2.7.3/share/hadoop/common/lib/jsr305 3.0.0.jar:/home/hadoop/hadoop/hadoop-2.7.3/share/hadoop/common/lib/commons-cli-1.2.jar:/home/hadoop/hadoop/hadoop-2.7.3/share/hadoop/common/lib/commons-math3-3.1.1.jar:/home/hadoop/hadoop/hadoop-2.7.3/share/hadoop/common/lib/xmlenc-0.52.jar:/home/hadoop/hadoop/hadoop-2.7.3/share/hadoop/common/lib/commons-httpclient-3.1.jar:/home/hadoop/hadoop/hadoop-2.7.3/share/hadoop/common/lib/commons-logging-1.1.3.jar:/home/hadoop/hadoop/hadoop-2.7.3/share/hadoop/common/lib/commons-codec-1.4.jar:/home/hadoop/hadoop/hadoop-2.7.3/share/hadoop/common/lib/commons-io-2.4.jar:/home/hadoop/hadoop/hadoop-2.7.3/share/hadoop/common/lib/commons-net-3.1.jar:/home/hadoop/hadoop/hadoop-2.7.3/share/hadoop/common/lib/commons-collections-3.2.2.jar:/home/hadoop/hadoop/hadoop-2.7.3/share/hadoop/common/lib/servlet-api-2.5.jar:/home/hadoop/hadoop/hadoop-2.7.3/share/hadoop/common/lib/jetty-6.1.26.jar:/home/hadoop/hadoop/hadoop-2.7.3/share/hadoop/common/lib/jetty-util-6.1.26.jar:/home/hadoop/hadoop/hadoop-2.7.3/share/hadoop/common/lib/jsp-api-2.1.jar:/home/hadoop/hadoop/hadoop-2.7.3/share/hadoop/common/lib/jersey-core-1.9.jar:/home/hadoop/hadoop/hadoop-2.7.3/share/hadoop/common/lib/jersey-json-1.9.jar:/home/hadoop/hadoop/hadoop-2.7.3/share/hadoop/common/lib/jettison-1.1.jar:/home/hadoop/hadoop/hadoop-2.7.3/share/hadoop/common-2.7.3.jar:/home/hadoop/hadoop/hadoop-2.7.3/share/hadoop/common/hadoop-common-2.7.3-tests.jar:/home/hadoop/hadoop/hadoop-2.7.3/share/hadoop/common/hadoop-nfs-2.7.3.jar:/home/hadoop/hadoop/hadoop-2.7.3/share/hadoop/hdfs:/home/hadoop/hadoop/hadoop-2.7.3/share/hadoop/hdfs/lib/commons-codec-1.4.jar:/home/hadoop/hadoop/hadoop-2.7.3/share/hadoop/hdfs/lib/log4j-1.2.17.jar:/home/hadoop/hadoop/hadoop-2.7.3/share/hadoop/hdfs/lib/commons-logging-1.1.3.jar:/home/hadoop/hadoop/hadoop-2.7.3/share/hadoop/hdfs/lib/netty-3.6.2.Final.jar:/home/hadoop/hadoop/hadoop-2.7.3/share/hadoop/hdfs/lib/guava-11.0.2.jar:/home/hadoop/hadoop/hadoop-2.7.3/share/hadoop/hdfs/lib/jsr305-3.0.0.jar:/home/hadoop/hadoop/hadoop-2.7.3/share/hadoop/hdfs/lib/commons-cli-1.2.jar:/home/hadoop/hadoop/hadoop-2.7.3/share/hadoop/hdfs/lib/xmlenc-0.52.jar:/home/hadoop/hadoop/hadoop-2.7.3/share/hadoop/hdfs/lib/commons-io-2.4.jar:/home/hadoop/hadoop/hadoop-2.7.3/share/hadoop/hdfs/lib/servlet-api-2.5.jar:/home/hadoop/hadoop/hadoop-2.7.3/share/hadoop/hdfs/lib/jetty-6.1.26.jar:/home/hadoop/hadoop/hadoop-2.7.3/share/hadoop/hdfs/lib/jetty-util-6.1.26.jar:/home/hadoop/hadoop/hadoop-2.7.3/share/hadoop/hdfs/lib/jersey-core-1.9.jar:/home/hadoop/hadoop/hadoop-2.7.3/share/hadoop/hdfs/lib/jackson-core-asl-1.9.13.jar:/home/hadoop/hadoop/hadoop-2.7.3/share/hadoop/hdfs/lib/jackson-mapper-asl-1.9.13.jar:/home/hadoop/hadoop/hadoop-2.7.3/share/hadoop/hdfs/lib/jersey-server-1.9.jar:/home/hadoop/hadoop/hadoop-2.7.3/share/hadoop/hdfs/lib/asm-3.2.jar:/home/hadoop/hadoop/hadoop-2.7.3/share/hadoop/hdfs/lib/commons-lang-2.6.jar:/home/hadoop/hadoop/hadoop-2.7.3/share/hadoop/hdfs/lib/protobuf-java-2.5.0.jar:/home/hadoop/hadoop/hadoop-2.7.3/share/hadoop/hdfs/lib/htrace-core-3.1.0-incubating.jar:/home/hadoop/hadoop/hadoop-2.7.3/share/hadoop/hdfs/lib/commons-daemon-1.0.13.jar:/home/hadoop/hadoop/hadoop-2.7.3/share/hadoop/hdfs/lib/netty-all-4.0.23.Final.jar:/home/Hadoop/hadoop/hadoop-2.7.3/share/hadoop/hdfs/lib/xercesImpl-2.9.1.jar:/home/hadoop/hadoop/hadoop-2.7.3/share/hadoop/hdfs/lib/xml-apis-1.3.04.jar:/home/hadoop/hadoop/hadoop-2.7.3/share/hadoop/hdfs/lib/leveldbjni-all-1.8.jar:/home/hadoop/hadoop/hadoop-2.7.3/share/hadoop/hdfs/hadoop-hdfs-2.7.3.jar:/home/hadoop/hadoop/hadoop-2.7.3/share/hadoop/hdfs/hadoop-hdfs-2.7.3-tests.jar:/home/hadoop/hadoop/hadoop-2.7.3/share/hadoop/hdfs/hadoop-hdfs-nfs-2.7.3.jar:/home/hadoop/hadoop/hadoop-2.7.3/share/hadoop/yarn/lib/zookeeper-3.4.6-tests.jar:/home/hadoop/hadoop/hadoop-2.7.3/share/hadoop/yarn/lib/commons-lang-2.6.jar:/home/hadoop/hadoop/hadoop-2.7.3/share/hadoop/yarn/lib/guava-11.0.2.jar:/home/hadoop/hadoop/hadoop-2.7.3/share/hadoop/yarn/lib/jsr305-3.0.0.jar:/home/hadoop/hadoop/hadoop-2.7.3/share/hadoop/yarn/lib/commons-logging-1.1.3.jar:/home/hadoop/hadoop/hadoop-2.7.3/share/hadoop/yarn/lib/protobuf-java-2.5.0.jar:/home/hadoop/hadoop/hadoop-2.7.3/share/hadoop/yarn/lib/log4j-1.2.17.jar:/home/hadoop/hadoop/hadoop-2.7.3/share/hadoop/yarn/lib/commons-cli-1.2.jar:/home/hadoop/hadoop/hadoop-2.7.3/share/hadoop/yarn/lib/jaxb-api-2.2.2.jar:/home/hadoop/hadoop/hadoop-2.7.3/share/hadoop/yarn/lib/stax-api-1.0-2.jar:/home/hadoop/hadoop/hadoop-2.7.3/share/hadoop/yarn/lib/activation-1.1.jar:/home/hadoop/hadoop/hadoop-2.7.3/share/hadoop/yarn/lib/commons-compress-1.4.1.jar:/home/hadoop/hadoop/hadoop-2.7.3/share/hadoop/yarn/lib/xz-1.0.jar:/home/hadoop/hadoop/hadoop-2.7

项目2 搭建高可用Hadoop整合平台

```
.3/share/hadoop/yarn/lib/servlet-api-2.5.jar:/home/hadoop/hadoop/hadoop-2.7.3/share/hadoop/yarn/lib/
commons-codec-1.4.jar:/home/hadoop/hadoop/hadoop-2.7.3/share/hadoop/yarn/lib/jetty-util-6.1.26.jar:/
home/hadoop/hadoop/hadoop-2.7.3/share/hadoop/yarn/lib/jersey-core-1.9.jar:/home/hadoop/hadoop
-2.7.3/share/hadoop/yarn/lib/jersey-client-1.9.jar:/home/hadoop/hadoop/hadoop-2.7.3/share/hadoop/yar
n/lib/jackson-core-asl-1.9.13.jar:/home/hadoop/hadoop/hadoop-2.7.3/share/hadoop/yarn/lib/jackson-map
per-asl-1.9.13.jar:/home/hadoop/hadoop/hadoop-2.7.3/share/hadoop/yarn/lib/jackson-jaxrs-1.9.13.jar:/
home/hadoop/hadoop/hadoop-2.7.3/share/hadoop/yarn/lib/jackson-xc-1.9.13.jar:/home/hadoop/hado
op-2.7.3/share/hadoop/yarn/lib/guice-servlet-3.0.jar:/home/hadoop/hadoop/hadoop-2.7.3/share/hadoop/y
arn/lib/guice-3.0.jar:/home/hadoop/hadoop/hadoop-2.7.3/share/hadoop/yarn/lib/javax.inject-1.jar:/hom
e/hadoop/hadoop/hadoop-2.7.3/share/hadoop/yarn/lib/aopalliance-1.0.jar:/home/hadoop/hadoop-2.
7.3/share/hadoop/yarn/lib/commons-io-2.4.jar:/home/hadoop/hadoop/hadoop-2.7.3/share/hadoop/yarn/lib/
jersey-server-1.9.jar:/home/hadoop/hadoop/hadoop-2.7.3/share/hadoop/yarn/lib/asm-3.2.jar:/home/hadoo
p/hadoop/hadoop-2.7.3/share/hadoop/yarn/lib/jersey-json-1.9.jar:/home/hadoop/hadoop/yarn/lib/sha
re/hadoop/yarn/lib/jettison-1.1.jar:/home/hadoop/hadoop/hadoop-2.7.3/share/hadoop/yarn/lib/jaxb-impl
-2.2.3-1.jar:/home/hadoop/hadoop/hadoop-2.7.3/share/hadoop/yarn/lib/jersey-guice-1.9.jar:/home/hadoo
p/hadoop/hadoop-2.7.3/share/hadoop/yarn/lib/zookeeper-3.4.6.jar:/home/hadoop/hadoop/share/hadoop/sha
re/hadoop/yarn/lib/netty-3.6.2.Final.jar:/home/hadoop/hadoop/hadoop-2.7.3/share/hadoop/yarn/lib/leve
ldbjni-all-1.8.jar:/home/hadoop/hadoop/hadoop-2.7.3/share/hadoop/yarn/lib/commons-collections-3.2.2.
jar:/home/hadoop/hadoop/hadoop-2.7.3/share/hadoop/yarn/lib/jetty-6.1.26.jar:/home/hadoop/hado
op-2.7.3/share/hadoop/yarn/hadoop-yarn-api-2.7.3.jar:/home/hadoop/hadoop/hadoop-2.7.3/share/hadoop/y
arn/hadoop-yarn-common-2.7.3.jar:/home/hadoop/hadoop-2.7.3/share/hadoop/yarn/hadoop-yarn-serv
er-common-2.7.3.jar:/home/hadoop/hadoop/hadoop-2.7.3/share/hadoop/yarn/hadoop-yarn-server-nodemanage
r-2.7.3.jar:/home/hadoop/hadoop/hadoop-2.7.3/share/hadoop/yarn/hadoop-yarn-server-web-proxy-2.7.3.ja
r:/home/hadoop/hadoop/hadoop-2.7.3/share/hadoop/yarn/hadoop-yarn-server-applicationhistoryservice-2.
7.3.jar:/home/hadoop/hadoop/hadoop-2.7.3/share/hadoop/yarn/hadoop-yarn-server-resourcemanager-2.7.3.
jar:/home/hadoop/hadoop/hadoop-2.7.3/share/hadoop/yarn/hadoop-yarn-server-tests-2.7.3.jar:/home/hado
op/hadoop/hadoop-2.7.3/share/hadoop/yarn/hadoop-yarn-client-2.7.3.jar:/home/hadoop/hadoop/hadoop-2.7
.3/share/hadoop/yarn/hadoop-yarn-server-sharedcachemanager-2.7.3.jar:/home/hadoop/hadoop/hadoop-2.7.
3/share/hadoop/yarn/hadoop-yarn-applications-distributedshell-2.7.3.jar:/home/hadoop/hadoop/hadoop-2
.7.3/share/hadoop/yarn/hadoop-yarn-applications-unmanaged-am-launcher-2.7.3.jar:/home/hadoop/hadoop/
hadoop-2.7.3/share/hadoop/yarn/hadoop-yarn-registry-2.7.3.jar:/home/hadoop/hadoop/hadoop-2.7.3/share
/hadoop/mapreduce/lib/protobuf-java-2.5.0.jar:/home/hadoop/hadoop/hadoop-2.7.3/share/hadoop/mapreduc
e/lib/avro-1.7.4.jar:/home/hadoop/hadoop/hadoop-2.7.3/share/hadoop/mapreduce/lib/jackson-core-asl-1.
9.13.jar:/home/hadoop/hadoop/hadoop-2.7.3/share/hadoop/mapreduce/lib/jackson-mapper-asl-1.9.13.jar:/
home/hadoop/hadoop/hadoop-2.7.3/share/hadoop/mapreduce/lib/paranamer-2.3.jar:/home/hadoop/hadoop/had
oop-2.7.3/share/hadoop/mapreduce/lib/snappy-java-1.0.4.1.jar:/home/hadoop/hadoop/hadoop-2.7.3/share
```

```
hadoop/mapreduce/lib/commons-compress-1.4.1.jar:/home/hadoop/hadoop/hadoop-2.7.3/share/hadoop/mapred
uce/lib/xz-1.0.jar:/home/hadoop/hadoop/hadoop-2.7.3/share/hadoop/mapreduce/lib/hadoop-annotations-2.
7.3.jar:/home/hadoop/hadoop/hadoop-2.7.3/share/hadoop/mapreduce/lib/commons-io-2.4.jar:/home/hadoop/
hadoop/hadoop-2.7.3/share/hadoop/mapreduce/lib/jersey-core-1.9.jar:/home/hadoop/hadoop/hadoop-2.7.3/
share/hadoop/mapreduce/lib/jersey-server-1.9.jar:/home/hadoop/hadoop/hadoop-2.7.3/share/hadoop/mapre
duce/lib/asm-3.2.jar:/home/hadoop/hadoop/hadoop-2.7.3/share/hadoop/mapreduce/lib/log4j-1.2.17.jar:/h
ome/hadoop/hadoop/hadoop-2.7.3/share/hadoop/mapreduce/lib/netty-3.6.2.Final.jar:/home/hadoop/hadoop/
hadoop-2.7.3/share/hadoop/mapreduce/lib/leveldbjni-all-1.8.jar:/home/hadoop/hadoop/hadoop-2.7.3/shar
e/hadoop/mapreduce/lib/guice-3.0.jar:/home/hadoop/hadoop/hadoop-2.7.3/share/hadoop/mapreduce/lib/jav
ax.inject-1.jar:/home/hadoop/hadoop/hadoop-2.7.3/share/hadoop/mapreduce/lib/aopalliance-1.0.jar:/hom
e/hadoop/hadoop/hadoop-2.7.3/share/hadoop/mapreduce/lib/jersey-guice-1.9.jar:/home/hadoop/hadoop/had
oop-2.7.3/share/hadoop/mapreduce/lib/guice-servlet-3.0.jar:/home/hadoop/hadoop/hadoop-2.7.3/share/ha
doop/mapreduce/lib/junit-4.11.jar:/home/hadoop/hadoop/hadoop-2.7.3/share/hadoop/mapreduce/lib/hamcre
st-core-1.3.jar:/home/hadoop/hadoop/hadoop-2.7.3/share/hadoop/mapreduce/hadoop-mapreduce-client-core
-2.7.3.jar:/home/hadoop/hadoop/hadoop-2.7.3/share/hadoop/mapreduce/hadoop-mapreduce-client-common-2.
7.3.jar:/home/hadoop/hadoop/hadoop-2.7.3/share/hadoop/mapreduce/hadoop-mapreduce-client-shuffle-2.7.
3.jar:/home/hadoop/hadoop/hadoop-2.7.3/share/hadoop/mapreduce/hadoop-mapreduce-client-app-2.7.3.jar:
/home/hadoop/hadoop/hadoop-2.7.3/share/hadoop/mapreduce/hadoop-mapreduce-client-hs-2.7.3.jar:/home/h
adoop/hadoop/hadoop-2.7.3/share/hadoop/mapreduce/hadoop-mapreduce-client-jobclient-2.7.3.jar:/home/h
adoop/hadoop/hadoop-2.7.3/share/hadoop/mapreduce/hadoop-mapreduce-client-hs-plugins-2.7.3.jar:/home/
hadoop/hadoop/hadoop-2.7.3/share/hadoop/mapreduce/hadoop-mapreduce-examples-2.7.3.jar:/home/h
adoop/hadoop/hadoop-2.7.3/share/hadoop/mapreduce/hadoop-mapreduce-client-jobclient-2.7.3-tests.jar:/home/ha
doop/hadoop/hadoop-2.7.3/contrib/capacity-scheduler/*.jar:/home/hadoop/hadoop/hadoop-2.7.3/contrib/c
apacity-scheduler/*.jar
```

```
STARTUP_MSG:   build = https://git-wip-us.apache.org/repos/asf/hadoop.git -r baa91f7c6bc9cb92be5982d
e4719c1c8af91ccff; compiled by 'root' on 2016-08-18T01:41Z
STARTUP_MSG:   java = 1.8.0_131
************************************************************/
18/06/26 22:51:56 INFO namenode.NameNode: registered UNIX signal handlers for [TERM, HUP, INT]
18/06/26 22:51:58 INFO namenode.NameNode: createNameNode [-format]
18/06/26 22:51:58 WARN common.Util: Path /home/hadoop/hadoop/name should be specified as a URI in co
nfiguration files. Please update hdfs configuration.
18/06/26 22:51:58 WARN common.Util: Path /home/hadoop/hadoop/name should be specified as a URI in co
nfiguration files. Please update hdfs configuration.
Formatting using clusterid: CID-b99ba26b-c7dc-4ad2-89b0-f3708389ca30
18/06/26 22:51:58 INFO namenode.FSNamesystem: No KeyProvider found.
18/06/26 22:51:58 INFO namenode.FSNamesystem: fsLock is fair:true
18/06/26 22:51:58 INFO blockmanagement.DatanodeManager: dfs.block.invalidate.limit=1000
18/06/26 22:51:58 INFO blockmanagement.DatanodeManager: dfs.namenode.datanode.registration.ip-hostna
```

```
me-check=true
18/06/26 22:51:58 INFO blockmanagement.BlockManager: dfs.namenode.startup.delay.block.deletion.sec i
s set to 000:00:00:00.000
18/06/26 22:51:58 INFO blockmanagement.BlockManager: The block deletion will start around 2018 Jun 2
6 22:51:58
18/06/26 22:51:58 INFO util.GSet: Computing capacity for map BlocksMap
18/06/26 22:51:58 INFO util.GSet: VM type       = 64-bit
18/06/26 22:51:58 INFO util.GSet: 2.0% max memory 966.7 MB = 19.3 MB
18/06/26 22:51:58 INFO util.GSet: capacity      = 2^21 = 2097152 entries
18/06/26 22:51:58 INFO blockmanagement.BlockManager: dfs.block.access.token.enable=false
18/06/26 22:51:58 INFO blockmanagement.BlockManager: defaultReplication     = 3
18/06/26 22:51:58 INFO blockmanagement.BlockManager: maxReplication         = 512

18/06/26 22:51:58 INFO blockmanagement.BlockManager: minReplication           = 1
18/06/26 22:51:58 INFO blockmanagement.BlockManager: maxReplicationStreams    = 2
18/06/26 22:51:58 INFO blockmanagement.BlockManager: replicationRecheckInterval = 3000
18/06/26 22:51:58 INFO blockmanagement.BlockManager: encryptDataTransfer      = false
18/06/26 22:51:58 INFO blockmanagement.BlockManager: maxNumBlocksToLog        = 1000
18/06/26 22:51:58 INFO namenode.FSNamesystem: fsOwner             = hadoop (auth:SIMPLE)
18/06/26 22:51:58 INFO namenode.FSNamesystem: supergroup          = supergroup
18/06/26 22:51:58 INFO namenode.FSNamesystem: isPermissionEnabled = true
18/06/26 22:51:58 INFO namenode.FSNamesystem: Determined nameservice ID: hadoop-ha
18/06/26 22:51:58 INFO namenode.FSNamesystem: HA Enabled: true
18/06/26 22:51:58 INFO namenode.FSNamesystem: Append Enabled: true
18/06/26 22:51:59 INFO util.GSet: Computing capacity for map INodeMap
18/06/26 22:51:59 INFO util.GSet: VM type       = 64-bit
18/06/26 22:51:59 INFO util.GSet: 1.0% max memory 966.7 MB = 9.7 MB
18/06/26 22:51:59 INFO util.GSet: capacity      = 2^20 = 1048576 entries
18/06/26 22:51:59 INFO namenode.FSDirectory: ACLs enabled? false
18/06/26 22:51:59 INFO namenode.FSDirectory: XAttrs enabled? true
18/06/26 22:51:59 INFO namenode.FSDirectory: Maximum size of an xattr: 16384
18/06/26 22:51:59 INFO namenode.NameNode: Caching file names occuring more than 10 times
18/06/26 22:51:59 INFO util.GSet: Computing capacity for map cachedBlocks
18/06/26 22:51:59 INFO util.GSet: VM type       = 64-bit
18/06/26 22:51:59 INFO util.GSet: 0.25% max memory 966.7 MB = 2.4 MB
18/06/26 22:51:59 INFO util.GSet: capacity      = 2^18 = 262144 entries
18/06/26 22:51:59 INFO namenode.FSNamesystem: dfs.namenode.safemode.threshold-pct = 0.99900001287460
33
18/06/26 22:51:59 INFO namenode.FSNamesystem: dfs.namenode.safemode.min.datanodes = 0
18/06/26 22:51:59 INFO namenode.FSNamesystem: dfs.namenode.safemode.extension     = 30000
18/06/26 22:51:59 INFO metrics.TopMetrics: NNTop conf: dfs.namenode.top.window.num.buckets = 10
18/06/26 22:51:59 INFO metrics.TopMetrics: NNTop conf: dfs.namenode.top.num.users = 10
18/06/26 22:51:59 INFO metrics.TopMetrics: NNTop conf: dfs.namenode.top.windows.minutes = 1,5,25
18/06/26 22:51:59 INFO namenode.FSNamesystem: Retry cache on namenode is enabled
18/06/26 22:51:59 INFO namenode.FSNamesystem: Retry cache will use 0.03 of total heap and retry cach
e entry expiry time is 600000 millis
18/06/26 22:51:59 INFO util.GSet: Computing capacity for map NameNodeRetryCache
18/06/26 22:51:59 INFO util.GSet: VM type       = 64-bit
18/06/26 22:51:59 INFO util.GSet: 0.029999999329447746% max memory 966.7 MB = 297.0 KB

18/06/26 22:51:59 INFO util.GSet: capacity      = 2^15 = 32768 entries
Re-format filesystem in QJM to [192.168.60.132:8485, 192.168.60.133:8485, 192.168.60.134:8485] ? (Y
or N) Y
18/06/26 22:52:07 INFO namenode.FSImage: Allocated new BlockPoolId: BP-1439782799-192.168.60.130-153
0067927781
18/06/26 22:52:07 INFO common.Storage: Storage directory /home/hadoop/hadoop/name has been successfu
lly formatted.
18/06/26 22:52:08 INFO namenode.FSImageFormatProtobuf: Saving image file /home/hadoop/hadoop/name/cu
rrent/fsimage.ckpt_0000000000000000000 using no compression
18/06/26 22:52:10 INFO namenode.FSImageFormatProtobuf: Image file /home/hadoop/hadoop/name/current/f
simage.ckpt_0000000000000000000 of size 353 bytes saved in 1 seconds.
18/06/26 22:52:10 INFO namenode.NNStorageRetentionManager: Going to retain 1 images with txid >= 0
18/06/26 22:52:10 INFO util.ExitUtil: Exiting with status 0
18/06/26 22:52:10 INFO namenode.NameNode: SHUTDOWN_MSG:
/************************************************************
SHUTDOWN_MSG: Shutting down NameNode at Cluster-01/192.168.60.130
************************************************************/
```

图 2-5-8 对 HDFS 文件系统格式化

（3）HDFS 文件系统的格式化操作完成之后，使用命令"scp ~/hadoop/name 用户名@目标主机名或 IP 地址：/home/用户名/hadoop/"将"hadoop"目录下的元数据文件的存放目录"name"发送到集群中所有备用的运行 HDFS 的

NameNode 服务的主机的对应目录下。

（4）在主节点上使用命令"hdfs zkfc -formatZK"对 Hadoop 集群在 Zookeeper 中的服务切换控制信息进行格式化，如图 2-5-9 所示，若格式化执行过程中没有出现报错提示信息，则表示格式化过程成功完成。

```
[hadoop@Cluster-01 hadoop]$ hdfs zkfc -formatZK
18/06/26 23:18:46 INFO tools.DFSZKFailoverController: Failover controller configured for NameNode Na
meNode at Cluster-01/192.168.60.130:9000
18/06/26 23:18:47 INFO zookeeper.ZooKeeper: Client environment:zookeeper.version=3.4.6-1569965, buil
t on 02/20/2014 09:09 GMT
18/06/26 23:18:47 INFO zookeeper.ZooKeeper: Client environment:host.name=Cluster-01
18/06/26 23:18:47 INFO zookeeper.ZooKeeper: Client environment:java.version=1.8.0_131
18/06/26 23:18:47 INFO zookeeper.ZooKeeper: Client environment:java.vendor=Oracle Corporation
18/06/26 23:18:47 INFO zookeeper.ZooKeeper: Client environment:java.home=/home/hadoop/java/jdk1.8.0_
131/jre
18/06/26 23:18:47 INFO zookeeper.ZooKeeper: Client environment:java.class.path=/home/hadoop/hadoop/h
adoop-2.7.3/etc/hadoop:/home/hadoop/hadoop/hadoop-2.7.3/share/hadoop/common/lib/jaxb-impl-2.2.3-1.ja
r:/home/hadoop/hadoop/hadoop-2.7.3/share/hadoop/common/lib/jaxb-api-2.2.2.jar:/home/hadoop/hadoop/ha
doop-2.7.3/share/hadoop/common/lib/stax-api-1.0-2.jar:/home/hadoop/hadoop/hadoop-2.7.3/share/hadoop/
common/lib/activation-1.1.jar:/home/hadoop/hadoop/hadoop-2.7.3/share/hadoop/common/lib/jackson-core-
asl-1.9.13.jar:/home/hadoop/hadoop/hadoop-2.7.3/share/hadoop/common/lib/jackson-mapper-asl-1.9.13.ja
r:/home/hadoop/hadoop/hadoop-2.7.3/share/hadoop/common/lib/jackson-jaxrs-1.9.13.jar:/home/hadoop/had
oop/hadoop-2.7.3/share/hadoop/common/lib/jackson-xc-1.9.13.jar:/home/hadoop/hadoop/hadoop-2.7.3/shar
e/hadoop/common/lib/jersey-server-1.9.jar:/home/hadoop/hadoop/hadoop-2.7.3/share/hadoop/common/lib/a
sm-3.2.jar:/home/hadoop/hadoop/hadoop-2.7.3/share/hadoop/common/lib/log4j-1.2.17.jar:/home/hadoop/ha
doop/hadoop-2.7.3/share/hadoop/common/lib/jets3t-0.9.0.jar:/home/hadoop/hadoop/hadoop-2.7.3/share/ha
doop/common/lib/httpclient-4.2.5.jar:/home/hadoop/hadoop/hadoop-2.7.3/share/hadoop/common/lib/httpco
re-4.2.5.jar:/home/hadoop/hadoop/hadoop-2.7.3/share/hadoop/common/lib/java-xmlbuilder-0.4.jar:/home/
hadoop/hadoop/hadoop-2.7.3/share/hadoop/common/lib/commons-lang-2.6.jar:/home/hadoop/hadoop/hadoop-2
.7.3/share/hadoop/common/lib/commons-configuration-1.6.jar:/home/hadoop/hadoop/hadoop-2.7.3/share/ha
doop/common/lib/commons-digester-1.8.jar:/home/hadoop/hadoop/hadoop-2.7.3/share/hadoop/common/lib/co
mmons-beanutils-1.7.0.jar:/home/hadoop/hadoop/hadoop-2.7.3/share/hadoop/common/lib/commons-beanutils
-core-1.8.0.jar:/home/hadoop/hadoop/hadoop-2.7.3/share/hadoop/common/lib/slf4j-api-1.7.10.jar:/home/
hadoop/hadoop/hadoop-2.7.3/share/hadoop/common/lib/slf4j-log4j12-1.7.10.jar:/home/hadoop/hadoop/hado
op-2.7.3/share/hadoop/common/lib/avro-1.7.4.jar:/home/hadoop/hadoop/hadoop-2.7.3/share/hadoop/common
/lib/paranamer-2.3.jar:/home/hadoop/hadoop/hadoop-2.7.3/share/hadoop/common/lib/snappy-java-1.0.4.1.
jar:/home/hadoop/hadoop/hadoop-2.7.3/share/hadoop/common/lib/commons-compress-1.4.1.jar:/home/hadoop
/hadoop/hadoop-2.7.3/share/hadoop/common/lib/xz-1.0.jar:/home/hadoop/hadoop/hadoop-2.7.3/share/hadoo
p/common/lib/protobuf-java-2.5.0.jar:/home/hadoop/hadoop/hadoop-2.7.3/share/hadoop/common/lib/gson-2
.2.4.jar:/home/hadoop/hadoop/hadoop-2.7.3/share/hadoop/common/lib/hadoop-auth-2.7.3.jar:/home/hadoop
/hadoop/hadoop-2.7.3/share/hadoop/common/lib/apacheds-kerberos-codec-2.0.0-M15.jar:/home/hadoop/hado
op/hadoop-2.7.3/share/hadoop/common/lib/apacheds-i18n-2.0.0-M15.jar:/home/hadoop/hadoop/hadoop-2.7.3
/share/hadoop/common/lib/api-asn1-api-1.0.0-M20.jar:/home/hadoop/hadoop/hadoop-2.7.3/share/hadoop/co
mmon/lib/api-util-1.0.0-M20.jar:/home/hadoop/hadoop/hadoop-2.7.3/share/hadoop/common/lib/zookeeper-3
.4.6.jar:/home/hadoop/hadoop/hadoop-2.7.3/share/hadoop/common/lib/netty-3.6.2.Final.jar:/home/hadoop
/hadoop/hadoop-2.7.3/share/hadoop/common/lib/curator-framework-2.7.1.jar:/home/hadoop/hadoop/hadoop-
2.7.3/share/hadoop/common/lib/curator-client-2.7.1.jar:/home/hadoop/hadoop/hadoop-2.7.3/share/hadoop
/common/lib/jsch-0.1.42.jar:/home/hadoop/hadoop/hadoop-2.7.3/share/hadoop/common/lib/curator-recipes
-2.7.1.jar:/home/hadoop/hadoop/hadoop-2.7.3/share/hadoop/common/lib/htrace-core-3.1.0-incubating.jar
:/home/hadoop/hadoop/hadoop-2.7.3/share/hadoop/common/lib/junit-4.11.jar:/home/hadoop/hadoop/hadoop-
2.7.3/share/hadoop/common/lib/hamcrest-core-1.3.jar:/home/hadoop/hadoop/hadoop-2.7.3/share/hadoop/co
mmon/lib/mockito-all-1.8.5.jar:/home/hadoop/hadoop/hadoop-2.7.3/share/hadoop/common/lib/hadoop-annot
ations-2.7.3.jar:/home/hadoop/hadoop/hadoop-2.7.3/share/hadoop/common/lib/guava-11.0.2.jar:/home/had
oop/hadoop/hadoop-2.7.3/share/hadoop/common/lib/jsr305-3.0.0.jar:/home/hadoop/hadoop/hadoop-2.7.3/sh
are/hadoop/common/lib/commons-cli-1.2.jar:/home/hadoop/hadoop/hadoop-2.7.3/share/hadoop/common/lib/c
ommons-math3-3.1.1.jar:/home/hadoop/hadoop/hadoop-2.7.3/share/hadoop/common/lib/xmlenc-0.52.jar:/hom
e/hadoop/hadoop/hadoop-2.7.3/share/hadoop/common/lib/commons-httpclient-3.1.jar:/home/hadoop/hadoop/
hadoop-2.7.3/share/hadoop/common/lib/commons-logging-1.1.3.jar:/home/hadoop/hadoop/hadoop-2.7.3/shar
e/hadoop/common/lib/commons-codec-1.4.jar:/home/hadoop/hadoop/hadoop-2.7.3/share/hadoop/common/lib/c
ommons-io-2.4.jar:/home/hadoop/hadoop/hadoop-2.7.3/share/hadoop/common/lib/commons-net-3.1.jar:/home
/hadoop/hadoop/hadoop-2.7.3/share/hadoop/common/lib/commons-collections-3.2.2.jar:/home/hadoop/hadoo
p/hadoop-2.7.3/share/hadoop/common/lib/servlet-api-2.5.jar:/home/hadoop/hadoop/hadoop-2.7.3/share/ha
doop/common/lib/jetty-6.1.26.jar:/home/hadoop/hadoop/hadoop-2.7.3/share/hadoop/common/lib/jetty-util
-6.1.26.jar:/home/hadoop/hadoop/hadoop-2.7.3/share/hadoop/common/lib/jsp-api-2.1.jar:/home/hadoop/ha
doop/hadoop-2.7.3/share/hadoop/common/lib/jersey-core-1.9.jar:/home/hadoop/hadoop/hadoop-2.7.3/share
/hadoop/common/lib/jersey-json-1.9.jar:/home/hadoop/hadoop/hadoop-2.7.3/share/hadoop/common/lib/jett
ison-1.1.jar:/home/hadoop/hadoop/hadoop-2.7.3/share/hadoop/common/lib/hadoop-common-2.7.3.jar:/home/hado
op/hadoop/hadoop-2.7.3/share/hadoop/common/hadoop-common-2.7.3-tests.jar:/home/hadoop/hadoop/hadoop-
2.7.3/share/hadoop/common/hadoop-nfs-2.7.3.jar:/home/hadoop/hadoop/hadoop-2.7.3/share/hadoop/hdfs:/h
```

ome/hadoop/hadoop/hadoop-2.7.3/share/hadoop/hdfs/lib/commons-codec-1.4.jar:/home/hadoop/hadoop/hadoop-2.7.3/share/hadoop/hdfs/lib/log4j-1.2.17.jar:/home/hadoop/hadoop/hadoop-2.7.3/share/hadoop/hdfs/lib/commons-logging-1.1.3.jar:/home/hadoop/hadoop/hadoop-2.7.3/share/hadoop/hdfs/lib/netty-3.6.2.Final.jar:/home/hadoop/hadoop/hadoop-2.7.3/share/hadoop/hdfs/lib/guava-11.0.2.jar:/home/hadoop/hadoop/hadoop-2.7.3/share/hadoop/hdfs/lib/jsr305-3.0.0.jar:/home/hadoop/hadoop/hadoop-2.7.3/share/hadoop/hdfs/lib/commons-cli-1.2.jar:/home/hadoop/hadoop/hadoop-2.7.3/share/hadoop/hdfs/lib/xmlenc-0.52.jar:/home/hadoop/hadoop/hadoop-2.7.3/share/hadoop/hdfs/lib/commons-io-2.4.jar:/home/hadoop/hadoop/hadoop-2.7.3/share/hadoop/hdfs/lib/servlet-api-2.5.jar:/home/hadoop/hadoop/hadoop-2.7.3/share/hadoop/hdfs/lib/jetty-6.1.26.jar:/home/hadoop/hadoop/hadoop-2.7.3/share/hadoop/hdfs/lib/jetty-util-6.1.26.jar:/home/hadoop/hadoop/hadoop-2.7.3/share/hadoop/hdfs/lib/jersey-core-1.9.jar:/home/hadoop/hadoop/hadoop-2.7.3/share/hadoop/hdfs/lib/jackson-core-asl-1.9.13.jar:/home/hadoop/hadoop/hadoop-2.7.3/share/hadoop/hdfs/lib/jackson-mapper-asl-1.9.13.jar:/home/hadoop/hadoop/hadoop-2.7.3/share/hadoop/hdfs/lib/jersey-server-1.9.jar:/home/hadoop/hadoop/hadoop-2.7.3/share/hadoop/hdfs/lib/asm-3.2.jar:/home/hadoop/hadoop/hadoop-2.7.3/share/hadoop/hdfs/lib/commons-lang-2.6.jar:/home/hadoop/hadoop/hadoop-2.7.3/share/hadoop/hdfs/lib/protobuf-java-2.5.0.jar:/home/hadoop/hadoop/hadoop-2.7.3/share/hadoop/hdfs/lib/htrace-core-3.1.0-incubating.jar:/home/hadoop/hadoop/hadoop-2.7.3/share/hadoop/hdfs/lib/commons-daemon-1.0.13.jar:/home/hadoop/hadoop/hadoop-2.7.3/share/hadoop/hdfs/lib/netty-all-4.0.23.Final.jar:/home/hadoop/hadoop/hadoop-2.7.3/share/hadoop/hdfs/lib/xercesImpl-2.9.1.jar:/home/hadoop/hadoop/hadoop-2.7.3/share/hadoop/hdfs/lib/xml-apis-1.3.04.jar:/home/hadoop/hadoop/hadoop-2.7.3/share/hadoop/hdfs/lib/leveldbjni-all-1.8.jar:/home/hadoop/hadoop/hadoop-2.7.3/share/hadoop/hdfs/hadoop-hdfs-2.7.3.jar:/home/hadoop/hadoop/hadoop-2.7.3/share/hadoop/hdfs/hadoop-hdfs-2.7.3-tests.jar:/home/hadoop/hadoop/hadoop-2.7.3/share/hadoop/hdfs/hadoop-hdfs-nfs-2.7.3.jar:/home/hadoop/hadoop/hadoop-2.7.3/share/hadoop/yarn/lib/zookeeper-3.4.6-tests.jar:/home/hadoop/hadoop/hadoop-2.7.3/share/hadoop/yarn/lib/commons-lang-2.6.jar:/home/hadoop/hadoop/hadoop-2.7.3/share/hadoop/yarn/lib/guava-11.0.2.jar:/home/hadoop/hadoop/hadoop-2.7.3/share/hadoop/yarn/lib/jsr305-3.0.0.jar:/home/hadoop/hadoop/hadoop-2.7.3/share/hadoop/yarn/lib/commons-logging-1.1.3.jar:/home/hadoop/hadoop/hadoop-2.7.3/share/hadoop/yarn/lib/protobuf-java-2.5.0.jar:/home/hadoop/hadoop/hadoop-2.7.3/share/hadoop/yarn/lib/commons-cli-1.2.jar:/home/hadoop/hadoop/hadoop-2.7.3/share/hadoop/yarn/lib/log4j-1.2.17.jar:/home/hadoop/hadoop/hadoop-2.7.3/share/hadoop/yarn/lib/jaxb-api-2.2.2.jar:/home/hadoop/hadoop/hadoop-2.7.3/share/hadoop/yarn/lib/stax-api-1.0-2.jar:/home/hadoop/hadoop/hadoop-2.7.3/share/hadoop/yarn/lib/activation-1.1.jar:/home/hadoop/hadoop/hadoop-2.7.3/share/hadoop/yarn/lib/commons-compress-1.4.1.jar:/home/hadoop/hadoop/hadoop-2.7.3/share/hadoop/yarn/lib/xz-1.0.jar:/home/hadoop/hadoop/hadoop-2.7.3/share/hadoop/yarn/lib/servlet-api-2.5.jar:/home/hadoop/hadoop/hadoop-2.7.3/share/hadoop/yarn/lib/commons-codec-1.4.jar:/home/hadoop/hadoop/hadoop-2.7.3/share/hadoop/yarn/lib/jetty-util-6.1.26.jar:/home/hadoop/hadoop/hadoop-2.7.3/share/hadoop/yarn/lib/jersey-core-1.9.jar:/home/hadoop/hadoop/hadoop-2.7.3/share/hadoop/yarn/lib/jersey-client-1.9.jar:/home/hadoop/hadoop/hadoop-2.7.3/share/hadoop/yarn/lib/jackson-core-asl-1.9.13.jar:/home/hadoop/hadoop/hadoop-2.7.3/share/hadoop/yarn/lib/jackson-mapper-asl-1.9.13.jar:/home/hadoop/hadoop/hadoop-2.7.3/share/hadoop/yarn/lib/jackson-jaxrs-1.9.13.jar:/home/hadoop/hadoop/hadoop-2.7.3/share/hadoop/yarn/lib/jackson-xc-1.9.13.jar:/home/hadoop/hadoop/hadoop-2.7.3/share/hadoop/yarn/lib/guice-servlet-3.0.jar:/home/hadoop/hadoop/hadoop-2.7.3/share/hadoop/yarn/lib/guice-3.0.jar:/home/hadoop/hadoop/hadoop-2.7.3/share/hadoop/yarn/lib/javax.inject-1.jar:/home/hadoop/hadoop/hadoop-2.7.3/share/hadoop/yarn/lib/aopalliance-1.0.jar:/home/hadoop/hadoop/hadoop-2.7.3/share/hadoop/yarn/lib/commons-io-2.4.jar:/home/hadoop/hadoop/hadoop-2.7.3/share/hadoop/yarn/lib/jersey-server-1.9.jar:/home/hadoop/hadoop/hadoop-2.7.3/share/hadoop/yarn/lib/asm-3.2.jar:/home/hadoop/hadoop/hadoop-2.7.3/share/hadoop/yarn/lib/jersey-json-1.9.jar:/home/hadoop/hadoop/hadoop-2.7.3/share/hadoop/yarn/lib/jettison-1.1.jar:/home/hadoop/hadoop/hadoop-2.7.3/share/hadoop/yarn/lib/jaxb-impl-2.2.3-1.jar:/home/hadoop/hadoop/hadoop-2.7.3/share/hadoop/yarn/lib/jersey-guice-1.9.jar:/home/hadoop/hadoop/hadoop-2.7.3/share/hadoop/yarn/lib/zookeeper-3.4.6.jar:/home/hadoop/hadoop/hadoop-2.7.3/share/hadoop/yarn/lib/netty-3.6.2.Final.jar:/home/hadoop/hadoop/hadoop-2.7.3/share/hadoop/yarn/lib/leveldbjni-all-1.8.jar:/home/hadoop/hadoop/hadoop-2.7.3/share/hadoop/yarn/lib/commons-collections-3.2.2.jar:/home/hadoop/hadoop/hadoop-2.7.3/share/hadoop/yarn/lib/jetty-6.1.26.jar:/home/hadoop/hadoop/hadoop-2.7.3/share/hadoop/yarn/hadoop-yarn-api-2.7.3.jar:/home/hadoop/hadoop/hadoop-2.7.3/share/hadoop/yarn/hadoop-yarn-common-2.7.3.jar:/home/hadoop/hadoop/hadoop-2.7.3/share/hadoop/yarn/hadoop-yarn-server-common-2.7.3.jar:/home/hadoop/hadoop/hadoop-2.7.3/share/hadoop/yarn/hadoop-yarn-server-nodemanager-2.7.3.jar:/home/hadoop/hadoop/hadoop-2.7.3/share/hadoop/yarn/hadoop-yarn-server-web-proxy-2.7.3.jar:/home/hadoop/hadoop/hadoop-2.7.3/share/hadoop/yarn/hadoop-yarn-server-applicationhistoryservice-2.7.3.jar:/home/hadoop/hadoop/hadoop-2.7.3/share/hadoop/yarn/hadoop-yarn-server-resourcemanager-2.7.3.jar:/home/hadoop/hadoop/hadoop-2.7.3/share/hadoop/yarn/hadoop-yarn-server-tests-2.7.3.jar:/home/hadoop/hadoop/hadoop-2.7.3/share/hadoop/yarn/hadoop-yarn-client-2.7.3.jar:/home/hadoop/hadoop/hadoop-2.7.3/share/hadoop/yarn/hadoop-yarn-server-sharedcachemanager-2.7.3.jar:/home/hadoop/hadoop/hadoop-2.7.3/share/hadoop/yarn/hadoop-yarn-applications-distributedshell-2.7.3.jar:/home/hadoop/hadoop/hadoop-2.7.3/share/hadoop/yarn/hadoop-yarn-applications-unmanaged-am-launcher-2.7.3.jar:/home/hadoop/hadoop/hadoop-2.7.3/share/hadoop/yarn/hadoop-yarn-registry-2.7.3.jar:/home/hadoop/hadoop/hadoop-2.7.3/share/hadoop/mapreduce/lib/protobuf-java-2.5.0.jar:/home/hadoop/hadoop/hadoop-2.7.3/share/hadoop/mapreduce/lib/avro-1.7.4.jar:/home/hadoop/hadoop/hadoop-2.7.3/share/hadoop/mapreduce/lib/jackson-core-asl-1.9.13.jar:/home/hadoop/hadoop/hadoop-2.7.3/share/hadoop/mapreduce/lib/jackson-mapper-asl-1.9.13.jar:/home/hadoop/hadoop/hadoop-2.7.3/share/hadoop/mapreduce/lib/paranamer-2.3.jar:/home/hadoop/hadoop/hadoop-2.7.3/share/hadoop/mapreduce/lib/snappy-java-1.0.4.1.jar:/home/hadoop/hadoop/hadoop-2.7.3/share/hadoop/mapreduce/lib/commons-compress-1.4.1.jar:/home/hadoop/hadoop/hadoop-2.7.3/share/hadoop/mapreduce/lib/xz-1.0.jar:/home/hadoop/hadoop/hadoop-2.7.3/share/hadoop/mapreduce/lib/hadoop-annotations-2.7.3.jar:/home/hadoop/hadoop/hadoop-2.7.3/share/hadoop/mapreduce/lib/commons-io-2.4.jar:/home/hadoop/hadoop/hadoop-2.7.3/share/hadoop/mapreduce/lib/jersey-core-1.9.jar:/home/hadoop/hadoop/hadoop-2.7.3/share/hadoop/mapreduce/lib/jersey-server-1.9.jar:/home/hadoop/hadoop/hadoop-2.7.3/share/hadoop/mapreduce/lib/asm-3.2.jar:/home/hadoop/hadoop/hadoop-

图 2-5-9 对服务切换控制信息格式化

（5）使用 Zookeeper 的客户端工具连接到 Zookeeper 集群，然后查看 Zookeeper 中的目录节点信息，如图 2-5-10 所示，可以看到其中新增加了一个名为"hadoop-ha"的目录节点，同时该目录节点下还包含有子目录节点。这便是高可用模式的 Hadoop 平台在 Zookeeper 中保存的服务切换控制信息。

（6）数据同步通信服务 JournalNode 会在 Hadoop 高可用模式集群启动时自动启动，所以为了完整的验证 Hadoop 高可用模式集群的启动是否正确，需要先关闭之前启动的所有 JournalNode 服务。在所有提供数据同步通信服务的主机（Cluster-03、Cluster-04、Cluster-05）上使用命令"hadoop-daemon.sh stop journalnode"可以关闭同步通信服务。

（7）在主节点上使用命令"start-all.sh"来启动 Hadoop 高可用模式的集群，如图 2-5-11 所示。

（8）Hadoop 高可用模式集群的启动命令的执行过程中，只会启动当前主机上 YARN 的 ResourceManager 服务，而其他所有备用的运行 ResourceManager 服务的主机都需要手动进行服务的启动操作，在这些主机上使用命令"yarn-daemon.sh start resourcemanager"来启动 YARN 的 ResourceManager 服务，如图 2-5-12 所示。

图 2-5-10　查看目录节点信息

图 2-5-11　启动高可用模式的集群

```
[hadoop@Cluster-02 hadoop]$ yarn-daemon.sh start resourcemanager
starting resourcemanager, logging to /home/hadoop/hadoop/hadoop-2.7.3/logs/yarn-hadoop-resourcemanag
er-Cluster-02.out
```

图 2-5-12　启动 ResourceManager 服务

（9）在查看集群中各个主机的 Java 进程信息，如图 2-5-13 所示，若满足以下的主机与进程名的对应关系，则表示 Hadoop 完全分布模式的集群启动成功。

Cluster-01	NameNode、ResourceManager、DFSZKFailoverController
Cluster-02	NameNode、ResourceManager、DFSZKFailoverController
Cluster-03	DataNode、NodeManager、JournalNode
Cluster-04	DataNode、NodeManager、JournalNode
Cluster-05	DataNode、NodeManager、JournalNode

```
[hadoop@Cluster-01 logs]$ jps
4839 QuorumPeerMain
10714 DFSZKFailoverController
10796 ResourceManager
10413 NameNode
11134 Jps

[hadoop@Cluster-02 hadoop]$ jps
5186 DFSZKFailoverController
3293 QuorumPeerMain
5357 ResourceManager
5389 Jps
5086 NameNode

[hadoop@Cluster-03 hadoop]$ jps
5265 Jps
3402 QuorumPeerMain
4923 DataNode
5003 JournalNode
5119 NodeManager

[hadoop@Cluster-04 hadoop]$ jps
2448 QuorumPeerMain
5476 DataNode
5829 Jps
5575 JournalNode
5674 NodeManager

[hadoop@Cluster-05 hadoop]$ jps
5906 Jps
2436 QuorumPeerMain
5558 DataNode
5638 JournalNode
5756 NodeManager
```

图 2-5-13　查看各个主机的 Java 进程信息

（10）第一次启动高可用模式的 Hadoop 平台，并且高可用模式的 Hadoop 平台的所有服务进程启动成功之后，使用 Zookeeper 的客户端工具连接到 Zookeeper 服务。然后查看 Zookeeper 中的目录节点信息，如图 2-5-14 所示，可以看到其中新增加了一个名为"yarn-leader"的目录节点，同时该目录节点下还有很多子目录节点。这便是高可用模式的 YARN 组件在 Zookeeper 中保存的服务状态信息以及服务状态切换控制信息。

```
[zk: Cluster-01:2181(CONNECTED) 0] ls /
[zookeeper, yarn-leader-election, hadoop-ha]
[zk: Cluster-01:2181(CONNECTED) 1] get /yarn-leader-election

cZxid = 0x5000000b1
ctime = Wed Jul 25 11:59:14 EDT 2018
mZxid = 0x5000000b1
mtime = Wed Jul 25 11:59:14 EDT 2018
pZxid = 0x5000000b2
cversion = 1
dataVersion = 0
aclVersion = 0
ephemeralOwner = 0x0
dataLength = 0
numChildren = 1
[zk: Cluster-01:2181(CONNECTED) 2] ls /yarn-leader-election
[yarn-ha]
[zk: Cluster-01:2181(CONNECTED) 3] ls /yarn-leader-election/yarn-ha
[ActiveBreadCrumb, ActiveStandbyElectorLock]
```

图 2-5-14　查看 Zookeeper 中的目录节点信息

2.5.6　安装和配置的常见问题及解决方法

高可用模式的 Hadoop 的安装与配置同完全分布模式的 Hadoop 的安装和配置的常见问题基本相同，主要就是服务节点的服务进程启动失败。

查找问题原因与解决问题方式也基本相似，主要就是检查配置文件中各个配置项的名称和值是否书写正确，或通过查看 Hadoop 的日志文件来确定问题和错误的详细信息。日志文件也同样是每个 Hadoop 服务节点的主机上只有自身所提供的 Hadoop 服务所对应的日志文件。需要注意的是在高可用模式下，不再有 SecondaryNameNode 服务节点，也就没有了相应的日志。但同时增加了数据同步通信服务 JournalNode 和 NameNode 服务节点状态自动切换服务 ZKFailoverController，其对应的日志文件的名称格式分别为"hadoop-用户名-journalnode-主机名.log"和"hadoop-用户名-zkfc-主机名.log"。

同样，若是修改了主机名、文件目录路径等方面的内容，建议将创建的四个功能目录"tmp""name""data""journal"中的内容清空,并重新将 HDFS 文件系统进行格式化之后,再启动 Hadoop 平台服务。

任务 2.6　安装 HBase 完全分布模式和高可用模式

2.6.1　HBase 完全分布模式和高可用模式安装规划

1. 环境要求

集群中所有主机的本地磁盘剩余空间 400 MB 以上;
集群中所有主机已安装 CentOS 7 1611 64 位操作系统;
集群中所有主机已完成基础网络环境配置;
集群中所有主机已安装 JDK 的 1.8.0_131 版本;
已完成 Zookeeper 的 3.4.9 版本的完全分布模式的安装和部署;
已完成 Hadoop 的 2.7.3 版本的完全分布模式或高可用模式的安装和部署。

2. 软件版本

与单节点模式 Hadoop 整合平台搭建中相同,选用 HBase 的 1.2.3 版本,软件包名为 hbase-1.2.3-bin.tar.gz。

3. 服务规划

HBase 的服务包括 Master 节点和 Region 节点共两个类型的服务节点,完全分布模式下要求有一台单独的主机作为 Master 节点,同时还需要使用两台及以上的主机作为 Region 节点来满足数据具备多个备份和分布式并行处理的需求,具体规划如表 2-6-1 所示。

表 2-6-1　节点规划

主机名	IP 地址	服务描述
Cluster-01	192.168.60.130	Master 节点（HMaster）
Cluster-03	192.168.60.132	Region 节点（HRegionServer）
Cluster-04	192.168.60.133	Region 节点（HRegionServer）
Cluster-05	192.168.60.134	Region 节点（HRegionServer）

HBase 的高可用模式只需要在完全分布模式的基础上,增加一台及以上的主机作为备用的 Master 节点即可,具体规划如表 2-6-2 所示。

表 2-6-2　HBase 的高可用模式节点规划

主机名	IP 地址	服务描述
Cluster-01	192.168.60.130	Master 节点（HMaster）
Cluster-02	192.168.60.131	备用 Master 节点（HMaster）
Cluster-03	192.168.60.132	Region 节点（HRegionServer）
Cluster-04	192.168.60.133	Region 节点（HRegionServer）
Cluster-05	192.168.60.134	Region 节点（HRegionServer）

2.6.2　同步集群主机时间

由于 HBase 数据库的数据模型中需要使用到时间戳来作为一条数据的版本标识，而时间戳的值是通过当前主机上操作系统的系统时间获取。所以在分布式环境下，HBase 会自动检查集群中各个服务节点之间的系统时间，若对于系统时间差异较大的主机，其上运行的 HBase 服务会被自动停止。

所以在搭建集群模式的 HBase 平台时，需要对集群中的所有主机的系统时间进行同步。当然要做到所有主机的系统时间完全一样比较困难，所以 HBase 中的时间同步检测允许集群中各个主机之间的系统时间有一个误差，该误差值默认为 30 s，并且可以通过 HBase 的配置文件 "hbase-site.xml" 中的配置项 "hbase.master.maxclockskew" 进行设定，设定值的单位为毫秒（ms）。下面提供两种对集群中主机进行时间同步的方法。

第一种方法是直接手动设定集群中主机的系统时间。使用 "root" 用户登录到操作系统，然后使用命令 "date -s '年-月-日 时：分：秒'" 可以对系统时间进行设置。系统时间设置完成之后，使用命令 "hwclock -w" 将设置的系统时间同步到硬件时钟，这样可以使新设置的系统时间永久生效，否则重新启动计算机之后，系统时间又会恢复到与原有的硬件时钟相同。

手动设定时间的方式不是太精确，集群中主机之间的系统时间可能会存在较大的差异（一般在 1 min 以内）。并且该方式的设置是一次性的，随着集群中主机的运行时间变长，主机之间的时间差也会逐渐变大，没有办法做到实时的时间同步。所以如果采用这种方式进行时间同步的话，建议将配置项 "hbase.master.maxclockskew" 的值改为 60 s 或更高。

第二种方法是利用 NTP（Network Time Protocol）服务器来进行时间同步。该方法需要使用到 ntpdate 工具，该工具在 CentOS 7 的大部分安装模式中都没有默认安装，而是需要用户手动进行安装。该工具的安装软件包在 CentOS 7 1611 的安装光盘中的"Packages"目录下可以找到，软件包名称为"ntpdate-4.2.6p5-25.el7.centos.x86_64.rpm"。ntpdate 工具安装完成之后，便可以使用命令"ntpdate -u NTP 服务器主机名或 IP 地址"将当前主机的系统时间与指定的 NTP 服务器上的时间进行同步，选项"-u"的作用是绕过防火墙进行时间同步，这样可以避免防火墙的设置产生无法进行时间同步操作等影响。同时对操作系统的"/etc"目录下的"crontab"配置文件进行编辑，在该文件的末尾添加"* */1 * * * ntpdate -u NTP 服务器主机名或 IP 地址"这样的一行内容，将时间同步操作添加为每隔一个小时执行一次的系统定时任务。

利用 NTP 服务器来进行的时间同步的方式精确度很高，集群中主机之间的系统时间差异很小。而且可以很方便块添加系统定时任务，使得集群中主机的系统时间可以做到时实进行同步，有效避免了由于长时间运行导致集群中主机之间的系统时间差异越来越大的问题。但是这种方法需要使用到 NTP 服务器，而该服务器的来源有两种。一种是使用集群中的某一台主机或是与集群之间可以进行有效网络通信的某台主机来作为 NTP 服务器，这种 NTP 服务器需要自己手动进行搭建。在 CentOS 7 1611 的安装光盘中也提供了用于搭建 NTP 服务器的 NTP 服务工具的软件安装包，该软件安装包同样位于"Packages"目录下，软件包名称为"ntp-4.2.6p5-25.el7.centos.x86_64"。另一种是使用互联网上现成的 NTP 服务器，这些 NTP 服务器一般广泛用于接入互联网的所有计算机的系统时间同步，并且其中很多都得到了权威机构的认证和授权，其上的时间基本可以作为世界标准时间来使用。不过使用这些 NTP 服务器进行系统时间同步需要集群中主机都能够连接到互联网。下面列出了一些常用的互联网中的 NTP 服务器：

中国国家授时中心：cn.pool.ntp.org；

上海电信授时中心：ntp.api.bz；

复旦大学授时中心：ntp.fudan.edu.cn；

台警大授时中心（中国台湾）：asia.pool.ntp.org；

美国国家授时中心：time.nist.gov；

微软公司授时主机（美国）：time.windows.com。

2.6.3 HBase 完全分布模式配置

（1）在作为 Master 节点的主机上的 Hadoop 专用用户下完成 HBase 的基础安装和环境配置。安装配置方法与单节点 Hadoop 平台的搭建过程中 HBase 的基础安装和环境配置方法相同,解压 HBase 软件包并在用户环境变量文件中配置相关环境变量。

(2)进入 HBase 相关文件的目录,如图 2-6-1 所示,分别创建 HBase 的临时文件的存放目录"tmp"和 HBase 的日志文件的存放目录"logs"。

```
[hadoop@Cluster-01 hbase]$ ls -l
total 0
drwxrwxr-x. 7 hadoop hadoop 160 Jul 10 12:06 hbase-1.2.3
drwxrwxr-x. 2 hadoop hadoop   6 Jul 11 06:33 logs
drwxrwxr-x. 2 hadoop hadoop   6 Jul 11 06:33 tmp
```

图 2-6-1 创建目录

(3)进入 HBase 软件目录下的配置文件所在目录"conf"。

(4)编辑配置文件"hbase-env.sh",编辑的内容与伪分布模式相同,如图 2-6-2 所示。在"JAVA_HOME"配置项中指定本地文件系统中 JDK 软件目录的所在路径,在"HBASE_CLASSPATH"配置项中指定本地文件系统中 Hadoop 软件的配置文件目录的所在路径,在"HBASE_LOG_DIR"配置项中指定 HBase 的日志文件在本地文件系统中的存放路径,在"HBASE_MANAGES_ZK"配置项中指定 HBase 是否使用自带的 Zookeeper 组件。

```
 26 # The java implementation to use.  Java 1.7+ required.
 27 export JAVA_HOME=/home/hadoop/java/jdk1.8.0_131
 29 # Extra Java CLASSPATH elements.  Optional.
 30 export HBASE_CLASSPATH=/home/hadoop/hadoop/hadoop-2.7.3/etc/hadoop
104 # Where log files are stored.  $HBASE_HOME/logs by default.
105 export HBASE_LOG_DIR=/home/hadoop/hbase/logs
127 # Tell HBase whether it should manage it's own instance of Zookeeper or not.
128 export HBASE_MANAGES_ZK=false
```

图 2-6-2 编辑配置文件

(5)编辑配置文件"hbase-site.xml",在其中配置以下内容:
<configuration>
　<property>
　　<name>hbase.rootdir</name>
　　<value>hdfs：//hadoop-ha/user/hadoop/hbase</value>
　</property>
　<property>
　　<name>hbase.tmp.dir</name>
　　HBase 的临时文件的存放目录"tmp"的路径

```
    </property>
    <property>
      <name>hbase.cluster.distributed</name>
      <value>true</value>
    </property>
    <property>
      <name>hbase.zookeeper.quorum</name>
      <value>Cluster-01:2181,Cluster-02:2181,Cluster-03:2181,Cluster-04:2181,Cluster-05:2181</value>
    </property>
    <property>
      <name>hbase.master.maxclockskew</name>
      <value>60000</value>
    </property>
</configuration>
```

配置项说明：

hbase.rootdir——这里再指定 HBase 数据的存放路径时，根据所使用的 Hadoop 平台的模式不同，设置的内容上有一定区别。对应伪分布模式或完全分布模式的 Hadoop 平台，其设置的 HDFS 访问路径为 NameNode 服务节点的主机名或地址，同时再加上访问的端口号，例如"Cluster-01"。而对于高可用模式的 Hadoop 平台，其设置的 HDFS 访问路径组需要使用命名空间的逻辑名称，如"hadoop-ha"。

hbase.master.maxclockskew——指定 HBase 集群中各主机之间进行时间校准时所允许的最大时间差，单位为毫秒。由于 HBase 在分布式存取数据的时候会使用到时间戳，所以 HBase 会在对数据进行操作，特别是写入类操作时，会对所有使用到的服务节点的主机时间进行校准，以免由于不同主机之间的系统时间差异过大而导致数据有问题。而该配置项便是设定时间校准操作所能允许的时间误差范围。

（6）编辑配置文件"regionservers"，删除文件中的所有原有内容，然后在其中按照每行一个主机名的方式，添加 HBase 的 Region 节点的主机名"Cluster-03""Cluster-04""Cluster-05"。

（7）使用命令"scp ~/.bash_profile 用户名@目标主机名或 IP 地址:/home/用户名/"和"scp -r ~/hbase 用户名@目标主机名或 IP 地址：/home/用户名/"依次将新修改的用户环境变量设置文件和 HBase 的专用目录发送到集群中

所有其他主机的相应用户家目录下。

2.6.4 HBase 高可用模式配置

（1）HBase 的高可用模式和完全分布模式在配置上的区别不大，只需要在完全分布模式的配置的基础之上，在 HBase 软件的配置文件目录“conf”之中手动创建一个名为“backup-masters”的配置文件。然后编辑该配置文件，在其中按照每行一个主机名的方式，添加 HBase 的所有备用 Master 节点的主机名。当前只设计了一个备用 Master 节点，所以只需要在其中添加“Cluster-02”即可。

（2）使用“scp”命令将新创建的配置文件“backup-masters”发送到集群中所有其他主机的 HBase 软件的配置文件目录下。

2.6.5 HBase 完全分布模式和高可用模式的启动和验证

```
[hadoop@Cluster-01 ~]$ start-hbase.sh
starting master, logging to /home/hadoop/hbase/logs/hbase-hadoop-master-Cluster-01.out
Java HotSpot(TM) 64-Bit Server VM warning: ignoring option PermSize=128m; support was removed in 8.0
Java HotSpot(TM) 64-Bit Server VM warning: ignoring option MaxPermSize=128m; support was removed in
8.0
Cluster-05: starting regionserver, logging to /home/hadoop/hbase/logs/hbase-hadoop-regionserver-Clus
ter-05.out
Cluster-04: starting regionserver, logging to /home/hadoop/hbase/logs/hbase-hadoop-regionserver-Clus
ter-04.out
Cluster-03: starting regionserver, logging to /home/hadoop/hbase/logs/hbase-hadoop-regionserver-Clus
ter-03.out
Cluster-05: Java HotSpot(TM) 64-Bit Server VM warning: ignoring option PermSize=128m; support was re
moved in 8.0
Cluster-05: Java HotSpot(TM) 64-Bit Server VM warning: ignoring option MaxPermSize=128m; support was
 removed in 8.0
Cluster-04: Java HotSpot(TM) 64-Bit Server VM warning: ignoring option PermSize=128m; support was re
moved in 8.0
Cluster-04: Java HotSpot(TM) 64-Bit Server VM warning: ignoring option MaxPermSize=128m; support was
 removed in 8.0
Cluster-03: Java HotSpot(TM) 64-Bit Server VM warning: ignoring option PermSize=128m; support was re
moved in 8.0
Cluster-03: Java HotSpot(TM) 64-Bit Server VM warning: ignoring option MaxPermSize=128m; support was
 removed in 8.0
Cluster-02: starting master, logging to /home/hadoop/hbase/logs/hbase-hadoop-master-Cluster-02.out
Cluster-02: Java HotSpot(TM) 64-Bit Server VM warning: ignoring option PermSize=128m; support was re
moved in 8.0
Cluster-02: Java HotSpot(TM) 64-Bit Server VM warning: ignoring option MaxPermSize=128m; support was
 removed in 8.0
```

图 2-6-3　启动 HBase 集群

（1）在 Master 节点所在主机上使用命令“start-hbase.sh”启动 HBase 集群，如图 2-6-3 所示。

（2）在 Master 节点上查看 Java 进程信息，如图 2-6-4 所示，若存在名为“HMaster”的进程，则表示 HBase 集群的 Master 节点服务启动成功。

（3）若搭建的是 HBase 的高可用模式的集群，则在所有备用 Master 节点上查看 Java 进程信息，如图 2-6-5 所示，若存在同样名为“HMaster”的进程，则表示 HBase 的高可用模式集群的备用 Master 节点服务启动成功。

项目 2　搭建高可用 Hadoop 整合平台

```
[hadoop@Cluster-01 ~]$ jps
4736 DFSZKFailoverController
4839 ResourceManager
6919 Jps
3433 QuorumPeerMain
4441 NameNode
6654 HMaster
```

图 2-6-4　查看 Master 节点的 Java 进程信息

```
[hadoop@Cluster-02 ~]$ jps
3889 ResourceManager
6469 Jps
3319 QuorumPeerMain
6313 HMaster
3726 NameNode
3823 DFSZKFailoverController
```

图 2-6-5　查看备用 Master 节点上查看 Java 进程信息

（4）在所有 Region 节点上查看 Java 进程信息，如图 2-6-6 所示，若存在名为 "HRegionServer" 的进程，则表示 HBase 集群的 Region 服务节点启动成功。

```
[hadoop@Cluster-03 ~]$ jps
4433 HRegionServer
4595 Jps
3860 NodeManager
3769 JournalNode
3675 DataNode
3261 QuorumPeerMain
```

```
[hadoop@Cluster-04 ~]$ jps
3697 JournalNode
3603 DataNode
4549 Jps
3174 QuorumPeerMain
3788 NodeManager
4365 HRegionServer
```

```
[hadoop@Cluster-05 ~]$ jps
3169 QuorumPeerMain
3766 NodeManager
4374 HRegionServer
4550 Jps
3675 JournalNode
3581 DataNode
```

图 2-6-6　R 查看 egion 节点的 Java 进程信息

（5）在集群中任意一个节点的主机上使用命令"hbase shell"并利用 HBase 的客户端工具连接 HBase 服务，如图 2-6-7 所示，若进入 HBase 客户端程序的控制台界面且没有出现报错信息，则表示 HBase 服务连接正常。在 HBase 控制台界面中使用命令"exit"可以退出 HBase 客户端程序并返回到操作系统的命令行界面。

```
[hadoop@Cluster-01 ~]$ hbase shell
SLF4J: Class path contains multiple SLF4J bindings.
SLF4J: Found binding in [jar:file:/home/hadoop/hbase/hbase-1.2.3/lib/slf4j-log4j12-1.7.5.jar!/org/sl
f4j/impl/StaticLoggerBinder.class]
SLF4J: Found binding in [jar:file:/home/hadoop/hadoop/hadoop-2.7.3/share/hadoop/common/lib/slf4j-log
4j12-1.7.10.jar!/org/slf4j/impl/StaticLoggerBinder.class]
SLF4J: See http://www.slf4j.org/codes.html#multiple_bindings for an explanation.
SLF4J: Actual binding is of type [org.slf4j.impl.Log4jLoggerFactory]
HBase Shell; enter 'help<RETURN>' for list of supported commands.
Type "exit<RETURN>" to leave the HBase Shell
Version 1.2.3, rbd63744624a26dc3350137b564fe746df7a721a4, Mon Aug 29 15:13:42 PDT 2016

hbase(main):001:0>
```

图 2-6-7　连接 HBase 服务

（6）在 HBase 控制台界面中使用命令"create '表名', '列名 1', '列名 2', …"在 HBase 数据库中创建数据表。如图 2-6-8 所示，若没有出现报错信息，并且显示变更行数和执行时间的信息，则表示数据表创建成功，HBase 数据库可以正常使用。

```
hbase(main):001:0> create 'test','col1','col2','col3'
0 row(s) in 3.1770 seconds

=> Hbase::Table - test
```

图 2-6-8　在 HBase 数据库中创建数据表

2.6.6　安装和配置的常见问题及解决方法

集群模式的 HBase 的安装与配置的常见问题大部分与单节点模式的 HBase 的安装和配置中的常见问题相同，主要就是服务节点的服务进程启动失败或启动一段时间之后自动关闭、不能正常进入 HBase 控制台、不能正常创建数据库表等。不过集群模式的 HBase 的安装与配置中有个需要特别注意的地方，就是由集群中主机之间系统时间差异过大导致的问题。这些问题导致的现象和前面所描述过的常见问题现象也比较相似，主要就是服务进程启动一段时间之后自动关闭，以及不能正常创建数据库表。

查找问题原因与解决问题方式也与单节点模式的 HBase 的安装和配置中查

找问题原因与解决问题方式基本相同，主要就是检查配置文件中各个配置项的名称和值是否书写正确，或通过查看 HBase 的日志文件来确定问题和错误的详细信息。即便是由于集群中主机之间系统时间差异过大导致的问题在日志文件中也会有详细的错误描述。不过需要注意的是，与集群模式下 Hadoop 相同，每个 HBase 服务节点的主机上只有自身所提供的 HBase 服务所对应的日志文件，如 HMaster 服务节点上只有 HMaster 的日志文件"hbase-用户名-master-主机名.log"，HRegionServer 服务节点上只有 HRegionServer 的日志文件"hbase-用户名-regionserver-主机名.log"，所以当某个节点的安装与配置出现问题时，只需要查看该问题节点的主机上的对应日志文件即可。

同样，若是修改了主机名、文件目录路径等方面的内容，建议将 HBase 的临时文件的存放目录"tmp"中的内容清空之后，再重新启动 HBase 平台服务。

项目 3 Hadoop 整合平台的使用与管理

任务 3.1 Hadoop 的使用和管理

这里将使用 Hadoop 软件自带的词频统计程序来分析数据文件中的每个英文单词的出现次数。这里以小说《教父》(The Godfather) 作为示例数据文件，数据文件的大小约为 926 KB，文件内容全部为只包含英文单词、英文符号、空格、空行标准的英文文本，如图 3-1-1 所示。词频统计程序默认只能识别标准的文本文件，而不能识别 Word 文件或是 PDF 文件，所以这里所使用的数据文件为 txt 格式的文本文件。

```
[hadoop@Cluster-01 ~]$ ls -l The_Godfather.txt
-rwxr-xr-x. 1 hadoop hadoop 948494 Aug 14 11:01 The_Godfather.txt
```

```
For Anthony Cleri

Book One

Chapter 01

Behind every great fortune there is a crime.
BALZAC

Amerigo Bonasera sat in New York Criminal Court Number 3 and waited for justice; vengeance on the me
n who had so cruelly hurt his daughter, who had tried to dishonor her.

The judge, a formidably heavy-featured man, rolled up the sleeves of his black robe as if to physica
lly chastise the two young men standing before the bench. His face was cold with majestic contempt.
But there was something false in all this that Amerigo Bonasera sensed but did not yet understand.

"You acted like the worst kind of degenerates," the judge said harshly. Yes, yes, thought Amerigo Bo
nasera. Animals. Animals. The two young men, glossy hair crew cut, scrubbed clean-cut faces composed
 into humble contrition, bowed their heads in submission.

The judge went on. "You acted like wild beasts in a jungle and you are fortunate you did not sexuall
y molest that poor girl or I'd put you behind bars for twenty years." The judge paused, his eyes ben
eath impressively thick brows flickered slyly toward the sallow-faced Amerigo Bonasera, then lowered
 to a stack of probation reports before him. He frowned and shrugged as if convinced against his own
 natural desire. He spoke again.

"But because of your youth, your clean records, because of your fine families, and because the law i
n its majesty does not seek vengeance, I hereby sentence you to three years' confinement to the peni
tentiary. Sentence to be suspended."

Only forty years of professional mourning kept the overwhelming frustration and hatred from showing
on Amerigo Bonasera's face. His beautiful young daughter was still in the hospital with her broken j
aw wired together; and now these two animales went free? It had all been a farce. He watched the hap
py parents cluster around their darling sons. Oh, they were all happy now, they were smiling now.

@
"The_Godfather.txt" [converted][dos] 5546L, 948556C
```

图 3-1-1 示例数据文件

3.1.1 Hadoop 平台的启动与关闭

在 Hadoop 软件中提供了一些现成的脚本文件命令来帮助使用者更为方便地启动和关闭整个 Hadoop 平台。在之前的 Hadoop 整合平台的搭建环节中，在 Hadoop 平台的格式化与启动的部分已经介绍了 Hadoop 平台中常用的几个启动命令，其中包括用于单独启动 HDFS 文件系统的命令"start-dfs.sh"，用于单独启动 YARN 资源管理器的命令"start-yarn.sh"，以及用于同时启动包括 HDFS 文件系统和 YARN 资源管理器在内的整个 Hadoop 平台的命令"start-all.sh"。不过，"start-all.sh"命令在新版本的 Hadoop 软件中已经不推荐使用，在使用该命令的时候会出现"This script is Deprecated"的提示信息，如图 3-1-2 所示，并且会被自动替换为依次使用"start-dfs.sh"命令和"start-yarn.sh"命令来启动 Hadoop 平台。

```
[hadoop@Cluster-01 ~]$ start-all.sh
This script is Deprecated. Instead use start-dfs.sh and start-yarn.sh
```

图 3-1-2　命令不推荐使用

除了用于启动 Hadoop 平台的相关脚本文件命令之外，自然还有对应的用于关闭 Hadoop 平台的相关脚本文件命令。与单独启动 HDFS 文件系统命令对应的单独关闭 HDFS 文件系统的命令为"stop-dfs.sh"，如图 3-1-3 所示。与单独启动 YARN 资源管理器命令对应的单独关闭 YARN 资源管理器的命令为"stop-yarn.sh"，如图 3-1-4 所示。对于启动整个 Hadoop 平台的命令也有对应的命令"stop-all.sh"来关闭整个 Hadoop 平台，如图 3-1-5 所示，该命令和"start-all.sh"命令一样，在新版本的 Hadoop 软件中已经不推荐使用，在使用该命令的时候会出现"This script is Deprecated"的提示信息，并且会被自动替换为依次使用"stop-dfs.sh"命令和"stop-yarn.sh"命令来启动 Hadoop 平台。

```
[hadoop@Cluster-01 ~]$ stop-dfs.sh
Stopping namenodes on [Cluster-01 Cluster-02]
Cluster-01: stopping namenode
Cluster-02: stopping namenode
Cluster-04: stopping datanode
Cluster-03: stopping datanode
Cluster-05: stopping datanode
Stopping journal nodes [Cluster-03 Cluster-04 Cluster-05]
Cluster-03: stopping journalnode
Cluster-05: stopping journalnode
Cluster-04: stopping journalnode
Stopping ZK Failover Controllers on NN hosts [Cluster-01 Cluster-02]
Cluster-02: stopping zkfc
Cluster-01: stopping zkfc
```

图 3-1-3　单独关闭 HDFS 文件系统的命令

```
[hadoop@Cluster-01 ~]$ stop-yarn.sh
stopping yarn daemons
stopping resourcemanager
Cluster-03: stopping nodemanager
Cluster-04: stopping nodemanager
Cluster-05: stopping nodemanager
no proxyserver to stop
```

图 3-1-4　单独关闭 YARN 资源管理器的命令

```
[hadoop@Cluster-01 ~]$ stop-all.sh
This script is Deprecated. Instead use stop-dfs.sh and stop-yarn.sh
Stopping namenodes on [Cluster-01 Cluster-02]
Cluster-01: stopping namenode
Cluster-02: stopping namenode
Cluster-05: stopping datanode
Cluster-03: stopping datanode
Cluster-04: stopping datanode
Stopping journal nodes [Cluster-03 Cluster-04 Cluster-05]
Cluster-04: stopping journalnode
Cluster-05: stopping journalnode
Cluster-03: stopping journalnode
Stopping ZK Failover Controllers on NN hosts [Cluster-01 Cluster-02]
Cluster-01: stopping zkfc
Cluster-02: stopping zkfc
stopping yarn daemons
stopping resourcemanager
Cluster-05: stopping nodemanager
Cluster-04: stopping nodemanager
Cluster-03: stopping nodemanager
no proxyserver to stop
```

图 3-1-5　关闭整个 Hadoop 平台

需要注意的是，通过使用"start-dfs.sh"命令和"start-yarn.sh"命令分别启动 HDFS 文件系统和 YARN 资源管理器来启动 Hadoop 平台时，以及通过使用"stop-dfs.sh"命令和"stop-yarn.sh"命令分别关闭 HDFS 文件系统和 YARN 资源管理器来关闭 Hadoop 平台时，最好按照一定的顺序进行操作。由于在 YARN 资源管理器中运行 MapReduce 程序需要使用到 HDFS 文件系统，所以在启动 Hadoop 平台时应该先启动 HDFS 文件系统，然后再启动 YARN 资源管理器。而在关闭 Hadoop 平台时应该先关闭 YARN 资源管理器，然后再关闭 HDFS 文件系统。这样做可以尽量避免错误的发生。

除了上面介绍的常用 Hadoop 平台的启动和关闭命令之外，Hadoop 中的所有组件的各个服务还可以单独启动和关闭。如在前面的 Hadoop 平台的安装和配置的常见问题及解决方法中提到过的"hadoop　HDFS 服务名"命令和"yarn

YARN 服务名"命令就是分别对应单独启动 HDFS 文件系统中指定服务和单独启动 YARN 资源管理器中指定服务的操作。其中的服务名与启动之后查看到的 Java 进程名称基本一样，只不过全部是小写字母，如 namenode、secondarynamenode、datanode、journalnode、resourcemanager、nodemanager，唯一不同的是 DFSZKFailoverController 进程对应的服务名为 zkfc。但这两个命令是以前台进程的方式来启动 Hadoop 的指定服务，该服务进程会随着控制台的关闭而关闭，而在通常情况下，我们都希望 Hadoop 的所有服务进程能够以守护进程的方式启动并运行。所以在单独启动指定服务时，需要使用到脚本文件命令"hadoop-daemon.sh"和"yarn-daemon.sh"，两者分别对应 HDFS 文件系统和 YARN 资源管理器。这两个命令在之前的 Hadoop 高可用模式格式化和启动部分都有用过，其完整的启动和关闭命令格式如下，使用效果如图 3-1-6 和图 3-1-7 所示。

启动 HDFS 服务：hadoop-daemon.sh　start　HDFS 服务名
关闭 HDFS 服务：hadoop-daemon.sh　stop　HDFS 服务名
启动 YARN 服务：yarn-daemon.sh　start　YARN 服务名
关闭 YARN 服务：yarn-daemon.sh　stop　YARN 服务名

```
[hadoop@Cluster-01 ~]$ hadoop-daemon.sh start namenode
starting namenode, logging to /home/hadoop/hadoop/hadoop-2.7.3/logs/hadoop-hadoop-namenode-Cluster-0
1.out
```

图 3-1-6　启用 HDFS 服务

```
[hadoop@Cluster-01 ~]$ hadoop-daemon.sh stop namenode
stopping namenode
```

图 3-1-7　关闭 HDFS 服务

另外，如果通过单独启动每个服务来启动整个 Hadoop 平台，需要严格按照指定的顺序来启动。首先是组件的启动顺序，依然按照先启动 HDFS 文件系统，再启动 YARN 资源管理器的顺序；其次 HDFS 文件系统中服务的启动顺序为 namenode → datanode → journalnode → zkfc；最后 YARN 资源管理器中服务的启动顺序为 resoucemanager → nodemanager。而通过单独关闭每个服务来关闭整个 Hadoop 平台时，关闭顺序也是严格按照启动顺序的逆向顺序来关闭。如果启动和关闭的顺序不正确，有可能导致 Hadoop 平台启动失败，或者关闭时导致 MapReduce 任务出错或是数据丢失。总体来说，Hadoop 平台中服务的单独启动和关闭通常只是在一些特定场合使用，而不推荐用于启动和关闭整个 Hadoop 平台，因为其操作复杂且容易出错，而达到的效果和常用的 Hadoop 平台启动和关闭命令却是一样的。

3.1.2 向 HDFS 上传数据文件

Hadoop 平台在搭建完成之后，HDFS 文件系统中是没有任何数据的。而要往 HDFS 中存放数据，就需要使用到 Hadoop 客户端中的相关操作命令，或者是通过 HDFS 的相关 API 接口来编写程序实现。这里介绍的是使用 Hadoop 客户端中的相关操作命令的方式。

Hadoop 客户端中的命令包括了用户命令和管理命令两大类。前面提到过的单独启动指定服务的命令"hadoop HDFS 服务名"就是属于 Hadoop 命令中的管理命令，当然该命令除了单独启动指定服务之外，还可以通过添加一些选项来执行一些其他操作，如格式化 HDFS 文件系统的命令"hadoop namenode -format"。而 Hadoop 命令中文件操作相关的命令属于用户命令中的 FS 命令，包括了文件的上传、下载、新建、删除、复制、移动等一系列操作。而由于文件操作命令是 Hadoop 客户端中最常用的命令，所以 FS 命令又被称为 Shell 命令。FS 命令的常见使用格式有以下三种：

hadoop fs {args}
hadoop dfs {args}
hdfs dfs {args}

三种不同使用格式的区别如下：

（1）"hadoop fs"命令除了可以操作 HDFS 文件系统之外，还可以操作多种其他的文件系统，多用于操作系统的本地文件系统或是网络中的其他文件系统与 HDFS 文件系统的交互操作中，其使用最为广泛。

（2）"hadoop dfs"命令与"hdfs dfs"命令只能操作 HDFS 文件系统，并且前者只存在于比较老的 Hadoop 版本分支中，而后者则是存在于比较新的 Hadoop 版本分支用于替代前者。

（3）在新的 Hadoop 版本中"hadoop fs"命令和"hadoop dfs"命令已经基本相同，"hadoop dfs"命令也可以对一些 HDFS 文件系统之外的其他文件系统进行操作。

FS 命令中"{args}"部分的具体文件操作命令与 Linux 操作系统下的文件操作命令基本相似，甚至有一些命令与 Linux 操作系统下的文件操作命令完全一样。在后面介绍的 FS 命令中，和 Linux 操作系统中的命令同名的命令在功能和用法上也基本相同。

而在使用 FS 命令时，还需要注意一些路径书写上的问题。若是操作操作系统的本地文件系统中的文件,其访问路径为"file:///操作系统中目录或文件路径",并且不能使用文件相对路径。若是操作远程的 HDFS 文件系统中的文件,则访问路径为"hdfs://Hadoop 访问路径:端口号/HDFS 中的目录或文件路径",并且同样不能使用文件的相对路径。而如果 Hadoop 平台位于当前操作系统的本地,则访问路径可以直接使用"/HDFS 中的目录或文件路径"的方式。并且在操作 HDFS 文件系统中的文件时可以使用相对路径,而相对路径的起始点固定为当前操作系统用户位于 HDFS 文件系统中的对应用户目录,如操作系统用户"hadoop"在 HDFS 文件系统中的用户目录路径为"/user/hadoop"。

(4)使用 Hadoop 的专用用户登录到 Hadoop 平台中任意主机的操作系统,在 HDFS 文件系统的操作系统用户对应目录下创建"data"目录,用于存放之后用作词频统计分析的原数据的数据文件,这里使用的是相对路径的方式在 HDFS 文件系统中创建目录,具体命令如下:

hadoop   fs   -mkdir   data

(5)在 HDFS 文件系统下查看操作系统用户对应目录下的文件和目录信息,检查"data"目录是否创建成功,具体命令如下:

hadoop   fs   -ls

hadoop   fs   -ls   /user/hadoop

两个命令分别采用的是相对路径和绝对路径的方式,都可以正常进行查看,但如图 3-1-8 和图 3-1-9 所示,显示的结果会有一些区别。

```
[hadoop@Cluster-01 ~]$ hadoop fs -ls
Found 3 items
drwxr-xr-x   - hadoop supergroup          0 2018-08-14 11:26 data
drwxr-xr-x   - hadoop supergroup          0 2018-07-23 12:48 hbase
drwxrwxrwx   - hadoop supergroup          0 2018-07-12 09:57 hive
```

图 3-1-8   "hadoop   fs   -ls"命令

```
[hadoop@Cluster-01 ~]$ hadoop fs -ls /user/hadoop
Found 3 items
drwxr-xr-x   - hadoop supergroup          0 2018-08-14 11:26 /user/hadoop/data
drwxr-xr-x   - hadoop supergroup          0 2018-07-23 12:48 /user/hadoop/hbase
drwxrwxrwx   - hadoop supergroup          0 2018-07-12 09:57 /user/hadoop/hive
```

图 3-1-9   "hadoop   fs   -ls   /user/hadoop"命令

(6)将数据文件"The_Godfather.txt"上传至 HDFS 文件系统中,存放在之前在操作系统用户对应目录下创建的"data"目录中,可使用的具体命令如下:

hadoop   fs   -cp   file:///home/hadoop/The_Godfather.txt   /user/hadoop/data

hadoop   fs   -put   /home/hadoop/The_Godfather.txt   /user/hadoop/data hadoop fs -copyFromLocal /home/hadoop/The_Godfather.txt data

三个命令都可以实现从本地文件系统中将文件拷贝至 HDFS 文件系统之中的操作，任选其一使用即可。三者之间的区别在于"cp"命令可以实现在任何文件系统之间进行双向的文件拷贝操作，并且在对本地文件系统进行操作时需要使用完整的本地文件系统访问路径"file：///操作系统中目录或文件路径"。而"put"命令和"copyFromLocal"命令只能完成将本地文件系统中的文件拷贝至 HDFS 文件系统之中的单向文件拷贝操作，但只需要使用一般的本地文件系统访问路径"/操作系统中目录或文件路径"来操作本地文件系统即可，因为命令会自动在本地文件系统之中查找除最后一个路径参数之外的所有路径参数。"copyFromLocal"命令与"put"命令的区别在于前者限制源路径只能有一个，而后者可以拥有多个源路径，使用起来更加灵活。

（7）通过查看上传至 HDFS 文件系统上的文件的大小或内容，检查文件的上传操作是否成功，如图 3-1-10 所示，具体命令如下：

hadoop fs -du /user/hadoop/data/The_Godfather.txt

hadoop fs -cat data/The_Godfather.txt

（由于数据文本的内容过长，这里不使用截图展示"cat"命令的使用效果）

```
[hadoop@Cluster-01 ~]$ hadoop fs -du /user/hadoop/data/The_Godfather.txt
948494   /user/hadoop/data/The_Godfather.txt
```

图 3-1-10　检查上传文件是否成功

3.1.3　运行词频统计分析程序

（1）这里使用的词频统计程序是 Hadoop 软件自带的一个示例程序，该程序被打包在 Hadoop 软件中的 Jar 包"hadoop-mapreduce-examples-2.7.3.jar"内，该 Jar 包位于 Hadoop 软件下的"share\hadoop\mapreduce"目录中，如图 3-1-11 所示。

```
[hadoop@Cluster-01 mapreduce]$ pwd
/home/hadoop/hadoop/hadoop-2.7.3/share/hadoop/mapreduce
[hadoop@Cluster-01 mapreduce]$ ls -l
total 4972
-rw-r--r--. 1 hadoop hadoop  537521 Aug 17  2016 hadoop-mapreduce-client-app-2.7.3.jar
-rw-r--r--. 1 hadoop hadoop  773581 Aug 17  2016 hadoop-mapreduce-client-common-2.7.3.jar
-rw-r--r--. 1 hadoop hadoop 1554595 Aug 17  2016 hadoop-mapreduce-client-core-2.7.3.jar
-rw-r--r--. 1 hadoop hadoop  189714 Aug 17  2016 hadoop-mapreduce-client-hs-2.7.3.jar
-rw-r--r--. 1 hadoop hadoop   27598 Aug 17  2016 hadoop-mapreduce-client-hs-plugins-2.7.3.jar
-rw-r--r--. 1 hadoop hadoop   61745 Aug 17  2016 hadoop-mapreduce-client-jobclient-2.7.3.jar
-rw-r--r--. 1 hadoop hadoop 1551594 Aug 17  2016 hadoop-mapreduce-client-jobclient-2.7.3-tests.jar
-rw-r--r--. 1 hadoop hadoop   71310 Aug 17  2016 hadoop-mapreduce-client-shuffle-2.7.3.jar
-rw-r--r--. 1 hadoop hadoop  295812 Aug 17  2016 hadoop-mapreduce-examples-2.7.3.jar
drwxr-xr-x. 2 hadoop hadoop    4096 Aug 17  2016 lib
drwxr-xr-x. 2 hadoop hadoop      30 Aug 17  2016 lib-examples
drwxr-xr-x. 2 hadoop hadoop    4096 Aug 17  2016 sources
```

图 3-1-11　词频统计程序包位置

（2）在 Hadoop 的 2.7.3 版本的软件包 hadoop-2.7.3.tar.gz 已经不再带有 Hadoop 项目的源代码，但其可以从 Apache 的官方资源网站（http://archive.apache.org/dist/）上获取，如图 3-1-12 所示。在该资源网站中包含了所有 Apache 项目的各种历史版本的 Release 包、源代码包等，之前及之后所使用到 Hadoop 及其各种组件的软件，若在其对应官方网站页面无法下载到所需要的版本，都可以在该资源网站中进行下载。

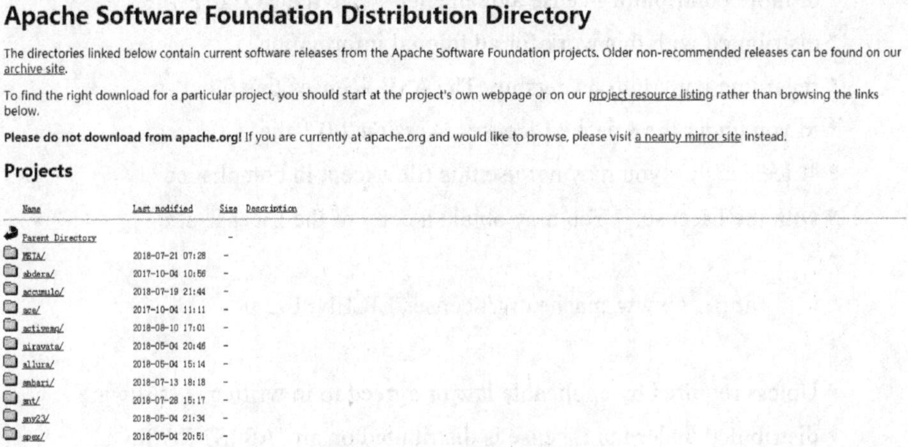
Apache Software Foundation Distribution Directory

The directories linked below contain current software releases from the Apache Software Foundation projects. Older non-recommended releases can be found on our archive site.

To find the right download for a particular project, you should start at the project's own webpage or on our project resource listing rather than browsing the links below.

Please do not download from apache.org! If you are currently at apache.org and would like to browse, please visit a nearby mirror site instead.

Projects

Name	Last modified	Size	Description
Parent Directory		-	
META/	2018-07-21 07:28	-	
abdera/	2017-10-04 10:56	-	
accumulo/	2018-07-19 21:44	-	
ace/	2017-10-04 11:11	-	
activemq/	2018-08-10 17:01	-	
airavata/	2018-05-04 20:46	-	
allura/	2018-05-04 15:14	-	
ambari/	2018-07-13 18:18	-	
ant/	2018-07-28 15:17	-	
any23/	2018-05-04 21:34	-	
apex/	2018-05-04 20:51	-	

图 3-1-12　获取资源网站

本书所使用的 Hadoop 的 2.7.3 版本所对应的所有资源在该资源网站中的存放路径为“hadoop/core/hadoop-2.7.3”，其中除了之前搭建 Hadoop 平台所使用到的 Release 包“hadoop-2.7.3.tar.gz”之外，另一个名为“hadoop-2.7.3-src.tar.gz”的软件包便是该版本 Hadoop 项目所对应的源代码包，如图 3-1-13 所示。

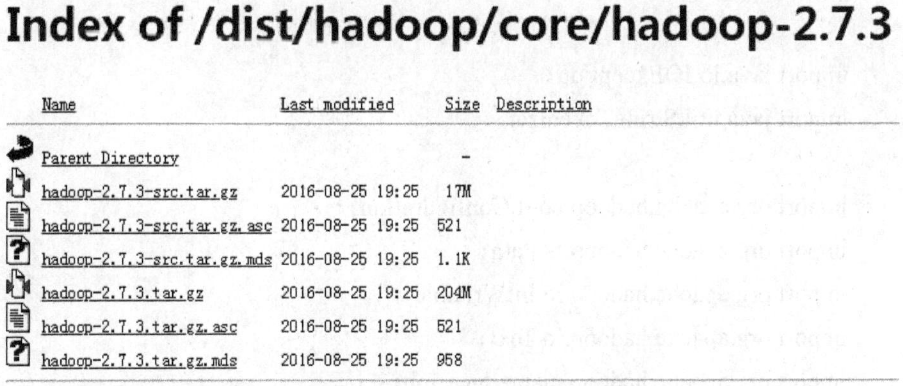
Index of /dist/hadoop/core/hadoop-2.7.3

Name	Last modified	Size	Description
Parent Directory		-	
hadoop-2.7.3-src.tar.gz	2016-08-25 19:25	17M	
hadoop-2.7.3-src.tar.gz.asc	2016-08-25 19:25	521	
hadoop-2.7.3-src.tar.gz.mds	2016-08-25 19:25	1.1K	
hadoop-2.7.3.tar.gz	2016-08-25 19:25	204M	
hadoop-2.7.3.tar.gz.asc	2016-08-25 19:25	521	
hadoop-2.7.3.tar.gz.mds	2016-08-25 19:25	958	

图 3-1-13　获取 hadoop-2.7.3 版本资源

（3）这里所使用的词频统计分析示例程序的源代码文件在源代码包中的存放位置为"hadoop-2.7.3-src\hadoop-mapreduce-project\hadoop-mapreduce- examples\src\main\java\org\apache\hadoop\examples"，文件名称为"WordCount.java"，程序的源代码如下：

```
/**
 * Licensed to the Apache Software Foundation（ASF）under one
 * or more contributor license agreements.  See the NOTICE file
 * distributed with this work for additional information
 * regarding copyright ownership. The ASF licenses this file
 * to you under the Apache License，Version 2.0（the
 * "License"）；you may not use this file except in compliance
 * with the License.  You may obtain a copy of the License at
 *
 *     http：//www.apache.org/licenses/LICENSE-2.0
 *
 * Unless required by applicable law or agreed to in writing，software
 * distributed under the License is distributed on an "AS IS" BASIS，
 * WITHOUT WARRANTIES OR CONDITIONS OF ANY KIND，either express or implied.
 * See the License for the specific language governing permissions and
 * limitations under the License.
 */
package org.apache.hadoop.examples；

import java.io.IOException；
import java.util.StringTokenizer；

import org.apache.hadoop.conf.Configuration；
import org.apache.hadoop.fs.Path；
import org.apache.hadoop.io.IntWritable；
import org.apache.hadoop.io.Text；
import org.apache.hadoop.mapreduce.Job；
import org.apache.hadoop.mapreduce.Mapper；
```

```java
import org.apache.hadoop.mapreduce.Reducer;
import org.apache.hadoop.mapreduce.lib.input.FileInputFormat;
import org.apache.hadoop.mapreduce.lib.output.FileOutputFormat;
import org.apache.hadoop.util.GenericOptionsParser;

public class WordCount {

  public static class TokenizerMapper
       extends Mapper<Object, Text, Text, IntWritable>{

    private final static IntWritable one = new IntWritable(1);
    private Text word = new Text();

    public void map(Object key, Text value, Context context
                    ) throws IOException, InterruptedException {
      StringTokenizer itr = new StringTokenizer(value.toString());
      while (itr.hasMoreTokens()) {
        word.set(itr.nextToken());
        context.write(word, one);
      }
    }
  }

  public static class IntSumReducer
       extends Reducer<Text, IntWritable, Text, IntWritable> {
    private IntWritable result = new IntWritable();

    public void reduce(Text key, Iterable<IntWritable> values,
                       Context context
                       ) throws IOException, InterruptedException {
      int sum = 0;
      for (IntWritable val: values) {
        sum += val.get();
```

```
        }
        result.set(sum);
        context.write(key, result);
    }
}

    public static void main(String[] args) throws Exception {
        Configuration conf = new Configuration();
        String[] otherArgs = new GenericOptionsParser(conf,args).getRemainingArgs();
        if (otherArgs.length < 2) {
            System.err.println("Usage: wordcount <in> [<in>...] <out>");
            System.exit(2);
        }
        Job job = Job.getInstance(conf, "word count");
        job.setJarByClass(WordCount.class);
        job.setMapperClass(TokenizerMapper.class);
        job.setCombinerClass(IntSumReducer.class);
        job.setReducerClass(IntSumReducer.class);
        job.setOutputKeyClass(Text.class);
        job.setOutputValueClass(IntWritable.class);
        for (int i = 0; i < otherArgs.length - 1; ++i) {
            FileInputFormat.addInputPath(job, new Path(otherArgs[i]));
        }
        FileOutputFormat.setOutputPath(job,
            new Path(otherArgs[otherArgs.length - 1]));
        System.exit(job.waitForCompletion(true) ? 0 : 1);
    }
}
```

（4）Hadoop 的 2.7.3 版本所提供的原始词频统计分析程序运行产生的结果可能不太符合正常的需求，因为它只是以" "（空格）、"\t"、"\n"这几个符号来分隔字符串，这会使得统计出来的单词并不是真正的英文单词，而是会包含有各种各样的符号或其他类型的文字，甚至会有"young,"这样的英文单词和标点符号在一起的字符串，并且这种字符串会被认为是和"young"不一样的另一个单词，如图 3-1-14 所示。而首字母大小写不同的同一个单词，也会被认为是不一样的两个单词。

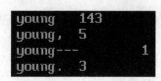

图 3-1-14　使用词频统计分析程序分离出单词

所以这里需要将原版的词频统计分析程序的代码进行一些优化，主要是对 map 函数中的分词方式相关代码进行一些修改，使其只从文本中提取出英文单词或字母，并且将所有单词全部转换成小写之后再进行统计。修改之后的 WordCount.java 文件的内容如下：

package org.apache.hadoop.examples;

import java.io.IOException;
import java.util.StringTokenizer;

import org.apache.hadoop.conf.Configuration;
import org.apache.hadoop.fs.Path;
import org.apache.hadoop.io.IntWritable;
import org.apache.hadoop.io.Text;
import org.apache.hadoop.mapreduce.Job;
import org.apache.hadoop.mapreduce.Mapper;
import org.apache.hadoop.mapreduce.Reducer;
import org.apache.hadoop.mapreduce.lib.input.FileInputFormat;
import org.apache.hadoop.mapreduce.lib.output.FileOutputFormat;
import org.apache.hadoop.util.GenericOptionsParser;

public class WordCount {

　　public static class WordMapper
　　　　extends Mapper<Object, Text, Text, IntWritable>{

```java
      private final static IntWritable one = new IntWritable(1);
      private Text word = new Text();

      public void map(Object key, Text value, Context context
                      ) throws IOException, InterruptedException {
        String[] items = value.toString().split("[^a-zA-Z]");
        for (String item: items) {
          if (item.length > 0) {
            word.set(item.toLowerCase());
            context.write(word, one);
          }
        }
      }
    }

    public static class WordReducer
          extends Reducer<Text, IntWritable, Text, IntWritable> {
      private IntWritable result = new IntWritable();

      public void reduce(Text key, Iterable<IntWritable> values,
                         Context context
                         ) throws IOException, InterruptedException {
        int sum = 0;
        for (IntWritable val: values) {
          sum += val.get();
        }
        result.set(sum);
        context.write(key, result);
      }
    }

    public static void main(String[] args) throws Exception {
      Configuration conf = new Configuration();
```

```
        String[] otherArgs = new GenericOptionsParser(conf, args).getRemaining
Args();
        if (otherArgs.length < 2) {
            System.err.println("Usage: wordcount <in> [<in>...] <out>");
            System.exit(2);
        }
        Job job = Job.getInstance(conf, "word count");
        job.setJarByClass(WordCount.class);
        job.setMapperClass(WordMapper.class);
        job.setCombinerClass(WordReducer.class);
        job.setReducerClass(WrodReducer.class);
        job.setOutputKeyClass(Text.class);
        job.setOutputValueClass(IntWritable.class);
        for (int i = 0; i < otherArgs.length - 1; ++i) {
            FileInputFormat.addInputPath(job, new Path(otherArgs[i]));
        }
        FileOutputFormat.setOutputPath(job,
            new Path(otherArgs[otherArgs.length - 1]));
        System.exit(job.waitForCompletion(true) ? 0 : 1);
    }
}
```

这里修改了 Mapper 类和 Reducer 类的类名，是为了避免在之后的打包运行时使用到 Hadoop 中的同名类，从而导致修改的内容失效。

（5）源代码修改完成之后，将"WordCount.java"文件单独导出为 Jar 包"WordCount.jar"，然后拷贝到 Hadoop 平台中的任意主机之上，如图 3-1-15 所示。

```
[hadoop@Cluster-01 ~]$ ls -l
total 932
drwxrwxr-x. 7 hadoop hadoop        76 Jul 11 05:32 hadoop
drwxrwxr-x. 5 hadoop hadoop        48 Jul 11 06:33 hbase
drwxrwxr-x. 3 hadoop hadoop        26 Jul 10 12:04 java
-rwxr-xr-x. 1 hadoop hadoop    948494 Aug 14 11:01 The_Godfather.txt
-rwxr-xr-x. 1 hadoop hadoop      3974 Aug 15 11:41 WordCount.jar
drwxrwxr-x. 5 hadoop hadoop        53 Jul 11 05:11 zookeeper
```

图 3-1-15 导出 Jar 包 "WordCount.jar"

（6）使用 Hadoop 的用户命令中的"jar"命令来运行修改之后的词频统计分析示例程序的具体命令如下：

hadoop jar WordCount.jar WordCount data output

该命令在之前的 Hadoop 平台搭建的验证过程中已经使用过，被用于运行计算 PI 值的示例程序。下面是对该命令中的各个参数的说明。

WordCount.jar——指定所要运行的程序的 Jar 包名称以及在本地文件系统中的存放路径，可以是相对路径也可以是绝对路径。

WordCount——指定程序运行的入口类（包含有 main 函数的主类）的类名。

data——指定用于进行词频统计分析的原数据文件在 HDFS 文件系统中的存放目录，可以有多个目录的路径。这个参数所代表的路径必须是一个目录，而不能是一个文件，这里使用的是相对路径。由于 MapReduce 程序会自动遍历该目录下的所有可以读取的文本文件，所以要保证该目录下没有其他与当前任务不相关的文件，避免运行之后的结果出现问题。从这个参数开始，以及后面的所有参数都是传递给入口类的 main 函数的参数。

outpu——指定程序运行之后输出的结果文件在 HDFS 文件系统中的存放目录的路径，这里使用的是相对路径。MapReduce 程序的运行结果目录必须是一个在当前 HDFS 文件系统中不存在的目录，即便是空目录也不行，否则程序运行会报错并中止。

不过直接使用上面的命令来运行示例程序会报错，错误信息如图 3-1-16 所示。错误的原因是命令找不到 WordCount 类。由于之前是直接在 Hadoop 的 2.7.3 版本的源代码项目中修改 WordCount 类，所以该类实际上是存在于"org.apache.hadoop.examples"这个包路径下面。可以通过两种方式来解决该问题，一种是在打包"WordCount.java"文件时，创建一个配置文件"MANIFEST.MF"，在其中的配置项"Main-Class"中指定入口类的完整路径为"org.apache.hadoop.examples.WordCount"。另一种方法是在命令中直接指定入口类的完整路径，即使用如下的命令来运行示例程序，运行过程如图 3-1-17 所示。

```
[hadoop@Cluster-01 ~]$ hadoop jar WordCount.jar WordCount data output
Exception in thread "main" java.lang.ClassNotFoundException: WordCount
        at java.net.URLClassLoader.findClass(URLClassLoader.java:381)
        at java.lang.ClassLoader.loadClass(ClassLoader.java:424)
        at java.lang.ClassLoader.loadClass(ClassLoader.java:357)
        at java.lang.Class.forName0(Native Method)
        at java.lang.Class.forName(Class.java:348)
        at org.apache.hadoop.util.RunJar.run(RunJar.java:214)
        at org.apache.hadoop.util.RunJar.main(RunJar.java:136)
```

图 3-1-16 错误信息

```
[hadoop@Cluster-01 ~]$ hadoop jar WordCount.jar org.apache.hadoop.examples.WordCount data output
18/08/15 12:14:21 INFO input.FileInputFormat: Total input paths to process : 1
18/08/15 12:14:21 INFO mapreduce.JobSubmitter: number of splits:1
18/08/15 12:14:22 INFO mapreduce.JobSubmitter: Submitting tokens for job: job_1534343693738_0006
18/08/15 12:14:22 INFO impl.YarnClientImpl: Submitted application application_1534343693738_0006
18/08/15 12:14:22 INFO mapreduce.Job: The url to track the job: http://Cluster-01:8088/proxy/applica
tion_1534343693738_0006/
18/08/15 12:14:22 INFO mapreduce.Job: Running job: job_1534343693738_0006
18/08/15 12:14:34 INFO mapreduce.Job: Job job_1534343693738_0006 running in uber mode : false
18/08/15 12:14:34 INFO mapreduce.Job:  map 0% reduce 0%
18/08/15 12:14:45 INFO mapreduce.Job:  map 100% reduce 0%
18/08/15 12:14:54 INFO mapreduce.Job:  map 100% reduce 100%
18/08/15 12:14:55 INFO mapreduce.Job: Job job_1534343693738_0006 completed successfully
18/08/15 12:14:55 INFO mapreduce.Job: Counters: 49
        File System Counters
                FILE: Number of bytes read=150719
                FILE: Number of bytes written=546907
                FILE: Number of read operations=0
                FILE: Number of large read operations=0
                FILE: Number of write operations=0
                HDFS: Number of bytes read=948610
                HDFS: Number of bytes written=109645
                HDFS: Number of read operations=6
                HDFS: Number of large read operations=0
                HDFS: Number of write operations=2
        Job Counters
                Launched map tasks=1
                Launched reduce tasks=1
                Data-local map tasks=1
                Total time spent by all maps in occupied slots (ms)=7896
                Total time spent by all reduces in occupied slots (ms)=6695
                Total time spent by all map tasks (ms)=7896
                Total time spent by all reduce tasks (ms)=6695
                Total vcore-milliseconds taken by all map tasks=7896
                Total vcore-milliseconds taken by all reduce tasks=6695
                Total megabyte-milliseconds taken by all map tasks=8085504
                Total megabyte-milliseconds taken by all reduce tasks=6855680
        Map-Reduce Framework
                Map input records=5546
                Map output records=201876
                Map output bytes=1745707
                Map output materialized bytes=150719
                Input split bytes=116
                Combine input records=201876
                Combine output records=10767
                Reduce input groups=10767
                Reduce shuffle bytes=150719
                Reduce input records=10767
                Reduce output records=10767
                Spilled Records=21534
                Shuffled Maps =1
                Failed Shuffles=0
                Merged Map outputs=1
                GC time elapsed (ms)=296
                CPU time spent (ms)=3860
                Physical memory (bytes) snapshot=306364416
                Virtual memory (bytes) snapshot=4199575552
                Total committed heap usage (bytes)=141852672
        Shuffle Errors
                BAD_ID=0
                CONNECTION=0
                IO_ERROR=0
                WRONG_LENGTH=0
                WRONG_MAP=0
                WRONG_REDUCE=0
        File Input Format Counters
                Bytes Read=948494
        File Output Format Counters
                Bytes Written=109645
```

图 3-1-17　在命令中指定类的完整路径

3.1.4 查看示例程序运行结果

（1）查看 HDFS 文件系统中操作系统用户对应目录下的文件和目录信息，可以看到新增的"output"目录，如图 3-1-18 所示。继续查看该目录中的文件，有"_SUCCESS"和"part-r-00000"两个文件，如图 3-1-19 所示。其中"_SUCCESS"文件为运行结果状态文件，里面没有任何内容，只是通过文件名来表示运行成功或者失败；而"part-r-00000"文件就是真正的运行结果文件，里面存放着词频统计分析程序对原数据文件中的英文单词的词频的统计分析结果。

```
[hadoop@Cluster-01 ~]$ hadoop fs -ls
Found 4 items
drwxr-xr-x   - hadoop supergroup          0 2018-08-14 12:13 data
drwxr-xr-x   - hadoop supergroup          0 2018-07-23 12:48 hbase
drwxrwxrwx   - hadoop supergroup          0 2018-07-12 09:57 hive
drwxr-xr-x   - hadoop supergroup          0 2018-08-15 12:14 output
```

图 3-1-18　查看新增"output"目录

```
[hadoop@Cluster-01 ~]$ hadoop fs -ls output
Found 2 items
-rw-r--r--   3 hadoop supergroup          0 2018-08-15 12:14 output/_SUCCESS
-rw-r--r--   3 hadoop supergroup     109645 2018-08-15 12:14 output/part-r-00000
```

图 3-1-19　查看"_SUCCESS"和"part-r-00000"文件

（2）由于结果文件中的内容较多，不方便使用 Hadoop 的客户端命令直接进行查看，所以将其下载到本地文件系统之中使用 Linux 操作系统的"more"命令或是 vi/vim 文本编辑器来进行查看。从 HDFS 文件系统下载文件到本地文件系统的具体命令如下：

　　hadoop　fs　-cp　output/part-r-00000　file：///home/hadoop/
　　hadoop　fs　-get　output/part-r-00000　/home/hadoop/
　　hadoop　fs　-copyToLocal　output/part-r-00000　/home/hadoop/

三个命令都可以实现从 HDFS 文件系统中将文件拷贝至本地文件系统之中的操作，任选其一使用即可。"cp"在前面已经进行过说明，可以在任何文件系统之间进行双向的文件拷贝操作，只是需要注意访问不同文件系统时的不同访问路径问题。而"get"命令和"copyToLocal"命令与之前的"put"命令和"copyFromLocal"命令相似，只是操作方向相反而已。这两个命令都只能完成将 HDFS 文件系统中的文件拷贝至本地文件系统之中的单向文件拷贝操作，对于命令中的最后一个参数会自动在本地文件系统之中进行查找，所以只需要使用一般的本地文件系统访问路径即可。而两个命令之间的区别也是前者限制源路径只能有一个，而后者可以拥有多个源路径，使用起来更加灵活。

（3）拷贝到本地文件系统中的词频统计分析程序运行的结果文件如图 3-1-20 所示。查看该结果文件，内容如图 3-1-21 所示（由于内容过长，这里只展示开头和结尾的内容）。

```
[hadoop@Cluster-01 ~]$ ls -l
total 1040
drwxrwxr-x. 7 hadoop hadoop       76 Jul 11 05:32 hadoop
drwxrwxr-x. 5 hadoop hadoop       48 Jul 11 06:33 hbase
drwxrwxr-x. 3 hadoop hadoop       26 Jul 10 12:04 java
-rw-r--r--. 1 hadoop hadoop   109645 Aug 15 12:41 part-r-00000
-rwxr-xr-x. 1 hadoop hadoop   948494 Aug 14 11:01 The_Godfather.txt
-rwxr-xr-x. 1 hadoop hadoop     3974 Aug 15 11:41 WordCount.jar
drwxrwxr-x. 5 hadoop hadoop       53 Jul 11 05:11 zookeeper
```

图 3-1-20　结果文件

```
a           3849
abandon 2
abbandanda      1
abbandando      32
abbandandos     1
abbandundo      1
abiding 1
abilities       1
ability 1
able    38
aborting        1
abortion        2
abortionist     3
abortions       3
about   440
above   14
abreast 1
abruptly        2
absence 2
absently        1
absentminded    3
absentmindedly  2
absolute        11
absolutely      22
absorb  1
abuse   3
abused  1
abusive 1
academy 12
accent  9
accented        2
accentiess      1
accept  15
acceptable      2
acceptance      2
accepted        13
"part-r-00000" 10076L, 103574C
```

```
yanked  1
yankee  3
yankees 2
yanking 1
yard    6
yards   2
yawned  1
yeah    38
year    79
years   151
yell    5
yelled  7
```

```
yellow  19
yes     33
yesterday       1
yet     73
yolk    1
yon     2
york    117
you     2129
young   158
younger 19
youngest        14
your    468
yours   12
yourself        27
yourselves      2
youth   16
youthful        3
youthfully      1
youths  2
yov     1
zaluchi 7
zero    1
zoom    1
zygoma  1
```

图 3-1-21 查看结果文件内容

（4）此时如果再次使用之前的命令来运行 Jar 包“WordCount.jar”中的示例程序，会出现错误信息并中止程序的运行，如图 3-1-22 所示。出错的原因是结果文件的存放目录“output”在上一次运行该示例程序时已经在 HDFS 文件系统中创建，而该参数必须是一个在 HDFS 文件系统中不存在的目录路径。避免该错误的方法是在第二次运行该示例程序时修改结果文件的存放目录的名称或路径，也可以删除 HDFS 文件系统中上次运行该示例程序产生的结果文件存放目录“output”，然后再运行示例程序。从 HDFS 文件系统中删除文件和目录的操作如图 3-1-23 所示，具体命令如下：

hadoop fs -rm -R output

在删除目录时必须添加命令选项“-R”，否则命令的运行会报错，如图 3-1-24 所示。

```
[hadoop@Cluster-01 ~]$ hadoop jar WordCount.jar org.apache.hadoop.examples.WordCount data output
Exception in thread "main" org.apache.hadoop.mapred.FileAlreadyExistsException: Output directory hdf
s://hadoop-ha/user/hadoop/output already exists
        at org.apache.hadoop.mapreduce.lib.output.FileOutputFormat.checkOutputSpecs(FileOutputFormat
.java:146)
        at org.apache.hadoop.mapreduce.JobSubmitter.checkSpecs(JobSubmitter.java:266)
        at org.apache.hadoop.mapreduce.JobSubmitter.submitJobInternal(JobSubmitter.java:139)
        at org.apache.hadoop.mapreduce.Job$10.run(Job.java:1290)
        at org.apache.hadoop.mapreduce.Job$10.run(Job.java:1287)
        at java.security.AccessController.doPrivileged(Native Method)
        at javax.security.auth.Subject.doAs(Subject.java:422)
        at org.apache.hadoop.security.UserGroupInformation.doAs(UserGroupInformation.java:1698)
        at org.apache.hadoop.mapreduce.Job.submit(Job.java:1287)
        at org.apache.hadoop.mapreduce.Job.waitForCompletion(Job.java:1308)
        at org.apache.hadoop.examples.WordCount.main(WordCount.java:83)
        at sun.reflect.NativeMethodAccessorImpl.invoke0(Native Method)
        at sun.reflect.NativeMethodAccessorImpl.invoke(NativeMethodAccessorImpl.java:62)
        at sun.reflect.DelegatingMethodAccessorImpl.invoke(DelegatingMethodAccessorImpl.java:43)
        at java.lang.reflect.Method.invoke(Method.java:498)
        at org.apache.hadoop.util.RunJar.run(RunJar.java:221)
        at org.apache.hadoop.util.RunJar.main(RunJar.java:136)
```

图 3-1-22 错误信息

```
[hadoop@Cluster-01 ~]$ hadoop fs -rm -R output
18/08/16 05:58:19 INFO fs.TrashPolicyDefault: Namenode trash configuration: Deletion interval = 0 mi
nutes, Emptier interval = 0 minutes.
Deleted output
```

图 3-1-23　删除文件和目录命令

```
[hadoop@Cluster-01 ~]$ hadoop fs -rm output
rm: `output': Is a directory
```

图 3-1-24　错误信息

3.1.5　其他 Hadoop 命令

1. 其他 FS 命令

touchz——创建一个空文本文件，用法和 Linux 操作系统中的"touch"命令相似。

mv——将文件或目录从源路径移动到目标路径，可以对多种文件系统进行操作，但不允许在不同的文件系统之间移动文件，用法和 Linux 操作系统中的"mv"命令相似。

chgrp——设置文件或目录的所属组，用法和 Linux 操作系统中的"chgrp"命令相似。

chown——设置文件或目录的所属用户，用法和 Linux 操作系统中的"chown"命令相似。

chmod——设置文件或目录的权限，用法和 Linux 操作系统中的"chmod"命令相似。

setrep——修改指定文件的副本系数。

getmerge——将 HDFS 文件系统中指定目录下的所有文件合并成为一个本地文件系统中的单个目标文件。

2. 其他用户命令

archive——归档命令，用于将 HDFS 中的多个小文件打包成为扩展名为"*.har"的归档文件，该归档文件中包含了原文件的所有元数据信息，本身就可以视作是一个小的文件系统。

distcp——用于集群内部或是集群与集群之间进行数据拷贝的工具。

fsck——用于检查 HDFS 文件系统的完整性和正确状况的工具，但不能主动恢复和备份缺失的数据块。

job——作业管理工具，可以查看指定作业的状态，也可以手动中止指定作业。

pipes——用于运行 C、C++等其他编程语言所编写的 MapReduce 程序的工具。通过将与应用逻辑相关的其他语言编写的代码放在单独的进程中运行，然后通过 Socket 让 Java 代码与其他语言代码进行通信，从而完成任务的运行。

3. 其他管理命令

balancer——用于在 Hadoop 平台运行过程中平衡集群负载的工具。主要是平衡集群中各个 DataNode 节点中的数据块分布，避免出现部分 DataNode 节点磁盘占用率过高的问题，通常这个问题可能导致该节点 CPU 使用率、内存使用率都会比其他服务器节点高。由于负载均衡时间花费较长，该命令可以在运行过程中简单地使用"Ctrl+C"来中止。

daemonlog——用于设置 Hadoop 相关服务进程的日志级别。

dfsadmin——管理员工具，可以用于获取 Hadoop 平台中各组件运行中的状态信息，以及执行升级、进入安全模式、设定配额等一系列管理操作。

3.1.6　Hadoop 平台的 Web 管理界面

Hadoop 平台中的很多组件都提供了 Web 客户端，不过一般来说这些自带的 Web 客户端都只能对对应的组件进行监控，而不能进行操作。

3.1.6.1　HDFS 的 Web 管理界面

Hadoop 平台中的 HDFS 文件系统的 Web 客户端的端口号为"50070"，访问地址为平台中任意 NameNode 服务节点主机的主机名或 IP 地址，如图 3-1-25 所示，访问的是处于 Standby 状态的 NameNode 服务节点主机，而访问处于 Active 状态的 NameNode 服务节点仅有"Overview"部分的内容有所不同，如图 3-1-26 所示。由于浏览器并不能直接解析 Hadoop 高可用模式平台中在配置文件"core-site.xml"的属性项"fs.defaultFS"中的所配置的命名空间，所以不能直接使用该命名空间的名称来访问 Web 客户端。HDFS 文件系统的 Web 客户端访问端口号可以通过配置文件手动进行配置。

在 HDFS 文件系统的 Web 客户端的主页中展示了 HDFS 平台的一些基本信息，包括平台整体概况、存储空间情况、DataNode 服务节点情况、Journal 服务节点情况、元数据存储位置等信息。而在首页最上面的一排链接中，"Datanodes"中展示了当前平台中所有 DataNode 服务节点的信息，如图 3-1-27 所示。"Datanode Volume Failures"是 Datanode 中的卷出错信息，"Snapshot"是当前平

项目 3　Hadoop 整合平台的使用与管理

台的快照信息，"Startup Progress"中是平台的启动信息，当前这些信息中都没有任何内容。

| Hadoop | Overview | Datanodes | Datanode Volume Failures | Snapshot | Startup Progress | Utilities |

Overview 'Cluster-01:9000' (standby)

Namespace:	hadoop-ha
Namenode ID:	namenode-01
Started:	Fri Aug 17 11:09:48 EDT 2018
Version:	2.7.3, rbaa91f7c6bc9cb92be5982de4719c1c8af91ccff
Compiled:	2016-08-18T01:41Z by root from branch-2.7.3
Cluster ID:	CID-17c31af1-d498-416b-9b9d-be94860f644e
Block Pool ID:	BP-1483832558-192.168.60.130-1531303249146

Summary

Security is off.
Safemode is off.

138 files and directories, 57 blocks = 195 total filesystem object(s).
Heap Memory used 58.49 MB of 102.09 MB Heap Memory. Max Heap Memory is 966.69 MB.
Non Heap Memory used 65.55 MB of 66.63 MB Commited Non Heap Memory. Max Non Heap Memory is -1 B.

Configured Capacity:	50.96 GB
DFS Used:	15.5 MB (0.03%)
Non DFS Used:	9.64 GB
DFS Remaining:	41.3 GB (81.04%)
Block Pool Used:	15.5 MB (0.03%)
DataNodes usages% (Min/Median/Max/stdDev):	0.03% / 0.03% / 0.03% / 0.00%
Live Nodes	3 (Decommissioned: 0)
Dead Nodes	0 (Decommissioned: 0)
Decommissioning Nodes	0
Total Datanode Volume Failures	0 (0 B)
Number of Under-Replicated Blocks	0
Number of Blocks Pending Deletion	0
Block Deletion Start Time	2018/8/17 下午11:09:48

NameNode Journal Status

Current transaction ID: 38360

Journal Manager	State
QJM to [192.168.60.132:8485, 192.168.60.133:8485, 192.168.60.134:8485]	open for read

NameNode Storage

Storage Directory	Type	State
/home/hadoop/hadoop/name	IMAGE_AND_EDITS	Active

Hadoop, 2016.

图 3-1-25　访问处于 Standby 状态的 NameNode 服务

Overview 'Cluster-02:9000' (active)

Namespace:	hadoop-ha
Namenode ID:	namenode-02
Started:	Fri Aug 17 11:09:47 EDT 2018
Version:	2.7.3, rbaa91f7c6bc9cb92be5982de4719c1c8af91ccff
Compiled:	2016-08-18T01:41Z by root from branch-2.7.3
Cluster ID:	CID-17c31af1-d498-416b-9b9d-be94860f644e
Block Pool ID:	BP-1483832558-192.168.60.130-1531303249146

图 3-1-26　访问 Active 状态的 NameNode 节点的 "Overview" 部分

而在最后的 "Utilities" 中包含了 HDFS 文件系统的 Web 客户端最重要的两个功能。一个功能是 "Browse the file system"，在该页面中可以直接浏览 HDFS 文件系统中的文件和目录，如图 3-1-28 所示，并且点击其中的文件还可以将文件从 HDFS 文件系统下载到本地文件系统之中，如图 3-1-29 所示。但需要注意的是只有访问处于 Active 状态的 NameNode 服务节点才能使用 "Browse the file system" 浏览 HDFS 文件系统中的文件和目录。另一个功能是 "Logs"，在该页

项目 3　Hadoop 整合平台的使用与管理

面中可以查看当前 HDFS 平台的日志文件，如图 3-1-30 所示。

Datanode Information

In operation

Node	Last contact	Admin State	Capacity	Used	Non DFS Used	Remaining	Blocks	Block pool used	Failed Volumes	Version
Cluster-04:50010 (192.168.60.133:50010)	0	In Service	16.99 GB	5.17 MB	3.22 GB	13.77 GB	53	5.17 MB (0.03%)	0	2.7.3
Cluster-05:50010 (192.168.60.134:50010)	0	In Service	16.99 GB	5.17 MB	3.21 GB	13.77 GB	53	5.17 MB (0.03%)	0	2.7.3
Cluster-03:50010 (192.168.60.132:50010)	2	In Service	16.99 GB	5.17 MB	3.21 GB	13.77 GB	53	5.17 MB (0.03%)	0	2.7.3

Decommissioning

Node	Last contact	Under replicated blocks	Blocks with no live replicas	Under Replicated Blocks In files under construction

Hadoop, 2016.

图 3-1-27　DataNode 服务节点信息

Browse Directory

/user/hadoop

Permission	Owner	Group	Size	Last Modified	Replication	Block Size	Name
drwxr-xr-x	hadoop	supergroup	0 B	2018/8/15 上午12:13:03	0	0 B	data
drwxr-xr-x	hadoop	supergroup	0 B	2018/8/20 下午11:35:46	0	0 B	hbase
drwxrwxrwx	hadoop	supergroup	0 B	2018/7/12 下午9:57:24	0	0 B	hive
drwxr-xr-x	hadoop	supergroup	0 B	2018/8/17 下午11:57:46	0	0 B	sqoop-data

Hadoop, 2016.

图 3-1-28　"Browse the file system" 功能

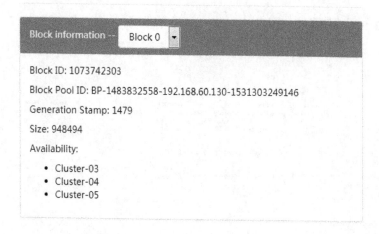

图 3-1-29　下载文件

Directory: /logs/

SecurityAuth-hadoop.audit	0 bytes	Jul 11, 2018 6:06:16 AM
hadoop-hadoop-namenode-Cluster-02.log	26150834 bytes	Aug 21, 2018 12:30:44 PM
hadoop-hadoop-namenode-Cluster-02.out	4965 bytes	Aug 21, 2018 12:11:18 PM
hadoop-hadoop-namenode-Cluster-02.out.1	716 bytes	Aug 17, 2018 10:31:18 AM
hadoop-hadoop-namenode-Cluster-02.out.2	716 bytes	Aug 17, 2018 7:31:50 AM
hadoop-hadoop-namenode-Cluster-02.out.3	716 bytes	Aug 15, 2018 10:34:17 AM
hadoop-hadoop-namenode-Cluster-02.out.4	716 bytes	Aug 14, 2018 11:18:14 AM
hadoop-hadoop-namenode-Cluster-02.out.5	716 bytes	Aug 10, 2018 11:49:25 AM
hadoop-hadoop-zkfc-Cluster-02.log	450993 bytes	Aug 17, 2018 11:10:11 AM
hadoop-hadoop-zkfc-Cluster-02.out	716 bytes	Aug 17, 2018 11:10:07 AM
hadoop-hadoop-zkfc-Cluster-02.out.1	716 bytes	Aug 17, 2018 10:31:40 AM
hadoop-hadoop-zkfc-Cluster-02.out.2	716 bytes	Aug 17, 2018 7:32:12 AM
hadoop-hadoop-zkfc-Cluster-02.out.3	716 bytes	Aug 15, 2018 10:34:40 AM
hadoop-hadoop-zkfc-Cluster-02.out.4	716 bytes	Aug 14, 2018 11:18:35 AM
hadoop-hadoop-zkfc-Cluster-02.out.5	716 bytes	Aug 10, 2018 11:49:47 AM
yarn-hadoop-resourcemanager-Cluster-02.log	932930 bytes	Aug 21, 2018 11:14:24 AM
yarn-hadoop-resourcemanager-Cluster-02.out	1531 bytes	Aug 17, 2018 11:14:18 AM
yarn-hadoop-resourcemanager-Cluster-02.out.1	1531 bytes	Aug 17, 2018 10:34:10 AM
yarn-hadoop-resourcemanager-Cluster-02.out.2	1524 bytes	Aug 17, 2018 7:35:53 AM
yarn-hadoop-resourcemanager-Cluster-02.out.3	1531 bytes	Aug 15, 2018 10:35:17 AM
yarn-hadoop-resourcemanager-Cluster-02.out.4	1531 bytes	Aug 10, 2018 11:51:22 AM
yarn-hadoop-resourcemanager-Cluster-02.out.5	1524 bytes	Aug 10, 2018 7:11:12 AM

项目 3 Hadoop 整合平台的使用与管理

图 3-1-30 查看日志文件

3.1.6.2 YARN 的 Web 管理界面

Hadoop 平台中的 YARN 资源管理器的 Web 客户端的端口号为 "8088"，访问地址为平台中 ResourceManager 服务节点主机的主机名或 IP 地址，如图 3-1-31 所示。

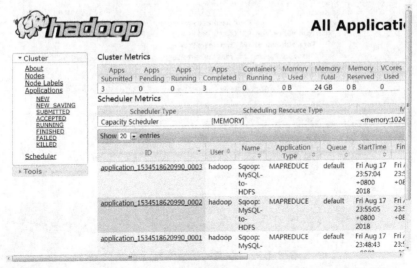

图 3-1-31 YARN 的 Web 管理界面

在 YARN 资源管理器的 Web 客户端的主页中展示了当前集群中的任务的一些概要信息，其中包括了运行中和已完成的所有任务。在概要信息中的任务列表中，每个任务都可以点击进入该任务的详细信息页面，如图 3-1-32 所示。在任务的详细信息页面有该任务的所有 Attempt 信息，在 Attempt 信息列表中可以点击进入该任务的详细信息页面，如图 3-1-33 所示，也可以查看该 Attempt 所在的 NodeManager 服务节点的详细信息，如图 3-1-34 所示，还可以查看对应的日志信息。

application_1534518620990_0003

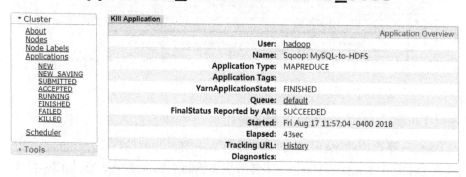

图 3-1-32　任务的详细信息页面

Application Attempt

appattempt_1534518620990_0003_000001

Application Attempt Overview

- **Application Attempt State:** FINISHED
- **AM Container:** container_1534518620990_0003_01_000001
- **Node:** Cluster-05:37210
- **Tracking URL:** History
- **Diagnostics Info:**
- **Blacklisted Nodes:** -

Total Allocated Containers: 5
Each table cell represents the number of NodeLocal/RackLocal/OffSwitch containers satisfied by NodeLocal/RackLocal/OffSwitch resource requests.

	Node Local Request	Rack Local Request	Off Switch Request
Num Node Local Containers (satisfied by)	0		
Num Rack Local Containers (satisfied by)	0	0	
Num Off Switch Containers (satisfied by)	0	0	5

图 3-1-33 Attempt 信息

图 3-1-34 NodeManager 服务节点信息

在 YARN 资源管理器的 Web 客户端的主页中的左侧有一系列链接，"Cluster"下面的链接是集群相关信息的监控页面。"About"中可以查看当前集群的一些概要信息，如图 3-1-35 所示。"Nodes"中可以查看当前集群中的所有

- 187 -

NodeManager 服务节点的列表信息，如图 3-1-36 所示，其中的每个 NodeManager 服务节点都可以点击进入该 NodeManager 服务节点的详细信息页面。"Node Label"中可以查看当前集群中的区域划分信息，由于当前没有进行区域划分，所以其中只有一个无名称的默认区域，如图 3-1-37 所示。而最后的"Applications"下面的连接可以查看当前集群中处于不同状态的任务的信息列表，这些任务状态包括"NEW""NEW SAVING""SUBMITTED""ACCEPTED""RUNNING""FINISHED""FAILED""KILLED"。

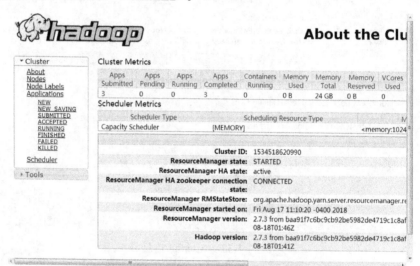

图 3-1-35　"About"页面信息

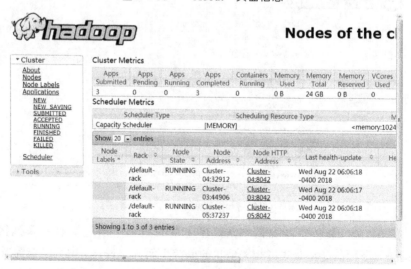

图 3-1-36　"Nodes"页面信息

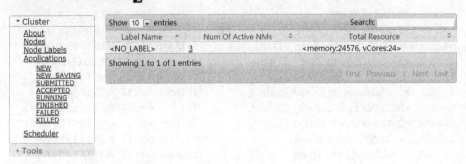

图 3-1-37 "Node Label" 页面

而在"Tools"下面的链接中,"Configuration"中可以查看当前集群的配置信息(见图 3-1-38),"Local Logs"中可以查看集群当前的日志信息(见图 3-1-39),"Server stacks"中可以查看集群的 Stacks 相关信息,"Server metrics"中可以查看集群的 Metrics 相关信息。

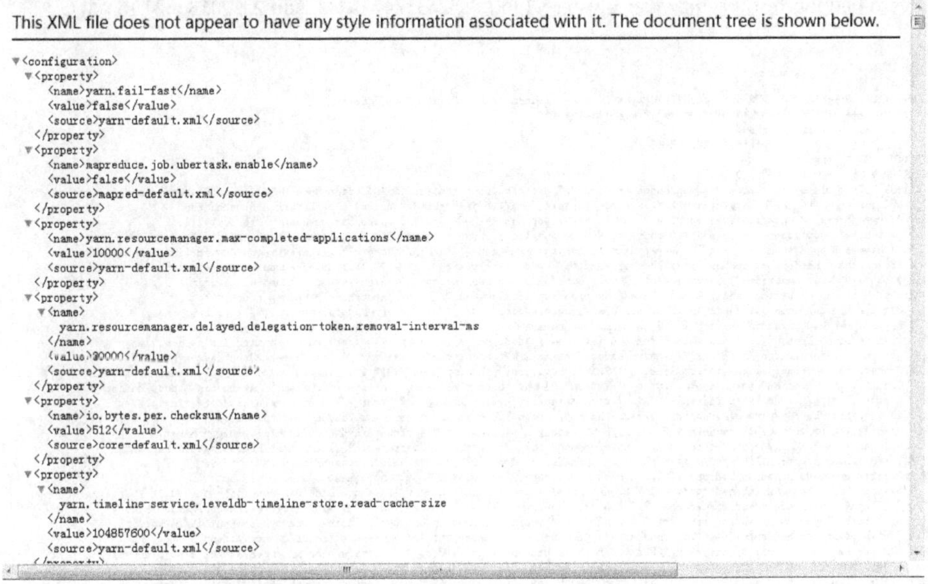

图 3-1-38 "Configuration" 页面信息

Directory: /logs/

SecurityAuth-hadoop.audit	0 bytes	Jul 11, 2018 6:14:15 AM
hadoop-hadoop-namenode-Cluster-01.log	31280123 bytes	Aug 22, 2018 6:27:35 AM
hadoop-hadoop-namenode-Cluster-01.out	4965 bytes	Aug 20, 2018 10:39:22 AM
hadoop-hadoop-namenode-Cluster-01.out.1	716 bytes	Aug 17, 2018 10:31:20 AM
hadoop-hadoop-namenode-Cluster-01.out.2	716 bytes	Aug 17, 2018 7:31:51 AM
hadoop-hadoop-namenode-Cluster-01.out.3	716 bytes	Aug 15, 2018 10:34:19 AM
hadoop-hadoop-namenode-Cluster-01.out.4	716 bytes	Aug 14, 2018 11:18:15 AM
hadoop-hadoop-namenode-Cluster-01.out.5	716 bytes	Aug 14, 2018 10:42:16 AM
hadoop-hadoop-zkfc-Cluster-01.log	473063 bytes	Aug 17, 2018 11:10:11 AM
hadoop-hadoop-zkfc-Cluster-01.out	716 bytes	Aug 17, 2018 11:10:09 AM
hadoop-hadoop-zkfc-Cluster-01.out.1	716 bytes	Aug 17, 2018 10:31:41 AM
hadoop-hadoop-zkfc-Cluster-01.out.2	716 bytes	Aug 17, 2018 7:32:14 AM
hadoop-hadoop-zkfc-Cluster-01.out.3	716 bytes	Aug 15, 2018 10:34:41 AM
hadoop-hadoop-zkfc-Cluster-01.out.4	716 bytes	Aug 14, 2018 11:18:37 AM
hadoop-hadoop-zkfc-Cluster-01.out.5	716 bytes	Aug 10, 2018 11:49:49 AM
mapred-hadoop-historyserver-Cluster-01.log	426626 bytes	Aug 22, 2018 6:26:47 AM
mapred-hadoop-historyserver-Cluster-01.out	1484 bytes	Aug 17, 2018 11:14:54 AM
mapred-hadoop-historyserver-Cluster-01.out.1	1484 bytes	Aug 17, 2018 10:36:05 AM
yarn-hadoop-resourcemanager-Cluster-01.log	2653743 bytes	Aug 22, 2018 6:24:16 AM

```
2018-07-11 06:14:15,280 INFO org.apache.hadoop.hdfs.server.namenode.NameNode: STARTUP_MSG:
/************************************************************
STARTUP_MSG: Starting NameNode
STARTUP_MSG:   host = Cluster-01/192.168.60.130
STARTUP_MSG:   args = []
STARTUP_MSG:   version = 2.7.3
STARTUP_MSG:   classpath = /home/hadoop/hadoop/hadoop-2.7.3/etc/hadoop:/home/hadoop/hadoop/hadoop-2.7.3/share/hadoop/common/lib/jaxb-impl-
2.2.3-1.jar:/home/hadoop/hadoop/hadoop-2.7.3/share/hadoop/common/lib/jaxb-api-2.2.2.jar:/home/hadoop/hadoop/hadoop-
2.7.3/share/hadoop/common/lib/stax-api-1.0-2.jar:/home/hadoop/hadoop/hadoop-2.7.3/share/hadoop/common/lib/activation-
1.1.jar:/home/hadoop/hadoop/hadoop-2.7.3/share/hadoop/common/lib/jackson-core-asl-1.9.13.jar:/home/hadoop/hadoop/hadoop-
2.7.3/share/hadoop/common/lib/jackson-mapper-asl-1.9.13.jar:/home/hadoop/hadoop/hadoop-2.7.3/share/hadoop/common/lib/jackson-jaxrs-
1.9.13.jar:/home/hadoop/hadoop/hadoop-2.7.3/share/hadoop/common/lib/jackson-xc-1.9.13.jar:/home/hadoop/hadoop/hadoop-
2.7.3/share/hadoop/common/lib/jersey-server-1.9.jar:/home/hadoop/hadoop/hadoop-2.7.3/share/hadoop/common/lib/asm-
3.2.jar:/home/hadoop/hadoop/hadoop-2.7.3/share/hadoop/common/lib/log4j-1.2.17.jar:/home/hadoop/hadoop/hadoop-
2.7.3/share/hadoop/common/lib/jets3t-0.9.0.jar:/home/hadoop/hadoop/hadoop-2.7.3/share/hadoop/common/lib/httpclient-
4.2.5.jar:/home/hadoop/hadoop/hadoop-2.7.3/share/hadoop/common/lib/httpcore-4.2.5.jar:/home/hadoop/hadoop/hadoop-
2.7.3/share/hadoop/common/lib/java-xmlbuilder-0.4.jar:/home/hadoop/hadoop/hadoop-2.7.3/share/hadoop/common/lib/commons-lang-
2.6.jar:/home/hadoop/hadoop/hadoop-2.7.3/share/hadoop/common/lib/commons-configuration-1.6.jar:/home/hadoop/hadoop/hadoop-
2.7.3/share/hadoop/common/lib/commons-digester-1.8.jar:/home/hadoop/hadoop/hadoop-2.7.3/share/hadoop/common/lib/commons-beanutils-
1.7.0.jar:/home/hadoop/hadoop/hadoop-2.7.3/share/hadoop/common/lib/commons-beanutils-core-1.8.0.jar:/home/hadoop/hadoop/hadoop-
2.7.3/share/hadoop/common/lib/slf4j-api-1.7.10.jar:/home/hadoop/hadoop/hadoop-2.7.3/share/hadoop/common/lib/slf4j-log4j12-
1.7.10.jar:/home/hadoop/hadoop/hadoop-2.7.3/share/hadoop/common/lib/avro-1.7.4.jar:/home/hadoop/hadoop/hadoop-
2.7.3/share/hadoop/common/lib/paranamer-2.3.jar:/home/hadoop/hadoop/hadoop-2.7.3/share/hadoop/common/lib/snappy-java-
1.0.4.1.jar:/home/hadoop/hadoop/hadoop-2.7.3/share/hadoop/common/lib/commons-compress-1.4.1.jar:/home/hadoop/hadoop/hadoop-
2.7.3/share/hadoop/common/lib/xz-1.0.jar:/home/hadoop/hadoop/hadoop-2.7.3/share/hadoop/common/lib/protobuf-java-
2.5.0.jar:/home/hadoop/hadoop/hadoop-2.7.3/share/hadoop/common/lib/gson-2.2.4.jar:/home/hadoop/hadoop/hadoop-
2.7.3/share/hadoop/common/lib/hadoop-auth-2.7.3.jar:/home/hadoop/hadoop/hadoop-2.7.3/share/hadoop/common/lib/apacheds-kerberos-codec-
2.0.0-M15.jar:/home/hadoop/hadoop/hadoop-2.7.3/share/hadoop/common/lib/apacheds-i18n-2.0.0-M15.jar:/home/hadoop/hadoop/hadoop-
2.7.3/share/hadoop/common/lib/api-asn1-api-1.0.0-M20.jar:/home/hadoop/hadoop/hadoop-2.7.3/share/hadoop/common/lib/api-util-1.0.0-
M20.jar:/home/hadoop/hadoop/hadoop-2.7.3/share/hadoop/common/lib/zookeeper-3.4.6.jar:/home/hadoop/hadoop/hadoop-
2.7.3/share/hadoop/common/lib/netty-3.6.2.Final.jar:/home/hadoop/hadoop/hadoop-2.7.3/share/hadoop/common/lib/curator-framework-
2.7.1.jar:/home/hadoop/hadoop/hadoop-2.7.3/share/hadoop/common/lib/curator-client-2.7.1.jar:/home/hadoop/hadoop/hadoop-
2.7.3/share/hadoop/common/lib/jsch-0.1.42.jar:/home/hadoop/hadoop/hadoop-2.7.3/share/hadoop/common/lib/curator-recipes-
2.7.1.jar:/home/hadoop/hadoop/hadoop-2.7.3/share/hadoop/common/lib/htrace-core-3.1.0-incubating.jar:/home/hadoop/hadoop/hadoop-
2.7.3/share/hadoop/common/lib/junit-4.11.jar:/home/hadoop/hadoop/hadoop-2.7.3/share/hadoop/common/lib/hamcrest-core-
1.3.jar:/home/hadoop/hadoop/hadoop-2.7.3/share/hadoop/common/lib/mockito-all-1.8.5.jar:/home/hadoop/hadoop/hadoop-
2.7.3/share/hadoop/common/lib/hadoop-annotations-2.7.3.jar:/home/hadoop/hadoop/hadoop-2.7.3/share/hadoop/common/lib/guava-
11.0.2.jar:/home/hadoop/hadoop/hadoop-2.7.3/share/hadoop/common/lib/jsr305-3.0.0.jar:/home/hadoop/hadoop/hadoop-
```

图 3-1-39 "Local Logs" 页面信息

任务 3.2　HBase 的使用和管理

这里使用一个的简单示例数据库来展示 HBase 的使用，该数据库是一个关系型数据库，只包含部门信息表（departments）和员工信息表（employees）这两个表，两个表通过部门信息中的部门编号（dept_no）进行关联。下面的表 3-2-1 和表 3-2-2 展示了这两个表的表结构，而表 3-2-3 和表 3-2-4 展示了这两个表中的数据。

表 3-2-1　部门信息表结构

部门信息表（departments）					
列名	数据类型	长度	允许空	主外键	说明
dept_no	CHAR	4	×	主	
dept_name	VARCHAR	40	×		

表 3-2-2　员工信息表结构

员工信息表（employees）					
列名	数据类型	长度	允许空	主外键	说明
emp_no	INT		×	主	
dept_no	CHAR	4	×	外	
birth_date	DATE		×		
first_name	VARCHAR	14	×		
last_name	VARCHAR	16	×		
gender	ENUM		×		枚举值：M、F
hire_date	DATE		×		

表 3-2-3　部门信息表数据

部门信息表数据	
dept_no	dept_name
D001	Marketing
D002	Finance
D003	Production
D004	Sales
D005	Customer Service

表 3-2-4 员工信息表数据

员工信息表数据						
emp_no	dept_no	birth_date	first_name	last_name	gender	hire_date
10001	D001	1953-09-02	Georgi	Facello	M	1986-06-26
10002	D001	1964-06-02	Bezalel	Simmel	F	1985-11-21
10003	D001	1959-12-03	Parto	Bamford	M	1986-08-28
10004	D001	1954-05-01	Chirstian	Koblick	M	1986-12-01
10005	D001	1955-01-21	Kyoichi	Maliniak	M	1989-09-12
10006	D002	1953-04-20	Anneke	Preusig	F	1989-06-02
10007	D002	1957-05-23	Tzvetan	Zielinski	F	1989-02-10
10008	D002	1958-02-19	Saniya	Kalloufi	M	1994-09-15
10009	D002	1952-04-19	Sumant	Peac	F	1985-02-18
10010	D002	1963-06-01	Duangkaew	Piveteau	F	1989-08-24

3.2.1 HBase 的表与关系型数据库的表

在前面的 Hadoop 平台搭建时已经介绍过 HBase 是一个基于列存储的非关系型数据库。HBase 的数据模型中虽然也有表、行等概念，但数据模型的结构和传统的关系型数据库中的基于行存储的数据模型的结构有着很大的区别。即便在各种非关系型数据库已经得到快速发展，相关技术已经基本成熟的今天，大部分 IT 从业人员依然是更加熟悉关系型数据库和基于行的数据模型。并且现今大部分的应用系统依然是使用的关系型数据库来存储数据，因为传统的关系型数据库在随机读写方面相比于新出现的非关系型数据库有着无法比拟的优势。而非关系型数据库的优势则多是在于对海量数据的存储与访问，它更多地被用于对海量数据的快速搜索和批量分析。所以在使用 HBase 之前，首先要了解如何在 HBase 中的数据模型和传统关系型数据库的数据模型之间进行转换。

3.2.1.1 单个关系表的转换

将单个关系表转换为 HBase 表的基本方式相对比较简单，可以简单地将关系表中的主键作为 HBase 表中的 Row Key，然后将其余的所有信息按照不同的属性转换为 Column Family 和 Column Key。这里以前面的示例数据库中的员工信息表（employees）为例，将其转换为 HBase 中的表之后其结构及数据如表 3-2-5 所示。

表 3-2-5 转换为 HBase 中的表

Row Key	Column Family "person_info"				Column Family "hire_info"	
emp_no	person_info: birth_date	person_info: first_name	person_info: last_name	person_info: gender	hire_info: hire_date	hire_info: dept_no
10001	1953-09-02	Georgi	Facello	M	1986-06-26	D001
10002	1964-06-02	Bezalel	Simmel	F	1985-11-21	D001
10006	1953-04-20	Anneke	Preusig	F	1989-06-02	D002
10007	1957-05-23	Tzvetan	Zielinski	M	1989-02-10	D002

在将关系表转换为 HBase 表时，用于 Row Key 的属性，Column Family 的建立，属于每个 Column Family 的 Column Key 等内容的设计和选择并不是唯一的，而是需要根据不同的需求和用途并结合 HBase 表的一些特性来进行设计和选择。比如 HBase 的表中的 Row Key 是按照字典顺序排列的，再比如 HBase 中一个表的数据文件是按照列族来划分的，并且在读取数据时会把一个 Column Family 中的所有 Column Key 的数据都取出来。所以在将关系表转换为 HBase 表时，可以将常用作排序的属性全部作为 Row Key，比如使用 dept_no+emp_no 来作为 Row Key，则可以将数据按照部门编号进行排序，而部门内的员工再按照员工编号进行排序，这样便可以省去在查询时去执行额外的排序操作。另外还可以将属于同一类型或者相似类型的属性归纳到一起放入同一个 Column Family 之中，这样可以有效减少执行查询操作时所读取的数据量，从而提高查询效率。

3.2.1.2 一对多关系表的转换

在前面给出的示例数据库中，部门信息表（departments）和员工信息表（employees）之间实际上是存在一对多关系的两个表，因为一个部门中会包含多个员工。而在将其转换为 HBase 中的表的时候，可能会有通过部门快速获取到部门中员工信息的需求。面对这种需求，将两个表分别转换为两个不同的 HBase 表是不合适的，因为在 HBase 里并没有传统关系型数据库那样的多表查询操作。所以只能将两个表的内容转换到同一个 HBase 表中，而转换的方式就是要将以前在传统关系型数据库中的"纵向延伸"的思想转变为在 HBase 数据库中的"横向延伸"的思想。具体来说就是针对同一个 Row Key 添加明细的 Column Key，对应到部门信息表（departments）和员工信息表（employees），就是把员工信息表（employees）中的每一个属性都作为 Column Family，而每一条数据的主键作为 Column Key 的结构来达到一对多的效果。经过转换之后生成的 HBase 表的结构和数据如表 3-2-6 所示。

表 3-2-6　两个关系表的转换

Row Key	Column Family "dept_info"	Column Family "first_name"		Column Family "last_name"		Column Family "gender"	
dept_no	dept_info: dept_name	first_name: 10001	first_name: 10006	last_name: 10001	last_name: 10006	gender: 10001	gender: 10006
D001	Marketing	Georgi		Facello		M	
D002	Finance		Anneke		Preusig		F

一般来说在 HBase 的表中，Column Family 下面可以有无限多个的 Column Key，但是 Column Family 最好不要超过 3 个，这是 HBase 的官方文档中给出的说明。所以在设计这种具备一对多的数据关系的表时，需要精确的根据需求来定位数据，如果所需要关联的数据项确实比较多，那么可以考虑将多个数据项合并成一项。比如上表中的属性项 "first_name" 和 "last_name" 就可以合并为 "name"，这样并不会影响根据部门获取部门中员工信息的需求。甚至 "gender" 也可以和 "first_name" "last_name" 进行合并，通过字符串拼接的方式直接组成 "emp_info"，只要不同数据之间用特定分隔符进行区分，就可以保证数据获取之后能够被处理。

3.2.2　创建表并添加数据

这里使用之前的表 3-2-5 中的由员工信息表（employees）及其数据转换得到的 HBase 数据库表及其数据为例来演示 HBase 数据库的基本操作。并将其中的部门编号（dept_no）属性在部门信息表（departments）中所对应的部门名称（dept_name）也添加到表中。

（1）进入 HBase 的控制台界面，本小节之后的所有步骤以及后面小节的所有步骤，若没有特殊说明，默认都是在 HBase 控制台界面之中进行操作。

（2）查看创建的员工信息表 "employees" 是否已经存在于 HBase 数据库中，如图 3-2-1 所示，具体命令如下：

exists 'employees'

```
hbase(main):001:0> exists 'employess'
Table employess does not exist
0 row(s) in 0.6270 seconds
```

图 3-2-1　查看员工信息表

（3）查看 HBase 数据库中当前的所有表信息，如图 3-2-2 所示，具体命令如下：

list（该命令不需要参数，直接使用）

其中的"test"数据库是在之前搭建 HBase 数据库平台时用于测试平台是否启动成功、功能是否正常所创建的测试用数据库。

```
hbase(main):002:0> list
TABLE
test
1 row(s) in 0.0810 seconds

=> ["test"]
```

图 3-2-2　查看所有表

（4）创建员工信息表"employees"，如图 3-2-3 所示，该表中包含了"person_info"和"hire_info"这两个列族（Column Family），具体命令如下：

create　'employees', 'person_info', 'hire_info'

create　'employees', {NAME=>'person_info'}, {NAME=>'hire_info'}

两条命令的含义和作用都是一样的，前一条是省略写法，后一条是完整写法。命令中第一个参数为表名，后面的全部为列族。如果需要对每个列族单独设定其他参数，则必须使用完整写法。

```
hbase(main):003:0> create 'employees','person_info','hire_info'
0 row(s) in 2.4200 seconds

=> Hbase::Table - employees
```

图 3-2-3　创建员工信息表

（5）查看创建的员工信息表"employees"的详细描述信息，如图 3-2-4 所示，具体命令如下：

describe　'employees'

```
hbase(main):004:0> describe 'employees'
Table employees is ENABLED
employees
COLUMN FAMILIES DESCRIPTION
{NAME => 'hire_info', BLOOMFILTER => 'ROW', VERSIONS => '1', IN_MEMORY => 'false', KEEP_DELETED_CELL
S => 'FALSE', DATA_BLOCK_ENCODING => 'NONE', TTL => 'FOREVER', COMPRESSION => 'NONE', MIN_VERSIONS
=> '0', BLOCKCACHE => 'true', BLOCKSIZE => '65536', REPLICATION_SCOPE => '0'}
{NAME => 'person_info', BLOOMFILTER => 'ROW', VERSIONS => '1', IN_MEMORY => 'false', KEEP_DELETED_CE
LLS => 'FALSE', DATA_BLOCK_ENCODING => 'NONE', TTL => 'FOREVER', COMPRESSION => 'NONE', MIN_VERSIONS
=> '0', BLOCKCACHE => 'true', BLOCKSIZE => '65536', REPLICATION_SCOPE => '0'}
2 row(s) in 0.1230 seconds
```

图 3-2-4　创建详细描述信息

（6）向员工信息表"employees"中添加数据，如图 3-2-5 所示。在 HBase 数据库中每次只能向表中一个行的一个列族下的一个列名（Column Key）进行数据的添加。而在该表中，列族"person_info"中包含了"birth_date""first_name""last_name""gender"这四个列，列族"hire_info"中包含了"hire_date""dept_no""dept_name"这三个列，所以需要多次执行数据添加操作，才能完成一行数据的添加。具体命令如下：

  put 'employees', '10001', 'person_info：birth_date', '1953-09-02'
  put 'employees', '10001', 'person_info：first_name', 'Georgi'
  put 'employees', '10001', 'person_info：last_name', 'Facello'
  put 'employees', '10001', 'person_info：gender', 'M'
  put 'employees', '10001', 'hire_info：hire_date', '1986-06-26'
  put 'employees', '10001', 'hire_info：dept_no', 'D001'
  put 'employees', '10001', 'hire_info：dept_name', 'Marketing'

命令中的第一个参数为表名，第二参数为行名（Row Key），即示例数据库中的员工编号（emp_no），第三个参数为列名，其格式为"Column Family：Column Key"，最后一个参数为具体的值，后面还可以个跟第五个参数，为手动指定该数据的时间戳。该命令还可以用来修改一个已存在的值，用法和格式与添加时一样。

```
hbase(main):005:0> put 'employees','10001','person_info:birth_date','1953-09-02'
0 row(s) in 0.6400 seconds

hbase(main):006:0> put 'employees','10001','person_info:first_name','Georgi'
0 row(s) in 0.0510 seconds

hbase(main):007:0> put 'employees','10001','person_info:last_name','Facello'
0 row(s) in 0.0300 seconds

hbase(main):008:0> put 'employees','10001','person_info:gender','M'
0 row(s) in 0.0250 seconds

hbase(main):009:0> put 'employees','10001','hire_info:hire_date','1986-06-26'
0 row(s) in 0.0550 seconds

hbase(main):010:0> put 'employees','10001','hire_info:dept_no','D001'
0 row(s) in 0.0300 seconds

hbase(main):011:0> put 'employees','10001','hire_info:dept_name','Marketing'
0 row(s) in 0.0230 seconds
```

图 3-2-5 向员工信息表添加数据

（7）再将几行示例数据库中的数据添加到 HBase 数据库中的员工信息表"employees"中，如图 3-2-6 所示。

```
hbase(main):012:0> put 'employees','10002','person_info:birth_date','1964-06-02'
0 row(s) in 0.0290 seconds

hbase(main):013:0> put 'employees','10002','person_info:first_name','Bezalel'
0 row(s) in 0.0210 seconds

hbase(main):014:0> put 'employees','10002','person_info:last_name','Simmel'
0 row(s) in 0.0420 seconds

hbase(main):015:0> put 'employees','10002','person_info:gender','F'
0 row(s) in 0.0460 seconds

hbase(main):016:0> put 'employees','10002','hire_info:hire_date','1985-11-21'
0 row(s) in 0.0250 seconds

hbase(main):017:0> put 'employees','10002','hire_info:dept_no','D001'
0 row(s) in 0.0210 seconds

hbase(main):018:0> put 'employees','10002','hire_info:dept_name','Marketing'
0 row(s) in 0.0170 seconds

hbase(main):019:0> put 'employees','10006','person_info:birth_date','1953-04-20'
0 row(s) in 0.0400 seconds

hbase(main):020:0> put 'employees','10006','person_info:first_name','Anneke'
0 row(s) in 0.0260 seconds

hbase(main):021:0> put 'employees','10006','person_info:last_name','Preusig'
0 row(s) in 0.0320 seconds

hbase(main):022:0> put 'employees','10006','person_info:gender','F'
0 row(s) in 0.0420 seconds

hbase(main):023:0> put 'employees','10006','hire_info:hire_date','1989-06-02'
0 row(s) in 0.0330 seconds

hbase(main):024:0> put 'employees','10006','hire_info:dept_no','D002'
0 row(s) in 0.0240 seconds

hbase(main):025:0> put 'employees','10006','hire_info:dept_name','Finance'
0 row(s) in 0.0150 seconds

hbase(main):026:0> put 'employees','10007','person_info:birth_date','1957-05-23'
0 row(s) in 0.0570 seconds

hbase(main):027:0> put 'employees','10007','person_info:first_name','Tzvetan'
0 row(s) in 0.0360 seconds

hbase(main):028:0> put 'employees','10007','person_info:last_name','Zielinski'
0 row(s) in 0.0460 seconds

hbase(main):029:0> put 'employees','10007','person_info:gender','M'
0 row(s) in 0.0340 seconds

hbase(main):030:0> put 'employees','10007','hire_info:hire_date','1989-02-10'
0 row(s) in 0.0430 seconds

hbase(main):031:0> put 'employees','10007','hire_info:dept_no','D002'
0 row(s) in 0.0260 seconds

hbase(main):032:0> put 'employees','10007','hire_info:dept_name','Finance'
0 row(s) in 0.0260 seconds
```

图 3-2-6　向员工信息表添加数据

3.2.3 查询数据

（1）扫描员工信息表"employees"中所有数据，如图 3-2-7 所示，具体命令如下：

scan 'employees'

图 3-2-7 扫描员工信息表数据

（2）扫描员工信息表"employees"中指定列的数据，可以只指定列族，也可以同时指定列族和列名，如图 3-2-8 和图 3-2-9 所示，具体命令如下：

scan 'employees', {COLUMNS=>'person_inf: first_name'}

scan 'employees', {COLUMNS=>['person_inf: first_name', 'person_inf: last_name']}

图 3-2-8 指定列族

```
hbase(main):041:0> scan 'employees',{COLUMNS=>['person_info:first_name','person_info:last_name']}
ROW                       COLUMN+CELL
 10001                    column=person_info:first_name, timestamp=1534782074952, value=Georgi
 10001                    column=person_info:last_name, timestamp=1534782098864, value=Facello
 10002                    column=person_info:first_name, timestamp=1534782355120, value=Bezalel
 10002                    column=person_info:last_name, timestamp=1534782370431, value=Simmel
 10006                    column=person_info:first_name, timestamp=1534782549516, value=Anneke
 10006                    column=person_info:last_name, timestamp=1534782570738, value=Preusig
 10007                    column=person_info:first_name, timestamp=1534782789238, value=Tzvetan
 10007                    column=person_info:last_name, timestamp=1534782810885, value=Zielinski
4 row(s) in 0.4510 seconds
```

图 3-2-9　指定列族和列名

（3）统计员工信息表"employees"中的数据记录数，如图 3-2-10 所示，具体命令如下：

count 'employees'

```
hbase(main):034:0> count 'employees'
4 row(s) in 1.3850 seconds

=> 4
```

图 3-2-10　统计数据记录数

（4）通过 Row Key 来获取数据记录中指定行的数据，如图 3-2-11 所示，具体命令如下：

get 'employees', '10006'

```
hbase(main):035:0> get 'employees','10006'
COLUMN                    CELL
 hire_info:dept_name      timestamp=1534782684932, value=Finance
 hire_info:dept_no        timestamp=1534782657424, value=D002
 hire_info:hire_date      timestamp=1534782641302, value=1989-06-02
 person_info:birth_date   timestamp=1534782520361, value=1953-04-20
 person_info:first_name   timestamp=1534782549516, value=Anneke
 person_info:gender       timestamp=1534782616191, value=F
 person_info:last_name    timestamp=1534782570738, value=Preusig
7 row(s) in 1.1940 seconds
```

图 3-2-11　通过 Row Key 获取指定行数据

（5）获取数据记录中指定行的指定列的数据，如图 3-2-12 所示，具体命令如下：

get 'employees', '10001', 'person_info: first_name', 'person_info: last_name'

get 'employees', '10001', ['person_info: first_name', 'person_info: last_name']

get 'employees', '10001', {COLLUMN=>['person_info: first_name', 'person_info: last_name']}

这三个命令的效果一样，只是写法上有一些区别。另外在只查询一个指定列的数据时，不需要书写"[]"。

```
hbase(main):042:0> get 'employees','10001','person_info:first_name','person_info:last_name'
COLUMN                    CELL
 person_info:first_name    timestamp=1534782074952, value=Georgi
 person_info:last_name     timestamp=1534782098064, value=Facello
2 row(s) in 0.1220 seconds
```

图 3-2-12　获取指定行的指定列的数据

（6）获取数据记录中指定行的指定时间范围内的数据，如图 3-2-13 所示，具体命令如下：

get 'employees', '10006', {TIMERANGE=>[1534782570000, 1534782580000]}"

这里的时间范围是使用的 HBase 的表中所记录的时间戳，这个时间戳会在添加数据的时候自动生成，也可以手动指定，HBase 表中的每个元素都有对应的时间戳，时间戳的单位为毫秒。"TIMERANGE"后的两个时间值分别代表时间范围的开始和结束，若是只指定一个时间，则表示查看指定时间的数据。

```
hbase(main):001:0> get 'employees','10006',{TIMERANGE=>[1534782570000,1534782580000]}
COLUMN                    CELL
 person_info:last_name     timestamp=1534782570738, value=Preusig
1 row(s) in 0.9280 seconds
```

图 3-2-13　获取指定行的指定范围内的数据

3.2.4　其他常用操作命令

（1）禁用指定表：disable　'表名'
（2）启动指定表：enable　'表名'
（3）查看指定表是否为启用状态：is_enabled　'表名'
（4）查看指定表是否为禁用状态：is_disabled　'表名'
（5）在指定表中添加列族：alter　'表名',{NAME=>'列族名'}
（6）删除指定表中的列族：alter　'表名',{NAME=>'列族名',METHOD=>'delete'}
（7）删除指定表：drop　'表名'
（8）删除指定表中指定行的指定列的数据：delete　'表名','行名','列族名:列名'
（9）删除指定表中指定行的数据：deleteall　'表名','行名'
（10）清空指定表中的数据：truncate　'表名'

修改表结构和删除表之前需要先将表设置为禁用状态。而修改表之后需要再将表设置为启用状态,才能执行添加数据、查询数据等操作。

3.2.5 HBase 的 Web 管理界面

HBase 数据库的 Web 客户端的端口号为"16010",访问地址为当前处于 Active 状态的 Master 服务节点主机的主机名或 IP 地址,如图 3-2-14 所示。HBase 数据库的 Web 客户端的访问端口号同样可以通过配置文件手动进行配置。

图 3-2-14 HBase 的 Web 管理界面

在 HBase 数据库的 Web 客户端首页中,"Master"后面显示的是当前的 Master 服务节点的主机。"Region Server"中显示的是 Region 服务节点的信息列表,默认显示的是基本信息,包括了服务器名称、启动时间、版本、每秒请求数、Region 数量,其他还有 Region 服务节点的存储空间信息(见图 3-2-15)、请求信息(见图 3-2-16)、存储文件信息(见图 3-2-17)、合并信息(见图 3-2-18)。而列表中的每一个 Region 服务节点都可以点击进入该 Region 服务节点对应的详细信息页面,如图 3-2-19 所示。

图 3-2-15 存储空间信息

Region Servers

Base Stats | Memory | **Requests** | Storefiles | Compactions

ServerName	Request Per Second	Read Request Count	Write Request Count
cluster-03,16020,1534779343134	0	28	28
cluster-04,16020,1534779343724	0	883	8
cluster-05,16020,1534779345394	0	4	0

图 3-2-16 请求信息

Region Servers

Base Stats | Memory | Requests | **Storefiles** | Compactions

ServerName	Num. Stores	Num. Storefiles	Storefile Size Uncompressed	Storefile Size	Index Size	Bloom Size
cluster-03,16020,1534779343134	2	2	0m	0mb	0k	0k
cluster-04,16020,1534779343724	1	1	0m	0mb	0k	0k
cluster-05,16020,1534779345394	4	1	0m	0mb	0k	0k

图 3-2-17 存储文件信息

Region Servers

Base Stats | Memory | Requests | Storefiles | **Compactions**

ServerName	Num. Compacting KVs	Num. Compacted KVs	Remaining KVs	Compaction Progress
cluster-03,16020,1534779343134	0	0	0	
cluster-04,16020,1534779343724	27	27	0	100.00%
cluster-05,16020,1534779345394	0	0	0	

图 3-2-18 合并信息

图 3-2-19 每个节点对应的详细信息页面

在"Region Server"的下面是"Backup Masters",显示的是备用 Master 服务节点的信息列表,如图 3-2-20 所示。同样,列表中的每一个备用 Master 服务节点都可以点击进入该备用 Master 服务节点对应的详细信息页面,如图 3-2-21 所示。

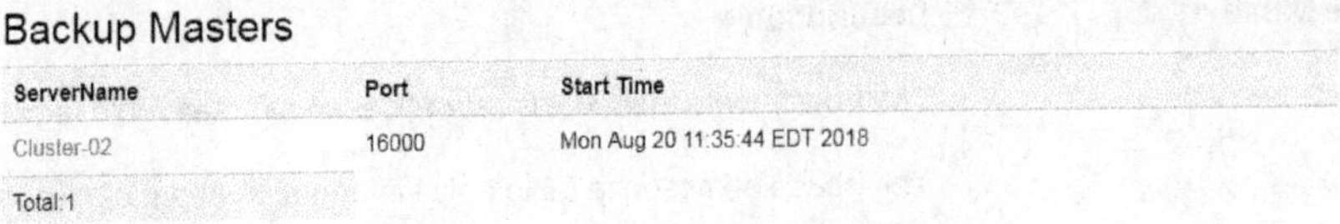

图 3-2-20　备用 Master 服务节点信息列表

图 3-2-21　备用节点对应的详细信息页面

在"Backup Masters"的下面是"Tables",显示的是当前数据库中表的信息列表,如图 3-2-22 所示。默认显示的是用户表的信息,其他还有系统表的信息(见图 3-2-23)以及快照的信息(当前没有建立快照,所以没有内容)。同样,列表中的每一个表都可以点击进入该表的详细信息页面,如图 3-2-24 所示。

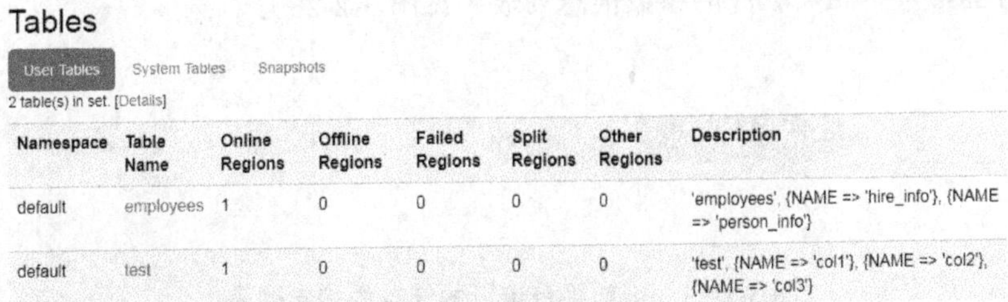

图 3-2-22　当前表的信息列表

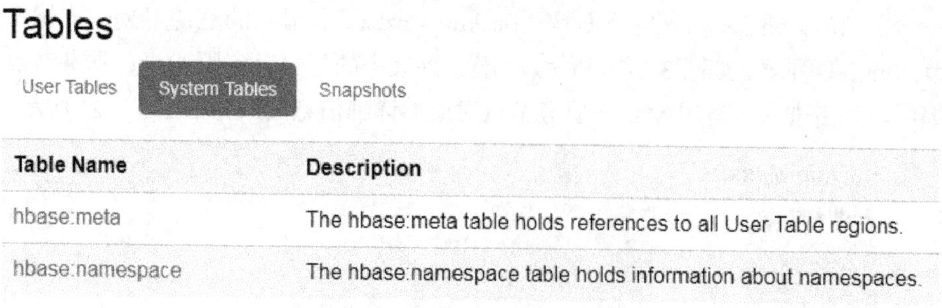

Tables

User Tables | **System Tables** | Snapshots

Table Name	Description
hbase:meta	The hbase:meta table holds references to all User Table regions.
hbase:namespace	The hbase:namespace table holds information about namespaces.

图 3-2-23　系统表的信息

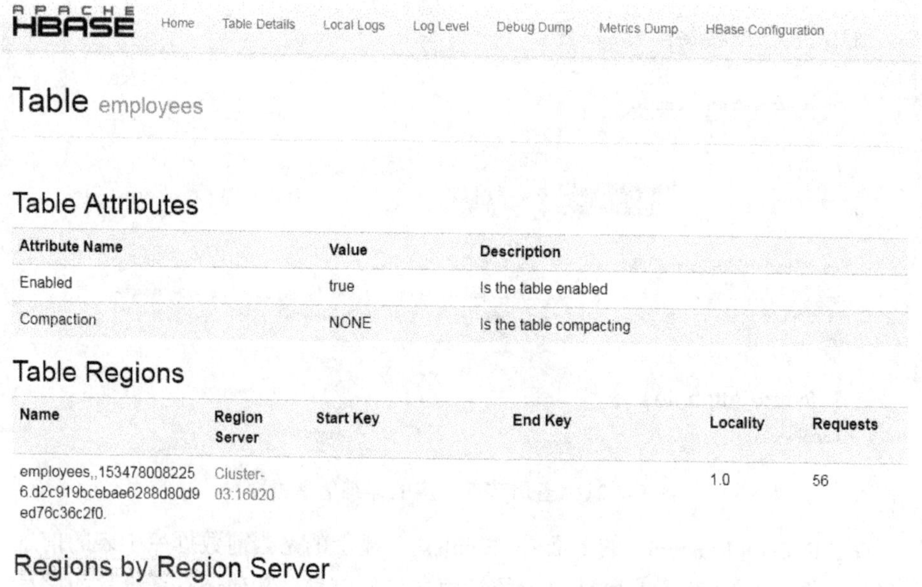

APACHE HBASE　　Home　Table Details　Local Logs　Log Level　Debug Dump　Metrics Dump　HBase Configuration

Table employees

Table Attributes

Attribute Name	Value	Description
Enabled	true	Is the table enabled
Compaction	NONE	Is the table compacting

Table Regions

Name	Region Server	Start Key	End Key	Locality	Requests
employees,,153478008225 6.d2c919bcebae6288d80d9 ed76c36c2f0.	Cluster-03:16020			1.0	56

Regions by Region Server

图 3-2-24　每个表对应的详细信息页面

在"Tables"之后是当前平台中运行的任务的信息列表"Task"（见图 3-2-25）以及 HBase 平台相关软件的参数信息列表（见图 3-2-26）。

Tasks

Show All Monitored Tasks　**Show non-RPC Tasks**　Show All RPC Handler Tasks　Show Active RPC Calls　Show Client Operations

View as JSON

No tasks currently running on this node.

图 3-2-25　当前平台运行的任务的信息列表

Software Attributes

Attribute Name	Value	Description
HBase Version	1.2.3, revision=bd63744624a26dc3350137b564fe746df7a721a4	HBase version and revision
HBase Compiled	Mon Aug 29 15:13:42 PDT 2016, stack	When HBase version was compiled and by whom
HBase Source Checksum	0ca49367ef6c3a680888bbc4f1485d18	HBase source MD5 checksum
Hadoop Version	2.5.1, revision=2e18d179e4a8065b6a9f29cf2de9451891265cce	Hadoop version and revision
Hadoop Compiled	2014-09-05T23:05Z, kasha	When Hadoop version was compiled and by whom
Hadoop Source Checksum	6424fcab95bfff8337780a181ad7c78	Hadoop source MD5 checksum
ZooKeeper Client Version	3.4.6, revision=1569965	ZooKeeper client version and revision
ZooKeeper Client Compiled	02/20/2014 09:09 GMT	When ZooKeeper client version was compiled
Zookeeper Quorum	Cluster-01:2181 Cluster-02:2181 Cluster-03:2181 Cluster-04:2181 Cluster-05:2181	Addresses of all registered ZK servers. For more, see zk dump.
Zookeeper Base Path	/hbase	Root node of this cluster in ZK.
HBase Root Directory	hdfs://hadoop-ha/user/hadoop/hbase	Location of HBase home directory
HMaster Start Time	Mon Aug 20 11:35:40 EDT 2018	Date stamp of when this HMaster was started
HMaster Active Time	Mon Aug 20 11:35:46 EDT 2018	Date stamp of when this HMaster became active
HBase Cluster ID	f6989344-fd6e-48c6-8ebd-fd5865922847	Unique identifier generated for each HBase cluster
Load average	1.33	Average number of regions per regionserver. Naive computation.
Coprocessors	[]	Coprocessors currently loaded by the master
LoadBalancer	org.apache.hadoop.hbase.master.balancer.StochasticLoadBalancer	LoadBalancer to be used in the Master

图 3-2-26　HBase 平台相关软件的参数信息列表

在 HBase 数据库的 Web 客户端首页的最上面一排链接中，"Table Details"中可以查看当前数据库中所有用户表的详细描述信息，如图 3-2-27 所示。"Local Logs"中可以查看当前 HBase 平台的日志文件，如图 3-2-28 所示。"Log Level"中可以查看和设置日志的级别，如图 3-2-29 所示。后面还有"Debug Dump"信息、"Metrics Dump"信息和"HBase Configuration"信息。

User Tables

2 table(s) in set.

Table	Description
employees	'employees', {NAME => 'hire_info', BLOOMFILTER => 'ROW', VERSIONS => '1', IN_MEMORY => 'false', KEEP_DELETED_CELLS => 'FALSE', DATA_BLOCK_ENCODING => 'NONE', TTL => 'FOREVER', COMPRESSION => 'NONE', MIN_VERSIONS => '0', BLOCKCACHE => 'true', BLOCKSIZE => '65536', REPLICATION_SCOPE => '0'}, {NAME => 'person_info', BLOOMFILTER => 'ROW', VERSIONS => '1', IN_MEMORY => 'false', KEEP_DELETED_CELLS => 'FALSE', DATA_BLOCK_ENCODING => 'NONE', TTL => 'FOREVER', COMPRESSION => 'NONE', MIN_VERSIONS => '0', BLOCKCACHE => 'true', BLOCKSIZE => '65536', REPLICATION_SCOPE => '0'}
test	'test', {NAME => 'col1', BLOOMFILTER => 'ROW', VERSIONS => '1', IN_MEMORY => 'false', KEEP_DELETED_CELLS => 'FALSE', DATA_BLOCK_ENCODING => 'NONE', TTL => 'FOREVER', COMPRESSION => 'NONE', MIN_VERSIONS => '0', BLOCKCACHE => 'true', BLOCKSIZE => '65536', REPLICATION_SCOPE => '0'}, {NAME => 'col2', BLOOMFILTER => 'ROW', VERSIONS => '1', IN_MEMORY => 'false', KEEP_DELETED_CELLS => 'FALSE', DATA_BLOCK_ENCODING => 'NONE', TTL => 'FOREVER', COMPRESSION => 'NONE', MIN_VERSIONS => '0', BLOCKCACHE => 'true', BLOCKSIZE => '65536', REPLICATION_SCOPE => '0'}, {NAME => 'col3', BLOOMFILTER => 'ROW', VERSIONS => '1', IN_MEMORY => 'false', KEEP_DELETED_CELLS => 'FALSE', DATA_BLOCK_ENCODING => 'NONE', TTL => 'FOREVER', COMPRESSION => 'NONE', MIN_VERSIONS => '0', BLOCKCACHE => 'true', BLOCKSIZE => '65536', REPLICATION_SCOPE => '0'}

图 3-2-27 "Table Details" 页面信息

Directory: /logs/

SecurityAuth.audit	15407 bytes	Aug 21, 2018 11:41:28 AM
hbase-hadoop-master-Cluster-01.log	330053 bytes	Aug 21, 2018 11:32:40 AM
hbase-hadoop-master-Cluster-01.out	700 bytes	Aug 20, 2018 11:35:40 AM
hbase-hadoop-master-Cluster-01.out.1	25238 bytes	Jul 25, 2018 11:28:44 AM
hbase-hadoop-master-Cluster-01.out.2	27741 bytes	Jul 16, 2018 3:11:15 AM
hbase-hadoop-master-Cluster-01.out.3	25238 bytes	Jul 11, 2018 9:48:18 AM

项目 3　Hadoop 整合平台的使用与管理

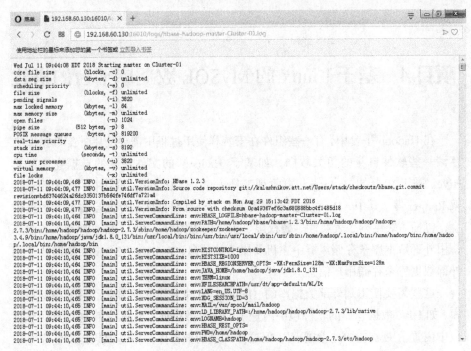

图 3-2-28　查看日志文件

Log Level

Get / Set

Log: [_____] Get Log Level

Log: [_____] Level: [_____] Set Log Level

Hadoop, 2018.

图 3-2-29　"LoyLevel"页面信息

项目 4　基于 Linux 的 MySQL 数据库平台的搭建

在 Hadoop 平台中，有一些组件在安装和使用过程中需要使用外部数据库来存储一些软件自身的相关数据，如基于 Hadoop 的文件系统的数据仓库工具 Hive、基于 Web 的 Hadoop 集群管理工具 Ambari、Hadoop 的集中式安全管理框架 Ranger 等。其中，Hive 数据仓库需要使用外部数据库来存储自身的元数据信息，包括了表结构、表属性、表权限、表分区等内容。而 Ambari 管理工具需要使用外部数据库来存储集群中主机及服务的信息，Ranger 安全框架则需要使用外部数据库来存储用户信息。

这些需要使用到外部数据库的 Hadoop 组件大部分都内置了一个小型数据库，如 Hive 数据仓库的软件包中内置了 Derby 数据库、Ambari 管理工具的软件包中内置了 PostgreSQL 数据库。但这些内置的数据库都不是原数据库的完整版本，而只是一个精简版本，其所能提供的功能都不够全面，在使用上会有很多限制，甚至在某些应用场景下的使用还会出现一些问题。另外像 Ambari 管理工具这种需要启动 Web 服务这种后台服务进程的组件，使用内置数据库甚至是本地的外部数据库（如 CentOS 7 操作系统自带的数据库 MariaDB）也会影响到服务本身的性能。所以一般选择使用独立的外部数据库，通过远程连接的方式进行使用。而 MySQL 作为使用最为广泛的开源数据库，这些需要使用到外部数据库的 Hadoop 组件默认都提供了远程连接的支持。

这里将介绍如何在 Linux 操作系统环境下搭建 MySQL 数据库平台。除了最常见的用于学习、开发、测试用的相对简易的单机模式 MySQL 数据库平台的搭建之外，还有可以提供对高可靠性、高可用性有一定需求的应对真实应用环境的集群模式 MySQL 数据库平台的搭建。

任务 4.1　安装基于 Linux 的 MySQL 单机模式

4.1.1　MySQL 数据库

MySQL 是一个关系型数据库管理系统（RDBMS，Relational Database

Management System），使用最常用的数据库管理语言 SQL（Structured Query Language，结构化查询语言）来进行数据库的管理。它最早由瑞典 MySQL AB 公司开发，目前属于 Oracle 公司的产品。MySQL 是业内非常流行的关系型数据库，大部分网站的开发，特别是中小型网站的开发都会选择 MySQL 作为数据库。MySQL 和大部分大型开源项目一样采用了双授权政策，分为免费开源的社区版和收费的商业版。

  MySQL 的产生最早可追溯至 1979 年，其雏形是由它的创始人之一 Monty Widenius 利用 BASIC 设计的一个报表工具。而其支持 SQL 的正式版本于 1996 年才发布，之后迅速得到广泛运用，被移植到了各个平台。之后 MySQL 关系型数据库管理系统的正式版本于 1998 年发布，提供了多线程的运行模式，以及对应各种编程语言的应用程序编程接口（API，Application Programming Interface）。1999 年，Monty Widenius 联合 MySQL 的其他几位创始人和主要开发人员创办了 MySQL AB 公司，专门致力于 MySQL 关系型数据管理系统，并与 Sleepycat 合作为 MySQL 添加了事务处理引擎。至此 MySQL 作为一个完整的关系型数据库管理系统已经基本成型。之后于 2008 年，MySQL 被 Sun 公司收购，而后随着 Sun 公司被 Oracle 公司收购 MySQL 成为了 Oracle 公司旗下的产品之一。

  MySQL 之所以会被业内广泛应用，是因为它具备以下的优势和特点：

  MySQL 是开放源代码的，并且没有太多版权制约。开发者和使用者不需要支付额外的费用，大幅降低了使用成本。

  MySQL 软件体积小，安装使用简单，并且易于维护，能够有效降低安装及维护成本。

  MySQL 支持大型数据库，能够高效的处理上千万条记录的数据。而且服务稳定，很少出现异常宕机。

  MySQL 使用标准的 SQL 数据语言形式，开发者、使用者、管理者都极易上手。

  MySQL 支持多种操作系统，提供多种 API 接口，支持多种开发语言，特别对现今最流行的 Web 开发语言 PHP 有着非常好的支持。

  MySQL 的自主性和可定制性强，其采用了 GPL 协议，开发者和使用者可以自由修改源码来开发满足自己功能需求的 MySQL 系统。

  MySQL 历史悠久，社区及用户都非常活跃，遇到问题可以非常方便地寻求到帮助。

  随着 MySQL 这么多年的发展，其整个系统也越来越完整和成熟，成千上万的网站依赖于 MySQL。并且对于许多人来说，它就是一个很好的解决方案。但是，适合许多人并不代表一定适合所有人，依然有很多用户的需求 MySQL 无法完全满足。如有些用户觉得现今的 MySQL 已经变得太过臃肿，提供了许多用户可能永远不会感兴趣的功能，牺牲了性能的简单性。或是有些用户认为 MySQL 并没有提供足够多的新功能，或者是添加新功能的速度太慢。他们可能认为 MySQL 没有跟上高可用性网站的目标市场的发展形势，而这些网站通常运行于具有大量内存的多核处理器之上。由于这些原因，MySQL 产生了很多分支，如 XtraDB、Percona、MariaDB、Drizzle 等。其中的 MariaDB 更是替代了 MySQL 在 Linux 操作系统中原有的地位，成为现今大部分 Linux 操作系统的默认安装的关系型数据库管理系统组件。

4.1.2　MySQL 单机模式的安装规划

1. 硬件和软件环境要求

本地磁盘剩余空间 3 GB 以上；
已安装 CentOS 7 1611 64 位操作系统；
已完成基础网络环境配置。

2. 软件版本

选用 MySQL Community Server 的 5.7.18 版本，软件包类型选择基于 Linux 版本的通用类型包 Generic，软件包名为 mysql-5.7.18-linux-glibc2.5-x86_64.tar.gz，该软件包可以在 MySQL 的官方网站（https://www.mysql.com）的开发者社区 DEVELOPER ZONE 的 Downloads 页面（https://dev.mysql.com/downloads/mysql）获取，如图 4-1-1 所示。

图 4-1-1 获取软件

3. 相关依赖软件

Linux 操作系统下 MySQL 的安装和使用需要依赖于软件 libaio，选用该软件的 0.3.109 版本，软件包名为 libaio-0.3.109-13.el7.x86_64.rpm，该软件为 Linux 操作系统的内核模块，如图 4-1-2 所示，可以在操作系统 CentOS 7 1611 的安装光盘中找到该软件包。

```
[root@Cluster-01 mnt]# ls -l Packages/libaio*
-rw-rw-r--. 2 root root 24744 Nov 25  2015 Packages/libaio-0.3.109-13.el7.x86_64.rpm
-rw-rw-r--. 2 root root 13176 Nov 25  2015 Packages/libaio-devel-0.3.109-13.el7.x86_64.rpm
```

图 4-1-2 相关依赖软件

4.1.3 卸载 MariaDB 数据库软件

MariaDB 也是和 MySQL 一样的关系型数据库管理系统。它实际上是 MySQL 数据库的发展分支之一，并且同样出 MySQL 的创始人 Michael Widenius 主导开发。其产生的主要原因是为了规避甲骨文公司收购 MySQL 之后可能产生的不开放源代码的风险。

在新版本的 Linux 操作系统中，大部分已经将其自身集成的数据库软件从之前的 MySQL 替换成了 MariaDB。而作为 MySQL 的分支之一，MariaDB 的一些内核组件和底层代码也是沿用的 MySQL 的对应内核组件或是代码实现方式。所以在自定义安装 MySQL 数据库软件时，需要将 Linux 操作系统自带的 MariaDB 数据库软件卸载，以避免与将要安装的 MySQL 数据库软件之间产生冲突的情况。

（1）使用"root"用户登录操作系统。MySQL 单机模式的整个安装和配置过程中的所有步骤，都需要使用 root 用户来进行操作。

（2）使用 RPM 的查询命令"rpm -qa | grep mariadb"或 YUM 的查询命令"yum list installed |grep mariadb"检查当前主机的操作系统中安装的 MariaDB 数据库软件的软件包名称，如图 4-1-3 所示。

```
[root@localhost ~]# rpm -qa | grep mariadb
mariadb-libs-5.5.52-1.el7.x86_64
```

图 4-1-3　检查软件包名称

（3）使用 RPM 命令"rpm -e --nodeps 软件包名"或是 YUM 命令"yum -y remove mariadb"将已经安装的 MariaDB 数据库软件包从操作系统中卸载。

4.1.4　卸载原有的 MySQL 数据库软件

对于未进行过重新安装操作系统的主机，需要检查其是否已经安装有其他老版本的 MySQL 数据库软件。在已安装有其他老版本的 MySQL 数据库软件的情况下，为了避免与将要安装的新版本的 MySQL 数据库软件之间产生冲突的情况，需要先将系统中已有的老版本的 MySQL 数据库软件进行卸载。可以使用 RPM 或 YUM 的查询命令通过关键字"mysql"来查看系统中是否存在已安装的 MySQL 数据库软件包。若存在已安装的 MySQL 数据库软件包，使用 RPM 或 YUM 的软件包删除命令可以将其从系统中卸载。

对于一些老版本的 Linux 操作系统，其自身集成了 MySQL 数据库软件，并且会在操作系统安装的过程中自动进行安装。对于这些版本的 Linux 操作系统，也需要将其自带的 MySQL 数据库软件卸载，以避免与将要安装的新版本的 MySQL 数据库软件之间产生冲突的情况。

若是操作系统中已经安装的 MySQL 数据库软件版本比当前选用的版本更新，并且可以正常使用，那么可以保留该版本的 MySQL 数据库软件，直接进行使用。

（1）使用 RPM 的查询命令"rpm -qa | grep mysql"或是 YUM 的查询命令"yum list installed | grep mysql"检查当前主机的操作系统中是否已经安装了其他版本的 MySQL 数据库软件。

（2）若已经安装有其他版本的 MySQL 数据库软件，使用 RPM 命令"rpm -e --nodeps 软件包名"或是 YUM 命令"yum -y remove mysql"将已经安装的 MySQL 数据库软件包从操作系统中卸载。

4.1.5 安装 Linux 操作系统的 libaio 模块

libaio 是 Linux 操作系统的异步磁盘 IO 模块，MySQL 数据库的磁盘读写功能会使用到该模块。该模块虽然是 Linux 操作系统的内核功能模块，但并不是默认的内核模块。在操作系统的安装过程中并不会自动安装到操作系统中，需要手动进行安装。不过作为操作系统的内核功能模块，其安装软件包一般在操作系统的安装光盘中就能找到。

（1）使用 RPM 的查询命令"rpm -qa | grep libaio"或 YUM 的查询命令"yum list installed | grep libaio"检查本机的操作系统中是否已经安装了 libaio 模块。如图 4-1-4 所示，若系统中已经安装有该模块，且版本与将要安装的 libaio 软件包版本相同或是更高，可以跳过接下来的 libaio 模块的安装步骤。

```
[root@localhost ~]# rpm -qa | grep libaio
libaio-0.3.109-13.el7.x86_64
```

图 4-1-4　检查是否安装了 libaio 模块

（2）若系统中已经安装有该模块，且版本比将要安装的 libaio 软件包版本低，则使用 RPM 命令"rpm -e --nodeps 软件包名"或是 YUM 命令"yum -y remove libaio"将已经安装的 libaio 模块软件包从操作系统中卸载。

（3）libaio 模块的软件安装包为 RPM 格式，可以使用命令"rpm -ivh 软件包路径"利用 RPM 软件包管理工具进行安装，如图 4-1-5 所示。安装完成后不需要进行任何配置。

```
[root@localhost ~]# rpm -ivh /software/libaio-0.3.109-13.el7.x86_64.rpm
warning: /software/libaio-0.3.109-13.el7.x86_64.rpm: Header V3 RSA/SHA256 Signature, key ID f4a80eb5: NOKEY
Preparing...                          ################################# [100%]
Updating / installing...
   1:libaio-0.3.109-13.el7            ################################# [100%]
```

图 4-1-5　安装软件包

4.1.6 创建专用用户和专用组

对于老版本的 MySQL 数据库软件，需要创建一个名为 mysql 的专用用户和一个名为 mysql 的专用组供 MySQL 数据库软件使用。但在新版本的 MySQL 数据库软件的官方说明文档中已经没有必须要创建 mysql 用户和 mysql 组的相关说明，所以创建 mysql 用户和 mysql 组的操作步骤可以不执行。不过如果没有创建 mysql 用户和 mysql 组，那么在后面初始化 MySQL 数据库的时候，指定的用户需要是 root 或者其他的拥有 MySQL 数据库软件所在目录权限的用户。

（1）使用命令"cat /etc/group | grep mysql"检查当前操作系统中是否已经存在名为 mysql 的组，若不存在则使用命令"groupadd mysql"创建 mysql 组。

（2）使用命令"cat /etc/passwd | grep mysql"检查当前操作系统中是否已经存在名为 mysql 的用户，若不存在则使用命令"useradd -r -g mysql mysql"创建 mysql 用户并加入 mysql 组中。其中选项"-r"表示该用户是内部用户，不允许外部登录，不使用此选项也同样可以。

（3）若当前操作系统中存在 mysql 用户以及 mysql 组，但 mysql 用户不属于 mysql 组，可以使用命令"usermod -g mysql mysql"将 mysql 用户的所属组修改为 mysql。

4.1.7　MySQL 单机模式的安装和配置

（1）在操作系统根目录下创建"mysql"目录用于存放 MySQL 的相关文件，该目录也可自行选择其他位置进行创建。创建完成后将当前的工作目录切换到该目录。

（2）使用命令"tar -xzf MySQL 安装包路径"将软件包解压解包到"mysql"目录下，如图 4-1-6 所示，解压解包出来的目录名为"mysql-5.7.18-linux-glibc2.5-x86_64"。

```
[root@localhost mysql]# ls -l
total 0
drwxr-xr-x. 9 root root 129 Jul  4 05:06 mysql-5.7.18-linux-glibc2.5-x86_64
```

图 4-1-6　解压安装包

（3）将当前的工作目录切换到操作系统的"/usr/local"目录下，并使用命令"ln -s /mysql/mysql-5.7.18-linux-glibc2.5-x86_64 mysql"在该目录下创建一个 MySQL 软件目录的链接，创建的链接如图 4-1-7 所示。然后将当前工作目录切换到该链接目录。

```
[root@localhost local]# ls -l mysql
lrwxrwxrwx. 1 root root 41 Jul  4 07:09 mysql -> /mysql/mysql-5.7.18-linux-glibc2.5-x86_64
```

图 4-1-7　创建链接

（4）创建 MySQL 数据文件的存放目录"data"，并使用命令"chmod 770

data"将该目录的权限更改为所属用户和所属组拥有所有权限，而其他用户没有任何权限，如图 4-1-8 所示。

```
[root@localhost mysql]# ls -dl data
drwxrwx---. 2 root root 6 Jul  4 07:12 data
```

图 4-1-8　创建存放目录"data"

（5）依次使用命令"chown -R mysql ."和"chgrp -R mysql ."将当前工作目录及其所有子目录和子文件的所属用户和所属组更改为 MySQL 数据库软件的专用用户和专用组，修改之后的目录和文件属性如图 4-1-9 所示。

```
[root@localhost mysql]# ls -l
total 0
drwxr-xr-x. 10 mysql mysql 141 Jul  4 07:12 mysql-5.7.18-linux-glibc2.5-x86_64
[root@localhost mysql]# ls -l
total 40
drwxr-xr-x.  2 mysql mysql  4096 Jul  4 05:06 bin
-rw-r--r--.  1 mysql mysql 17987 Mar 18  2017 COPYING
drwxrwx---.  5 mysql mysql  4096 Jul  6 04:46 data
drwxr-xr-x.  2 mysql mysql    55 Jul  4 05:06 docs
drwxr-xr-x.  3 mysql mysql  4096 Jul  4 05:06 include
drwxr-xr-x.  5 mysql mysql   229 Jul  4 05:06 lib
drwxr-xr-x.  4 mysql mysql    30 Jul  4 05:06 man
-rw-r--r--.  1 mysql mysql  2478 Mar 18  2017 README
drwxr-xr-x. 28 mysql mysql  4096 Jul  4 05:06 share
drwxr-xr-x.  2 mysql mysql    90 Jul  4 05:14 support-files
```

图 4-1-9　修改之后的目录和文件

（6）在操作系统的配置文件"/etc/profile"中配置 MySQL 相关的环境变量，如图 4-1-10 所示，在文件末尾添加以下内容：

\# mysql environment

MYSQL_HOME=/usr/local/mysql

PATH=$MYSQL_HOME/bin：$PATH

export　MYSQL_HOME　PATH

```
# mysql environment
MYSQL_HOME=/usr/local/mysql
PATH=$MYSQL_HOME/bin:$PATH
export MYSQL_HOME PATH
```

图 4-1-10　修改配置文件

（7）使用命令"source etc/profile"使新配置的环境变量立即生效。

（8）使用命令"echo $变量名"可以查看新添加和修改的环境变量的值是否正确，如图4-1-11所示。

```
[root@localhost ~]# echo $MYSQL_HOME
/usr/local/mysql
```

图 4-1-11　查看主环境变量

（9）使用命令"mysqld --initialize --user=mysql --basedir=/usr/local/mysql --datadir=/usr/local/mysql/data"对 MySQL 数据库的安装进行初始化，该命令执行过程完成之后会有与一些执行过程相关的输出信息，如图4-1-12所示。需要特别注意的是在所有输出信息中最后一行的包含"[Note]"内容的信息，该信息的内容格式如下：

[Note] A temporary password is generated for root@localhost: XXXXXXXXXXXXXX

信息末尾的"XXXXXXXXXXXXXX"是安装的初始化程序所生成的随机密码，也就是 MySQL 数据库的 root 用户的初始登录密码。在完成 MySQL 数据库软件的安装和配置之后，首次登录 MySQL 数据库时需要使用此密码进行登录。所以这段信息非常重要，最好通过复制或截图等方式将其保存下来。

```
[root@localhost ~]# mysqld --initialize --user=mysql --basedir=/usr/local/mysql --datadir=/usr/local/mysql/data
2018-07-04T15:24:30.414184Z 0 [Warning] TIMESTAMP with implicit DEFAULT value is deprecated. Please use --explicit_defaults_for_timestamp server option (see documentation for more details).
2018-07-04T15:24:31.294794Z 0 [Warning] InnoDB: New log files created, LSN=45790
2018-07-04T15:24:31.461783Z 0 [Warning] InnoDB: Creating foreign key constraint system tables.
2018-07-04T15:24:31.580317Z 0 [Warning] No existing UUID has been found, so we assume that this is the first time that this server has been started. Generating a new UUID: 595faca0-7f9e-11e8-9b11-aa4c7d8f54f7.
2018-07-04T15:24:31.581089Z 0 [Warning] Gtid table is not ready to be used. Table 'mysql.gtid_executed' cannot be opened.
2018-07-04T15:24:31.583103Z 1 [Note] A temporary password is generated for root@localhost: i4oFwqc%0KL)
```

图 4-1-12　对数据库初始化

MySQL 数据库的初始化程序所生成的 root 用户的随机初始登录密码有一定的时间限制，过期之后密码将失效。若忘记随机初始密码或者密码过期无法使用，可以清空 MySQL 数据文件的存放目录"data"中的所有内容，然后重新执行安装的初始化操作。

由于 MySQL 数据库的初始化程序所生成的 root 用户的随机初始登录密码中包含了数字、大小写字母、符号等内容，不仅不便于记忆，而且在密码输入的时候也容易出错。可以选择在 MySQL 数据库软件的安装初始化过程中不生成 root 用户的初始密码，只需要将初始化操作命令改为"mysqld --initialize-insecure --user=mysql --basedir=/usr/local/mysql --datadir=/usr/local/mysql/

data"即可。但若是采用这种方式执行安装初始化操作,数据库会变的不安全,所以一定要记得执行后面的修改 root 用户的登录密码的步骤。

(10)如图 4-1-13 示,使用命令"mysql_ssl_rsa_setup --basedir=/usr/local/mysql --datadir=/usr/local/mysql/data"进行 MySQL 数据库软件的安装。

```
[root@localhost ~]# mysql_ssl_rsa_setup --basedir=/usr/local/mysql --datadir=/usr/local/mysql/data
Generating a 2048 bit RSA private key
.........+++
..............+++
writing new private key to 'ca-key.pem'
-----
Generating a 2048 bit RSA private key
...............+++
............+++
writing new private key to 'server-key.pem'
-----
Generating a 2048 bit RSA private key
..+++
.......................+++
writing new private key to 'client-key.pem'
-----
```

图 4-1-13　进行数据库软件的安装

4.1.8　MySQL 单机模式的启动和验证

MySQL 数据库在安装过程中生成的 root 用户的随机初始密码一般较为复杂,难以记忆,通常都会将其重新设定为便于记忆的常用密码。而重新设置 root 用户的登录密码自然需要使用 root 用户身份登录到 MySQL 数据库进行修改,可以通过该操作来验证 MySQL 数据库是否能够正常使用。

(1)这里使用安全模式来启动 MySQL 数据库服务,对应的启动命令为"mysqld_safe　--user=mysql　--basedir=/usr/local/mysql　--datadir=/usr/local/mysql/data　&",如图 4-1-14 所示。

```
[root@localhost ~]# mysqld_safe --user=mysql --basedir=/usr/local/mysql --datadir=/usr/local/mysql/data &
[1] 2100
[root@localhost ~]# 2018-07-04T15:45:06.359246Z mysqld_safe Logging to '/usr/local/mysql/data/localhost.localdomain.err'.
2018-07-04T15:45:06.386901Z mysqld_safe Starting mysqld daemon with databases from /usr/local/mysql/data
```

图 4-1-14　使用安全模式来启动数据库服务

使用命令"mysqld"也可以启动 MySQL 数据库,并且后面的选项和参数也相同。但是更推荐使用命令"mysqld_safe",因为在安全模式下增加了一些保护机制和安全特性,如 MySQL 数据库服务挂掉的情况下自动进行重新启动的操作,以及在出现错误时向错误日志文件写入相关信息。

(2)使用命令"ps -ef | grep mysql"查看操作系统的进程信息。如图

4-1-15 所示，若存在进程信息中包含"mysql"关键字的进程则表示 MySQL 数据库启动成功。

图 4-1-15　查看进程信息

（3）使用命令"mysql -u root -p"登录 MySQL 数据库，这时会提示输入数据库的 root 用户的密码，如图 4-1-16 所示，该密码为 MySQL 数据库安装初始化过程中所显示的随机密码，正确输入密码之后便可以登录到 MySQL 数据库，并进入到 MySQL 的控制台界面。

图 4-1-16　登录 MySQL 数据库

（4）在 MySQL 的控制台界面使用命令"SET PASSWORD=PASSWORD ('****');"重新设置数据库的"root"用户的登录密码，其中"****"部分为自定义的新密码，如图 4-1-17 所示。

图 4-1-17　重新设置登录密码

（5）在 MySQL 的控制台可以使用命令"exit"退出 MySQL 控制台返回到操作系统的命令行界面。

4.1.9　配置 MySQL 的系统服务

MySQL 数据库的服务可以通过相应的命令手动进行启动，不过是这样的话，每次计算机或操作系统因某些原因需要重新启动的时候，都需要管理员手动操作来启动 MySQL 数据库服务。若是因为异常断电等一些特殊原因导致的计算机或操作系统的重新启动，管理员可能无法第一时间知晓或进行管理操作，这就会导致 MySQL 数据库服务的长时间中断。所以大部分时候都希望 MySQL 数据库这类的服务能够跟随操作系统的启动而自动启用，这就需要将 MySQL 数据库服务配置为系统服务。

（1）MySQL 数据库软件的系统服务脚本文件名为"mysql.server"，该文件位于 MySQL 软件目录下的"support-files"目录之中，如图 4-1-18 所示。将该服务脚本文件拷贝到操作系统的可控制服务目录"/etc/init.d"之中，并将拷贝的副本文件重新命名为"mysql"，如图 4-1-19 所示，该名称即对应系统服务的名称。

```
[root@localhost mysql]# ls -l support-files/
total 24
-rw-r--r--. 1 mysql mysql   773 Mar 18  2017 magic
-rwxr-xr-x. 1 mysql mysql  1061 Mar 18  2017 mysqld_multi.server
-rwxr-xr-x. 1 mysql mysql   894 Mar 18  2017 mysql-log-rotate
-rwxr-xr-x. 1 mysql mysql 10576 Mar 18  2017 mysql.server
```

图 4-1-18　"support-files"目录

```
[root@localhost etc]# ls -l init.d/
total 44
-rw-r--r--. 1 root root 15131 Sep 12  2016 functions
-rwxr-xr-x. 1 root root 10576 Jul  5 05:01 mysql
-rwxr-xr-x. 1 root root  2989 Sep 12  2016 netconsole
-rwxr-xr-x. 1 root root  6643 Sep 12  2016 network
-rw-r--r--. 1 root root  1160 Nov  6  2016 README
```

图 4-1-19　拷贝副本文件

（2）成功将 MySQL 数据库服务添加为系统的可控制服务之后，便可以使用命令"service mysql start"和"service mysql stop"来启动和关闭 MySQL 数据库服务，如图 4-1-20 和图 4-1-21 所示。通过使用命令"ps -ef | grep mysql"查看系统进程信息，可以确认 MySQL 数据库服务是否已经启动或者关闭。

```
[root@localhost ~]# service mysql start
Unit mysql.service could not be found.
Starting MySQL. SUCCESS!
```

图 4-1-20　启动数据库服务

```
[root@localhost ~]# service mysql stop
Shutting down MySQL.. SUCCESS!
```

图 4-1-21　关闭数据库服务

此处并没有将 MySQL 数据库服务注册到系统服务管理器命令 systemctl 下，所以不能使用 systemctl 命令来启动和关闭 MySQL 数据库服务。

（3）使用命令"chkconfig --add mysql"和"chkconfig --level 2345 mysql on"可以将 MySQL 服务设置为在操作系统启动时自动启动。

此处可以使用 systemctl 命令来设置 MySQL 数据库服务是否在操作系统启动时自动启动。虽然同上面的 service 命令一样，没有将 MySQL 数据库服务注册到系统服务管理器命令 systemctl 下，但系统会自动将 systemctl 命令替换为 chkconfig 命令来执行，如图 4-1-22 所示。

```
[root@localhost ~]# systemctl enable mysql
mysql.service is not a native service, redirecting to /sbin/chkconfig.
Executing /sbin/chkconfig mysql on
```

图 4-1-22　设置服务自动启动

（4）使用命令"reboot"重新启动操作系统，然后使用命令"ps -ef | grep mysql"查看系统进程信息，如图 4-1-23 所示，确认 MySQL 数据库服务是否能够自动启动。

```
[root@localhost ~]# ps -ef | grep mysql
root       945     1  0 05:27 ?        00:00:00 /bin/sh /usr/local/mysql/bin/mysqld_safe --datadir
=/usr/local/mysql/data --pid-file=/usr/local/mysql/data/localhost.localdomain.pid
mysql     1168   945  0 05:27 ?        00:00:01 /usr/local/mysql/bin/mysqld --basedir=/usr/local/m
ysql --datadir=/usr/local/mysql/data --plugin-dir=/usr/local/mysql/lib/plugin --user=mysql --log-err
or=/usr/local/mysql/data/localhost.localdomain.err --pid-file=/usr/local/mysql/data/localhost.locald
omain.pid
root      2235  2209  0 05:30 tty1     00:00:00 grep mysql
```

图 4-1-23　查看系统进程信息

4.1.10　配置 MySQL 的远程访问

CentOS 7 操作系统默认的防火墙服务不再是以前的 Iptables，而是替换成了新的防火墙 Firewall。Firewall 防火墙可以看作是 Iptables 防火墙的升级版本，两者之间的关系如图 4-1-24 所示。可以看出，Firewall 防火墙是在 Iptables 防火墙基础进行的升级和扩展，其底层核心部分依然使用的是 Iptables 防火墙。而 Firewall 防火墙相对于 Iptables 防火墙不仅是提供了更为丰富的功能，同时也提供更方便的操作和管理方式。

Linux 操作系统自带的防火墙在初始状态下的默认策略都是禁止所有外部

访问，所有 MySQL 数据库要提供远程访问的功能，需要先配置系统防火墙的相应端口策略，以开放远程访问 MySQL 数据库服务所需要使用到的网络端口。同时还需要在数据库中配置指定用户的权限信息，使该用户能够接收远程访问的请求信息。

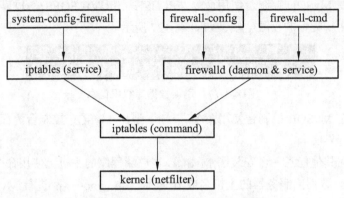

图 4-1-24 Firewall 与 Iptables 关系

（1）使用命令"firewall-cmd --zone=public --add-port=3306/tcp --permanent"添加系统防火墙的端口策略，对外开启 MySQL 数据库所使用的端口"3306"，若执行完成后显示信息"success"，则表示端口策略添加成功。

（2）使用命令"firewall-cmd --reload"重新启动操作系统的防火墙服务，使新添加的端口策略生效，若执行完成后显示信息"success"则表示防火墙重启成功。

（3）使用命令"mysql -u root -p"登录 MySQL 数据库，正确输入数据库的 root 用户的登录密码之后进入 MySQL 数据库的控制台界面。

（4）在 MySQL 控制台使用命令"USE mysql;"切换到"mysql"数据库，如图 4-1-25 所示。

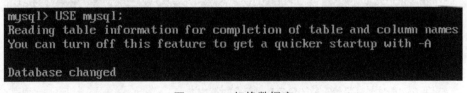

图 4-1-25 切换数据库

（5）在 MySQL 控制台使用命令"UPDATE user SET host='%' WHERE user='root';"修改数据库的 root 用户所接收的请求来源的范围，如图 4-1-26 所示。

```
mysql> UPDATE user SET host='%' WHERE user='root';
Query OK, 1 row affected (0.01 sec)
Rows matched: 1  Changed: 1  Warnings: 0
```

图 4-1-26　修改 root 用户接收请求来源的范围

（6）在 MySQL 控制台使用命令"FLUSH PRIVILEGES;"刷新数据库的权限信息使新配置的权限生效，如图 4-1-27 所示。

```
mysql> FLUSH PRIVILEGES;
Query OK, 0 rows affected (0.00 sec)
```

图 4-1-27　刷新数据库权限信息

（7）在 MySQL 控制台使用命令"exit"退出 MySQL 控制台界面，返回到系统命令行界面。

（8）在其他任意一台安装了 MySQL 客户端程序的主机上使用命令"mysql -h MySQL 数据库服务器的主机名或 IP 地址　-u　root　-p"进行 MySQL 数据库的远程登录，如图 4-1-28 所示。此时会提示输入数据库的 root 用户的登录密码，输入正确的密码之后便能远程连接上 MySQL 数据库。也可以在安装了 Windows 操作系统的主机上使用一些可视化的 MySQL 数据库客户端软件来测试远程连接。

```
[root@BackupHost network-scripts]# mysql -h 192.168.60.140 -u root -p
Enter password:
Welcome to the MySQL monitor.  Commands end with ; or \g.
Your MySQL connection id is 4
Server version: 5.7.18 MySQL Community Server (GPL)

Copyright (c) 2000, 2017, Oracle and/or its affiliates. All rights reserved.

Oracle is a registered trademark of Oracle Corporation and/or its
affiliates. Other names may be trademarks of their respective
owners.

Type 'help;' or '\h' for help. Type '\c' to clear the current input statement.

mysql>
```

图 4-1-28　进行数据库进程登录

任务 4.2　安装 MySQL 集群模式

4.2.1　MySQL 数据库的集群模式

MySQL Cluster 是 MySQL 适用于分布式计算环境的版本，使用了 NDB

Cluster 存储引擎，允许在 1 个 Cluster 中运行多个 MySQL 服务器。并且通过无共享架构（Share Nothing Architecture），整个系统可以通过使用廉价的硬件构成，对软硬件没有太多特殊要求。在 MyQL 5.0 及以上的二进制版本以及与最新的 Linux 版本兼容的 RPM 中也提供了该存储引擎。目前能够运行 MySQL Cluster 的操作系统有 Linux、Mac OS X 和 Solaris。

4.2.1.1 MySQL Cluster 架构

MySQL Cluster 采用了无共享架构（Share Nothing Architecture）的分布式节点架构的存储方案，其目的是提供高容错性和高性能。该技术允许在无共享的系统中部署运行于内存中的数据库的 Cluster。而且由于 Cluster 中每个节点都有自己的内存和磁盘，能够有效解决单点故障等问题。

无共享架构的对等节点使得某台服务器上的更新操作在其他服务器上立即可见。同时通过多个 MySQL 服务器来分担负载，最大限度地实现高性能。而且还可以通过在不同位置存储数据来保证高可用性和数据的冗余。

MySQL Cluster 包括了 MySQL 服务器、NDB Cluster 的数据节点、管理服务器、专门的数据访问程序，其各个组件之间的关系如图 4-2-1 所示。

1. NDB

NDB 是一种位于内存中的存储引擎，它具有可用性高和数据一致性好的特点。MySQL Cluster 的 NDB 存储引擎包含完整的数据集，并且能够使用多种故障切换和负载平衡选项来配置 NDB 存储引擎。目前，MySQL Cluster 的 Cluster 部分可独立于 MySQL 服务器来进行配置。在 MySQL Cluster 中，Cluster 的每个部分被视为 1 个节点。

2. 管理（MGM）节点

管理节点的作用是管理 MySQL Cluster 内的其他所有节点，如提供配置数据、启动并停止节点、运行备份等。它负责管理 Cluster 配置文件和 Cluster 日志，而 Cluster 中的每个节点都需要从管理节点检索配置数据。同时当数据节点内出现新的事件时，节点会将关于这类事件的信息传输到管理节点，然后由管理节点将这类信息写入 Cluster 日志。由于该节点负责管理其他节点的配置，所以应在启动其他节点之前首先启动该节点。

3. 数据节点

数据节点用于保存 Cluster 的数据。数据节点的数目与副本的数目相关，并且是片段的整数倍。例如，对于 2 个副本，每个副本有 2 个片段，那么就有 4 个数据节点。

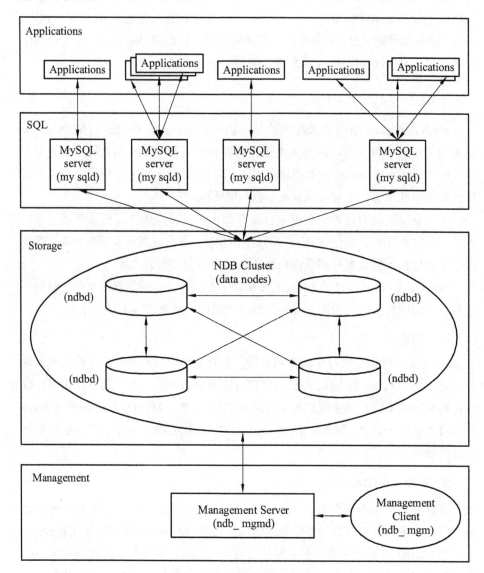

图 4-2-1　各组件的关系图

4. SQL 节点

SQL 节点是用来访问 Cluster 数据的节点。对于 MySQL Cluster 来说，客户端节点就是使用 NDB Cluster 存储引擎的传统 MySQL 服务器。

5. 客户端

MySQL Cluster 的客户端分为两种类型。一种是标准的 MySQL 客户端，它们与标准的非 Cluster 类型的 MySQL 客户端程序没有区别。另一种是管理客户端，这类客户端与管理服务器直接相连，并提供启动或停止节点、启动和停止消息跟踪、显示节点版本和状态、启动和停止备份等管理操作命令。

4.2.1.2 MySQL Cluster 的数据存储

MySQL Cluster 的数据节点组内的主从同步采用的是同步复制方式，数据更新使用读已提交隔离级别（Read-Committed Isolation）以及两阶段提交机制（Two-Phased Commit）来保证所有节点数据的一致性，其工作过程如下：

（1）Master 执行提交语句时，事务被发送到 Slave，Slave 开始准备事务的提交。

（2）每个 Slave 都要准备事务，然后向 Master 发送 OK（或 ABORT）消息，表明事务已经准备好（或者无法准备该事务）。

（3）Master 等待所有 Slave 发送 OK 或 ABORT 消息。

（4）如果 Master 收到所有 Slave 的 OK 消息，它就会向所有 Slave 发送提交消息，告诉 Slave 提交该事务。而如果 Master 收到来自任何一个 Slave 的 ABORT 消息，它就会向所有 Slave 发送 ABORT 消息，告诉 Slave 中止该事务。

（5）每个 Slave 等待来自 Master 的 OK 或 ABORT 消息。

（6）如果 Slave 收到提交请求，它们就会提交事务，并向 Master 发送事务已提交的确认消息。如果 Slave 收到取消请求，它们就会撤销所有改变并释放所占有的资源，从而中止事务，然后向 Master 送事务已中止的确认消息。

（7）当 Master 收到来自所有 Slave 的确认之后，就会报告该事务被提交（或中止），然后继续进行下一个事务的处理。

由于同步复制一共需要进行 4 次消息传送，所以 MySQL Cluster 的数据更新速度比单机模式的 MySQL 要慢。所以 MySQL Cluster 一般要求运行在千兆以上的局域网内，节点可以采用双网卡，节点组之间采用直连的方式。

另外，MySQL Cluster 将所有的索引列都保存在主存之中，而其他非索引列则可以存储在内存中或是通过建立表空间存储到磁盘上。如果数据发生改变（INSERT，UPDATE，DELETE 等操作），MySQL Cluster 会将发生改变的记录写入重做日志，然后通过检查点定期将数据持久化到磁盘上。由于重做日志是异步提交的，所以故障期间可能造成少量事务丢失。为了减少事务丢失发生的情况，MySQL Cluster 实现了延迟写入（默认延迟时间为 2 s，可手动进行配置），这样就可以在故障发生时完成检查点写入，而不会丢失最后一个检查点。一般单个数据节点故障不会导致任何数据丢失，因为集群内部会采用同步数据复制。

4.2.1.3　MySQL Cluster 的优缺点

1. 优点

99.999%的高可用性；
快速的自动失效切换；
灵活的分布式体系结构，没有单点故障问题；
高吞吐量和低延迟；
可扩展性强，支持在线的动态扩容。

2. 缺点

存在很多方面的限制，比如不支持外键等；
部署、管理、配置相对比较复杂；
占用的磁盘空间和内存空间较大；
备份和恢复的执行不方便；
重新启动 Cluster 的时候，数据节点将数据加载到内存中所需要花费的时间很长。

4.2.2　MySQL 集群模式的安装规划

1. 硬件和软件环境要求

集群中所有主机的本地磁盘剩余空间 4 GB 以上；
集群中所有主机已安装 CentOS 7 1611 64 位操作系统；
集群中所有主机已完成基础网络环境配置，包括 IP 地址、免密码登录、主机名、主机名与 IP 地址的映射关系的配置。

2. 软件版本

选用 MySQL Cluster 的 7.5.7 版本，软件包类型选择基于 Linux 的通用类型包 Generic，软件包名为 mysql-cluster-gpl-7.5.7-linux-glibc2.12-x86_64.tar.gz，该软件包可以在 MySQL 的官方网站（https://www.mysql.com）的开发者社区 DEVELOPER ZONE 的 Downloads 页面（https://dev.mysql.com/downloads/cluster）

获取，如图 4-2-2 所示。

图 4-2-2 获取软件资源

3. 依赖软件

Linux 操作系统下 MySQL 的安装和使用需要依赖于软件 libaio，选用该软件的 0.3.109 版本，软件包名为 libaio-0.3.109-13.el7.x86_64.rpm，该软件为 Linux 操作系统的内核模块，可以在操作系统 CentOS 7 1611 的安装光盘中找到该软件包。

4. 服务规划

MySQL 数据库的集群模式包括管理节点、数据节点、SQL 节点共三种类型的服务节点。

集群模式中的管理节点一般单独使用集群中的一台主机，而集群模式需要满足数据有备份和数据处理能够分布并行的基本要求，所以数据节点和 SQL 节点都需要两台或以上的主机，具体规划如表 4-2-1 所示。

表 4-2-1 集群模式节点规划

主机名	IP 地址	服务描述
MySQL-01	192.168.60.150	管理节点
MySQL-02	192.168.60.151	数据节点
MySQL-03	192.168.60.152	数据节点
MySQL-04	192.168.60.153	SQL 节点
MySQL-05	192.168.60.154	SQL 节点

4.2.3　MySQL 集群模式基础环境的安装和配置

MySQL 集群模式中 MySQL 数据库软件本身不需要使用免密码登录的功能，但也可以配置 MySQL 集群的主机之间的免密码登录，这可以为之后的同步 MySQL 集群模式配置以及管理操作提供便利。而 MySQL 集群模式中的免密码登录只需要配置管理节点能够免密码登录到其他所有数据节点和 SQL 节点即可。

MySQL 集群模式的基础环境的安装和配置的部分内容和 MySQL 单机模式的基础环境的安装和配置相同，依次卸载原有的 MariaDB 数据库软件和 MySQL 数据库软件，安装 Linux 操作系统的 libaio 模块，创建 MySQL 专用用户和专用组，不过需要注意的是这些步骤需要在集群中所有主机上进行操作。

4.2.4　关闭 Linux 操作系统的 SELinux 服务

SELinux（Security-Enhanced Linux）是 Linux 操作系统中的安全子系统，它所提供的访问控制体系可以限制进程执行过程中所能够访问文件和目录，从而最大限度地减小系统中服务进程可访问的资源。SELinux 的安全控制可能使得 MySQL 数据库服务因为无法访问某些文件而导致启动失败或是无法远程访问，这些问题在集群模式下尤为明显。虽然同样可以和配置防火墙一样通过 SELinux 的安全策略来解决此类问题，但这种方式需要对 SELinux 的安全策略以及 MySQL 所使用到的文件和目录有较为深入的了解，实施难度较大。这里选择相对较为简单的处理方式，直接永久关闭掉操作系统的 SELinux 服务。同时该操作步骤也需要在集群中所有主机上进行操作。

（1）使用"root"用户登录操作系统。MySQL 集群模式的整个安装和配置过程中的所有步骤，都需要使用 root 用户来进行操作。

（2）使用命令"sestatus -v"可以查看当前操作系统中 SELinux 服务的运行状态，如图 4-2-3 所示。其中第一行的参数"SELinux status"的所在行显示的内容为当前 SELinux 服务的启动状态，默认情况下操作系统会自动启动 SELinux 服务。

（3）CentOS 7 的 SELinux 服务的配置文件位于操作系统的"/etc/selinux"目录下，配置文件的文件名为"config"，如图 4-2-4 所示是。编辑该配置文件，找到配置项"SELINUX"的所在行，如图 4-2-5 所示，将其改为如下的内容：

SELINUX=disabled

项目 4　基于 Linux 的 MySQL 数据库平台的搭建

```
[root@MySQL-01 ~]# sestatus -v
SELinux status:                 enabled
SELinuxfs mount:                /sys/fs/selinux
SELinux root directory:         /etc/selinux
Loaded policy name:             targeted
Current mode:                   enforcing
Mode from config file:          enforcing
Policy MLS status:              enabled
Policy deny_unknown status:     allowed
Max kernel policy version:      28

Process contexts:
Current context:                unconfined_u:unconfined_r:unconfined_t:s0-s0:c0.c1023
Init context:                   system_u:system_r:init_t:s0
/usr/sbin/sshd                  system_u:system_r:sshd_t:s0-s0:c0.c1023

File contexts:
Controlling terminal:           unconfined_u:object_r:user_tty_device_t:s0
/etc/passwd                     system_u:object_r:passwd_file_t:s0
/etc/shadow                     system_u:object_r:shadow_t:s0
/bin/bash                       system_u:object_r:shell_exec_t:s0
/bin/login                      system_u:object_r:login_exec_t:s0
/bin/sh                         system_u:object_r:bin_t:s0 -> system_u:object_r:shell_exec_t:s0
/sbin/agetty                    system_u:object_r:getty_exec_t:s0
/sbin/init                      system_u:object_r:bin_t:s0 -> system_u:object_r:init_exec_t:s0
/usr/sbin/sshd                  system_u:object_r:sshd_exec_t:s0
```

图 4-2-3　查看服务运行状态

```
[root@MySQL-01 ~]# ls -l /etc/selinux/
total 8
-rw-r--r--. 1 root root  546 Jul  5 10:38 config
drwx------. 2 root root    6 Jul  3 10:38 final
-rw-r--r--. 1 root root 2321 Nov  6  2016 semanage.conf
drwxr-xr-x. 7 root root  215 Jul  3 10:38 targeted
drwxr-xr-x. 2 root root    6 Nov  6  2016 tmp
```

图 4-2-4　查询配置文件

```
2 # This file controls the state of SELinux on the system.
3 # SELINUX= can take one of these three values:
4 #     enforcing - SELinux security policy is enforced.
5 #     permissive - SELinux prints warnings instead of enforcing.
6 #     disabled - No SELinux policy is loaded.
7 SELINUX=disabled
```

图 4-2-5　编辑配置文件

（4）配置文件修改完成之后使用命令"reboot"重新启动操作系统。

（5）操作系统重新启动完成之后，再次使用命令"sestatus -v"查看当前系统中 SELinux 服务的运行状态，如图 4-2-6 所示，若参数"SELinux status"的所在行显示的内容为"disabled"，且不再有其他运行状态信息的内容，则表示 SELinux 服务未启动，已经成功关闭。

```
[root@MySQL-01 ~]# sestatus -v
SELinux status:                 disabled
```

图 4-2-6　SELinux 服务未启动

4.2.5 MySQL 集群模式的安装和配置

MySQL 集群模式的安装和配置过程与 MySQL 单机模式的安装和配置的过程大致上相同，唯一需要注意的就是解压出来的软件的目录名称和需要配置的系统环境变量的名称与 MySQL 单机模式的安装和配置过程中的内容有一些不同，以及不需要执行最后的初始化安装程序和安装注册的过程。

（1）在操作系统根目录下创建目录"mysql"，用于存放 MySQL Cluster 的相关文件，该目录也可自行选择其他位置进行创建。创建完成后将当前的工作目录切换到该目录。

（2）将 MySQL Cluster 的软件包解压解包到"mysql"目录下，如图 4-2-7 所示，解压解包出来的软件目录名称为"mysql-cluster-gpl-7.5.7-linux-glibc2.12-x86_64"。

```
[root@MySQL-01 mysql]# ls -l
total 0
drwxr-xr-x 10 root root 147 Jul  6 10:44 mysql-cluster-gpl-7.5.7-linux-glibc2.12-x86_64
```

图 4-2-7　解压软件包

（3）将当前的工作目录切换到操作系统的"/usr/local"目录下，并使用命令"ln -s /mysql/mysql-cluster-gpl-7.5.7-linux-glibc2.12-x86_64 mysql"在该目录下创建一个 MySQL Cluster 软件目录的链接，创建的链接如图 4-2-8 所示。然后将当前工作目录切换到该链接目录。

```
[root@MySQL-01 local]# ls -l mysql
lrwxrwxrwx 1 root root 54 Jul  6 10:46 mysql -> /mysql/mysql-cluster-gpl-7.5.7-linux-glibc2.12-x86_64
```

图 4-2-8　创建 MySQL Cluster 软件目录链接

（4）创建 MySQL 数据文件的存放目录"data"，并使用命令"chmod 770 data"将该目录的权限更改为所属用户和所属组拥有所有权限，而其他用户没有任何权限，修改之后的目录属性如图 4-2-9 所示。

```
[root@MySQL-01 mysql]# ls -dl data
drwxrwx--- 2 root root 6 Jul  6 10:47 data
```

图 4-2-9　更改目录权限

（5）依次使用命令"chown -R mysql ."和"chgrp -R mysql ."将当前工作目录及其所有子目录和子文件的所属用户和所属组更改为 MySQL 数据库软件的专用用户和专用组，如图 4-2-10 所示。

项目 4　基于 Linux 的 MySQL 数据库平台的搭建

```
[root@MySQL-01 mysql]# ls -l
total 0
drwxr-xr-x 11 mysql mysql 159 Jul  6 10:47 mysql-cluster-gpl-7.5.7-linux-glibc2.12-x86_64
```

```
[root@MySQL-01 mysql]# ls -l
total 36
drwxr-xr-x   2 mysql mysql  4096 Jul  6 10:44 bin
-rw-r--r--   1 mysql mysql 17987 Jun 23  2017 COPYING
drwxrwx---   2 mysql mysql     6 Jul  6 10:47 data
drwxr-xr-x   2 mysql mysql    55 Jul  6 10:44 docs
drwxr-xr-x   4 mysql mysql  4096 Jul  6 10:43 include
drwxr-xr-x   5 mysql mysql   331 Jul  6 10:44 lib
drwxr-xr-x   4 mysql mysql    30 Jul  6 10:44 man
drwxr-xr-x  10 mysql mysql   289 Jul  6 10:45 mysql-test
-rw-r--r--   1 mysql mysql  2478 Jun 23  2017 README
drwxr-xr-x  32 mysql mysql  4096 Jul  6 10:44 share
drwxr-xr-x   2 mysql mysql    90 Jul  6 10:43 support-files
```

图 4-2-10　更改目录及子文件的所属用户与组

（6）在操作系统的配置文件"/etc/profile"中配置 MySQL 相关的环境变量，如图 4-2-11 所示，在文件末尾添加以下内容：

\# mysql cluster environment

MYSQL_CLUSTER_HOME=/usr/local/mysql-cluster

PATH=$MYSQL_CLUSTER_HOME/bin：$PATH

export　MYSQL_CLUSTER_HOME　PATH

```
# mysql cluster environment
MYSQL_CLUSTER_HOME=/usr/local/mysql
PATH=$MYSQL_CLUSTER_HOME/bin:$PATH
export MYSQL_CLUSTER_HOME PATH
```

图 4-2-11　修改配置文件

（7）使用命令"source　/etc/profile"使新配置的环境变量立即生效。

（8）使用命令"echo　$变量名"可以查看新添加和修改的环境变量的值是否正确，如图 4-2-12 所示。

```
[root@MySQL-01 ~]# echo $MYSQL_CLUSTER_HOME
/usr/local/mysql
```

图 4-2-12　查看环境变量

（9）在集群中所有其他主机上执行以上操作步骤，或是使用"scp"命令依次将 MySQL 相关文件的"/mysql""/usr/local"目录下的链接目录"mysql-cluster"、系统环境变量配置文件"/etc/profile"发送到集群中所有其他主机的相应目录下。

4.2.6 管理节点 MySQL-01 的配置

（1）使用"root"用户登录到管理节点的主机"MySQL-01"，执行以下的操作步骤。

（2）进入"/usr/local"目录下的 MySQL Cluster 软件目录的链接目录"mysql"，创建 MySQL Cluster 的管理节点的配置文件存放目录"mysql-cluster"，并将当前工作目录切换到该目录。

（3）创建 MySQL Cluster 的管理节点的配置文件"config.ini"，然后编辑该配置文件，在其中配置以下内容，"#"号之后的内容为对应配置项或配置内容的说明，不需要写入到配置文件之中：

\# 管理节点的通用基本属性配置

[NDB_MGMD DEFAULT]

\# 管理节点的数据和日志文件的存放目录

DataDir=/usr/local/mysql/data

\# 数据节点的通用基本属性配置

[NDBD DEFAULT]

\# 每个数据节点的镜像数量

NoOfReplicas=2

\# 每个数据节点中给数据分配的内存

DataMemory=512M

\# 每个数据节点中给索引分配的内存

IndexMemory=256M

\# 数据节点的数据和日志文件的存放目录

DataDir=/usr/local/mysql/data

\# 管理节点的属性配置

[NDB_MGMD]

\# 节点主机在集群中的编号

NodeId=1

\# 节点主机的主机名或 IP 地址

HostName=MySQL-01

\# 数据节点的属性配置

[NDBD]
NodeId=2
HostName=MySQL-02
[NDBD]
NodeId=3
HostName=MySQL-03
# SQL 节点的属性配置
[MYSQLD]
NodeId=4
HostName=MySQL-04
[MYSQLD]
NodeId=5
HostName=MySQL-05

（4）依次使用命令"chown -R mysql ."和"chgrp -R mysql ."将当前的工作目录"mysql-cluster"及其所有子目录和子文件的所属用户和所属组更改为 MySQL 数据库软件的专用用户和专用组，如图 4-2-13 所示。

```
[root@MySQL-01 mysql]# ls -dl mysql-cluster
drwxr-xr-x 2 mysql mysql 24 Jul  6 11:10 mysql-cluster
[root@MySQL-01 mysql-cluster]# ls -l
total 4
-rw-r--r-- 1 mysql mysql 350 Jul  6 11:10 config.ini
```

图 4-2-13　更改目录及子文件的所属用户与组

（5）使用命令"ndb_mgmd -f /usr/local/mysql/mysql-cluster/config.ini --initial"初始化并启动 MySQL Cluster 的管理节点，如图 4-2-14 所示。

★ 初始化选项"--initial"只需要在首次执行该命令时添加，之后启动 MySQL Cluster 的管理节点的时候不再添加该参数，除非是在备份恢复数据、配置变化后重新启动管理节点时才需要再次添加该选项，否则管理节点中的数据会被清空。

```
[root@MySQL-01 ~]# ndb_mgmd -f /usr/local/mysql/mysql-cluster/config.ini --initial
MySQL Cluster Management Server mysql-5.7.19 ndb-7.5.7
```

图 4-2-14　初始化并启动管理节点

（6）使用命令"ps -ef | grep ndb_mgmd"查看系统进程信息，如图 4-2-15 所示，若存在进程信息中包含"ndb_mgmd"关键字的进程则表示 MySQL

Cluster 的管理节点启动成功。

```
[root@MySQL-01 ~]# ps -ef | grep ndb_mgmd
root       4027     1  1 05:19 ?        00:00:00 ndb_mgmd -f /usr/local/mysql/mysql-cluster/config.ini --initial
root       4042  2238  0 05:19 tty1     00:00:00 grep ndb_mgmd
```

图 4-2-15　查看系统进程信息

（7）使用命令"ndb_mgm"可以进入 MySQL Cluster 管理节点的控制台界面，如图 4-2-16 所示。

```
[root@MySQL-01 ~]# ndb_mgm
-- NDB Cluster -- Management Client --
ndb_mgm>
```

图 4-2-16　进入控制台界面

（8）在控制台界面中使用命令"show"可以查看当前 MySQL Cluster 集群中的节点信息，如图 4-2-17 所示。不过因为当前只是启动了 MySQL Cluster 的管理节点，所以只有管理节点的连接信息，还没有数据节点和 SQL 节点的连接信息。

```
ndb_mgm> show
Connected to Management Server at: localhost:1186
Cluster Configuration
---------------------
[ndbd(NDB)]     2 node(s)
id=2 (not connected, accepting connect from MySQL-02)
id=3 (not connected, accepting connect from MySQL-03)

[ndb_mgmd(MGM)] 1 node(s)
id=1    @192.168.60.150  (mysql-5.7.19 ndb-7.5.7)

[mysqld(API)]   2 node(s)
id=4 (not connected, accepting connect from MySQL-04)
id=5 (not connected, accepting connect from MySQL-05)
```

图 4-2-17　查看节点信息

（9）在控制台界面中使用命令"exit"可以退出控制台，如图 4-2-18 所示。

```
ndb_mgm> exit
[root@MySQL-01 ~]#
```

图 4-2-18　退出控制台

（10）MySQL Cluster 的管理节点默认使用的端口是 1186，配置防火墙端口策略开启此端口,配置方式与 MySQL 数据库的单机模式中防火墙端口策略的配置方式相同。

(11)配置管理节点的服务开机自动启动,编辑操作系统中的"/etc/rc.d/"目录下的配置文件"rc.local",如图 4-2-19 所示,在该配置文件中用户可以自定义在操作系统启动时需要执行的操作命令,在文件的末尾添加如下的内容:

/usr/local/mysql/bin/ndb_mgmd　-f　/usr/local/mysql-cluster/etc/config.ini

```
#!/bin/bash
# THIS FILE IS ADDED FOR COMPATIBILITY PURPOSES
#
# It is highly advisable to create own systemd services or udev rules
# to run scripts during boot instead of using this file.
#
# In contrast to previous versions due to parallel execution during boot
# this script will NOT be run after all other services.
#
# Please note that you must run 'chmod +x /etc/rc.d/rc.local' to ensure
# that this script will be executed during boot.

touch /var/lock/subsys/local

/usr/local/mysql/bin/ndb_mgmd -f /usr/local/mysql/mysql-cluster/config.ini
```

图 4-2-19　编辑配置文件

在 Centos 7 版本中,配置文件"/etc/rc.d/rc.local"的权限被降低了,默认情况下并不具有可执行权限,所以需要使用命令"chmod +x /etc/rc.d/rc.local"为该配置文件添加可执行权限。

(12)使用命令"reboot"重新启动操作系统,然后使用命令"ps -ef | grep ndb_mgmd"查看系统进程信息,如图 4-2-20 所示,确认 MySQL Cluster 的管理节点的服务是否能够自动启动。

```
[root@MySQL-01 ~]# ps -ef | grep ndb_mgmd
root       1214      1  0 12:32 ?        00:00:02 /usr/local/mysql/bin/ndb_mgmd -f /usr/local/mysql/mysql-cluster/config.ini
root       2200   2238  0 12:39 tty1     00:00:00 grep ndb_mgmd
```

图 4-2-20　查看系统进程信息

4.2.7　数据节点 MySQL-02、MySQL-03 的配置

(1)使用"root"用户依次登录到数据节点的主机"MySQL-02"和"MySQL-03",执行以下的操作步骤。

(2)进入操作系统的配置文件所在目录"etc",创建 MySQL Cluster 的数据节点的配置文件"my.cnf",然后编辑该配置文件,在其中配置以下内容,"#"

号之后的内容为对应配置项或配置内容的说明，不需要写入到配置文件之中：

[MYSQLD]

＃指定使用 NDB 集群存储引擎

ndbcluster

＃指定管理节点所在主机的主机名或 IP 地址，若有多个多个管理节点，每个主机名或 IP 地址之间用","分隔开

ndb-connectstring=MySQL-01

[MYSQL_CLUSTER]

ndb-connectstring=MySQL-01

（3）依次使用命令"chown mysql my.cnf"和"chgrp mysql my.cnf"将新创建的配置文件的所属用户和所属组更改为 MySQL 数据库软件的专用用户和专用组，如图 4-2-21 所示。

```
[root@MySQL-02 etc]# ls -l my.cnf
-rw-r--r-- 1 mysql mysql 90 Jul  6 12:43 my.cnf
```

图 4-2-21　更改所属用户和所属组

（4）使用命令"ndbd --initial"初始化并启动 MySQL Cluster 的数据节点，如图 4-2-22 所示。

同管理节点中的启动命令一样，初始化选项"--initial"只需要在首次执行该命令时添加，之后启动 MySQL Cluster 的数据节点的时候不再添加该参数，除非是在备份恢复数据以及配置变化后重新启动数据节点时才需要再次添加该选项，否则数据节点中的数据会被清空。

```
[root@MySQL-02 ~]# ndbd --initial
2018-07-09 04:45:50 [ndbd] INFO     -- Angel connected to 'MySQL-01:1186'
2018-07-09 04:45:50 [ndbd] INFO     -- Angel allocated nodeid: 2
```

图 4-2-22　初始化并启动数据节点

（5）使用命令"ps -ef | grep ndbd"查看系统进程信息，如图 4-2-23 所示，若存在信息中包含"ndbd"关键字的进程则表示 MySQL Cluster 的数据服务节点启动成功。

```
[root@MySQL-02 ~]# ps -ef | grep ndbd
root      2240     1  0 04:45 ?        00:00:00 ndbd --initial
root      2241  2240  1 04:45 ?        00:00:04 ndbd --initial
root      2305  2173  0 04:50 tty1     00:00:00 grep ndbd
```

图 4-2-23　查看系统进程信息

（6）进入 MySQL Cluster 的管理节点的控制台，在控制台界面中使用命令"show"查看当前 MySQL Cluster 集群中的节点信息，如图 4-2-24 所示，若有相应数据服务节点的连接信息，则表示 MySQL Cluster 的数据服务节点启动并连接成功。

```
ndb_mgm> show
Cluster Configuration
---------------------
[ndbd(NDB)]     2 node(s)
id=2    @192.168.60.151  (mysql-5.7.19 ndb-7.5.7, starting, Nodegroup: 0)
id=3    @192.168.60.152  (mysql-5.7.19 ndb-7.5.7, starting, Nodegroup: 0)

[ndb_mgmd(MGM)] 1 node(s)
id=1    @192.168.60.150  (mysql-5.7.19 ndb-7.5.7)

[mysqld(API)]   2 node(s)
id=4 (not connected, accepting connect from MySQL-04)
id=5 (not connected, accepting connect from MySQL-05)
```

图 4-2-24　查看节点信息

（7）配置操作系统的防火墙。在官方说明文档中给出的 MySQL Cluster 的数据节点所使用的默认端口是 2202，但实际上 MySQL Cluster 的数据节点一般都是使用的随机端口，并且在每次重新启动服务之后，所使用的随机端口号都会有所不同。所以推荐使用命令"systemctl stop firewalld"和"systemctl disable firewalld"关闭并禁用掉操作系统的防火墙，如图 4-2-25 所示。如果因为特殊原因一定需要启用防火墙服务，可以使用命令"netstat -anp | grep ndbd"来查看 MySQL Cluster 的数据节点服务所使用到的端口号，如图 4-2-26 所示，然后使用与 MySQL 数据库的单机模式中相同的方式配置防火墙的端口策略，开启所有使用到的端口。不过需要注意的是在每次重新启动数据节点服务之后，都需要针对新使用到的随机端口号重新配置防火墙的端口策略。

```
[root@MySQL-02 ~]# systemctl disable firewalld
Removed symlink /etc/systemd/system/dbus-org.fedoraproject.FirewallD1.service.
Removed symlink /etc/systemd/system/basic.target.wants/firewalld.service.
```

图 4-2-25　关闭操作系统防火墙

```
[root@MySQL-02 ~]# netstat -anp | grep ndbd
tcp    0    0 192.168.60.151:37772    0.0.0.0:*              LISTEN       1402/ndbd
tcp    0    0 192.168.60.151:41007    0.0.0.0:*              LISTEN       1402/ndbd
tcp    0    0 192.168.60.151:37848    0.0.0.0:*              LISTEN       1402/ndbd
tcp    0    0 192.168.60.151:36816    192.168.60.150:1186    ESTABLISHED  1398/ndbd
tcp    0    0 192.168.60.151:36818    192.168.60.150:1186    ESTABLISHED  1402/ndbd
unix   3    [ ]          STREAM      CONNECTED    19349              1402/ndbd
unix   3    [ ]          STREAM      CONNECTED    19350              1402/ndbd
```

图 4-2-26　查看端口号

（8）配置数据节点的服务开机自动启动，编辑操作系统中的"/etc/rc.d/"目录下的配置文件"rc.local"，如图 4-2-27 所示，在文件的末尾添加如下的内容：

/usr/local/mysql/bin/ndbd

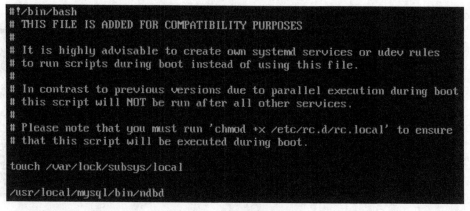

图 4-2-27　编辑配置文件

这里和管理节点的情况一样，同样需要使用命令"chmod +x /etc/rc.d/rc.local"为配置文件"/etc/rc.d/rc.local"添加可执行权限。

（9）使用命令"reboot"重新启动操作系统，然后使用命令"ps -ef | grep ndbd"查看系统进程信息，如图 4-2-28 所示，确认 MySQL Cluster 的数据节点的服务是否能够自动启动。

图 4-2-28　查看系统进程信息

4.2.8　SQL 节点 MySQL-04、MySQL-05 的安装和配置

（1）依次使用"root"用户登录到 SQL 节点的主机"MySQL-04"和"MySQL-05"，执行以下的操作步骤。

（2）进入操作系统的配置文件所在目录"etc"，创建 MySQL Cluster 的 SQL 节点的配置文件"my.cnf"，然后编辑该配置文件，在其中配置以下内容，"#"号之后的内容为对应配置项或配置内容的说明，不需要写入到配置文件之中：

[MYSQLD]

SQL 节点主机上的 MySQL Cluster 数据库软件所在目录

basedir=/usr/local/mysql
# SQL 节点的数据和日志文件的存放目录
datadir=/usr/local/mysql/data
# 指定使用 NDB 集群存储引擎
ndbcluster
指定管理节点所在主机的主机名或 IP 地址，若有多个管理节点，每个主机名或 IP 地址之间用","分隔开
ndb-connectstring=MySQL-01
[MYSQL_CLUSTER]
ndb-connectstring=MySQL-01

（3）依次使用命令"chown mysql my.cnf"和"chgrp mysql my.cnf"将新创建的配置文件的所属用户和所属组更改为 MySQL 数据库软件的专用用户和专用组，如图 4-2-29 所示。

```
[root@MySQL-04 etc]# ls -l my.cnf
-rw-r--r-- 1 mysql mysql 145 Jul  9 06:13 my.cnf
```

图 4-2-29　更改所属用户和用户组

（4）使用"mysqld --initialize --user=mysql"命令对 MySQL 数据库的 SQL 节点的安装进行初始化，如图 4-2-30 所示。同 MySQL 的单机模式的安装初始化过程相同，安装初始化程序会生成数据库的 root 用户的随机初始密码，将该密码记住或保存下来。

```
[root@MySQL-04 ~]# mysqld --initialize --user=mysql
2018-07-09T10:19:37.715744Z 0 [Warning] TIMESTAMP with implicit DEFAULT value is deprecated. Please use --explicit_defaults_for_timestamp server option (see documentation for more details).
2018-07-09T10:19:38.170595Z 0 [Warning] InnoDB: New log files created, LSN=45790
2018-07-09T10:19:38.222779Z 0 [Warning] InnoDB: Creating foreign key constraint system tables.
2018-07-09T10:19:38.370944Z 0 [Warning] No existing UUID has been found, so we assume that this is the first time that this server has been started. Generating a new UUID: 95d631a3-8361-11e8-9c3e-000c299f3077.
2018-07-09T10:19:38.376187Z 0 [Warning] Gtid table is not ready to be used. Table 'mysql.gtid_executed' cannot be opened.
2018-07-09T10:19:38.380297Z 1 [Note] A temporary password is generated for root@localhost: +*G6tdci(?j-
```

图 4-2-30　对 SQL 节点的安装进行初始化

同样若是忘记了数据库的 root 用户的随机初始密码，或者是随机初始密码过期无法使用，可以清空 SQL 节点的数据文件的存放目录"data"中的所有内容，然后重新执行安装的初始化操作。也可以选择使用"mysqld --initialize-insecure --user=mysql"命令来进行不生成 root 用户的初始密码的安装初始化操作。

（5）将 MySQL 数据库服务配置成为系统服务，同 MySQL 单机模式下的配置方式相同，将位于 MySQL 软件目录下的"support-files"目录中的系统服务脚本文件"mysql.server"拷贝到操作系统的可控制服务目录"/etc/init.d"之中，并将拷贝的副本文件命名为"mysql"。然后使用命令"chkconfig --add mysql"和"chkconfig --level 2345 mysql on"将 MySQL 服务配置为开机自动启动。

（6）使用命令"reboot"重新启动操作系统，然后使用命令"ps -ef | grep mysql"查看系统进程信息，如图 4-2-31 所示，若存在进程信息中包含"mysql"关键字的进程则表示 MySQL Cluster 的 SQL 节点能够成功的自动启动。

图 4-2-31　查看系统进程信息

（7）MySQL Cluster 的 SQL 服务节点默认使用的端口是 3306，配置防火墙端口策略开启此端口，配置方式与 MySQL 数据库的单机模式中防火墙端口策略的配置方式相同。

（8）使用数据库的 root 用户以及安装初始化过程中所显示的 root 用户的随机初始密码登录到本机的 MySQL 数据库。如图 4-2-32 所示，修改数据库的 root 用户的登录密码为自定义密码。然后配置接收远程访问，修改和配置方法与 MySQL 单机模式中的相同，如图 4-2-33 所示，修改配置完成后退出 MySQL 数据库的控制台界面返回操作系统的命令行界面。

图 4-2-32　修改登录密码

这里需要注意 MySQL Cluster 集群中所有 SQL 节点的数据库 root 用户的登录密码要使用相同的密码，否则该 SQL 节点会被视为不能使用的无效节点，无法进行数据的同步。

```
mysql> FLUSH PRIVILEGES;
Query OK, 0 rows affected (0.00 sec)
```

图 4-2-33　配置接收远程访问

（9）进入 MySQL Cluster 的管理节点的控制台，在控制台界面中使用命令"show"查看当前 MySQL Cluster 集群中的节点信息，如图 4-2-34 所示，若有相应 SQL 节点的连接信息，则表示 MySQL Cluster 的 SQL 节点启动并连接成功。

```
ndb_mgm> show
Cluster Configuration
---------------------
[ndbd(NDB)]     2 node(s)
id=2    @192.168.60.151  (mysql-5.7.19 ndb-7.5.7, Nodegroup: 0, *)
id=3    @192.168.60.152  (mysql-5.7.19 ndb-7.5.7, Nodegroup: 0)

[ndb_mgmd(MGM)] 1 node(s)
id=1    @192.168.60.150  (mysql-5.7.19 ndb-7.5.7)

[mysqld(API)]   2 node(s)
id=4    @192.168.60.153  (mysql-5.7.19 ndb-7.5.7)
id=5    @192.168.60.154  (mysql-5.7.19 ndb-7.5.7)
```

图 4-2-34　查看节点信息

4.2.9　MySQL 集群模式验证

（1）选择多个 SQL 节点中的任意一个，登录到其上的 MySQL 数据库，如图 4-2-35 所示。

```
[root@MySQL-01 ~]# mysql -h MySQL-04 -u root -p
Enter password:
Welcome to the MySQL monitor.  Commands end with ; or \g.
Your MySQL connection id is 6
Server version: 5.7.19-ndb-7.5.7-cluster-gpl MySQL Cluster Community Server (GPL)

Copyright (c) 2000, 2017, Oracle and/or its affiliates. All rights reserved.

Oracle is a registered trademark of Oracle Corporation and/or its
affiliates. Other names may be trademarks of their respective
owners.

Type 'help;' or '\h' for help. Type '\c' to clear the current input statement.

mysql>
```

图 4-2-35　登录节点数据库

（2）在 MySQL 数据库的控制台界面中使用 SQL 语句"CREATE DATABSE mydb;"新建一个测试用的数据库，如图 4-2-36 所示。然后使用 SQL 语句"USE mydb;"选择该数据库,如图4-2-37所示。

```
mysql> CREATE DATABASE mydb;
Query OK, 1 row affected (0.05 sec)
```

图 4-2-36　新建数据库

```
mysql> USE mydb;
Database changed
```

图 4-2-37　选择数据库

（3）在 MySQL 数据库的控制台界面中依次使用下面的 SQL 语句创建用户信息表并在其中添加一些测试数据,如图 4-2-38 所示。

```
mysql> CREATE TABLE user_info(
    -> id INT UNSIGNED NOT NULL AUTO_INCREMENT PRIMARY KEY,
    -> name VARCHAR(20) NOT NULL DEFAULT '')
    -> ENGINE=NDBCLUSTER DEFAULT CHARSET utf8;
Query OK, 0 rows affected (0.26 sec)

mysql> INSERT INTO user_info VALUES(1,'Jack');
Query OK, 1 row affected (0.04 sec)

mysql> INSERT INTO user_info VALUES(2,'Tom');
Query OK, 1 row affected (0.01 sec)

mysql> INSERT INTO user_info VALUES(3,'Adam');
Query OK, 1 row affected (0.00 sec)
```

图 4-2-38　创建用户信息表并添加数据

CREATE TABLE user_info（
id INT UNSIGNED NOT NULL AUTO_INCREMENT PRIMARY KEY,
name VARCHAR（20） NOT NULL DEFAULT ''）
ENGINE=NDBCLUSTER DEFAULT CHARSET utf8;
INSERT INTO user_info VALUES（1, 'Jack'）;
INSERT INTO user_info VALUES（2, 'Tom'）;
INSERT INTO user_info VALUES（3, 'Adam'）;

在 MySQL 的集群模式中,数据库表的存储引擎必须指定为 NDBCLUSTER,否则数据库表及其中的数据不会同步到其他 SQL 节点服务器上。

（4）登录到其他任意一个 SQL 节点上的 MySQL 数据库,如图 4-2-39 所示。

```
[root@MySQL-01 ~]# mysql -h MySQL-05 -u root -p
Enter password:
Welcome to the MySQL monitor.  Commands end with ; or \g.
Your MySQL connection id is 6
Server version: 5.7.19-ndb-7.5.7-cluster-gpl MySQL Cluster Community Server (GPL)

Copyright (c) 2000, 2017, Oracle and/or its affiliates. All rights reserved.

Oracle is a registered trademark of Oracle Corporation and/or its
affiliates. Other names may be trademarks of their respective
owners.

Type 'help;' or '\h' for help. Type '\c' to clear the current input statement.

mysql>
```

图 4-2-39　登录数据库

（5）在 MySQL 控制台界面中使用 SQL 语句"SHOW　DATABSES;"查看数据库的信息中是否存在名为"mydb"的数据库，如图 4-2-40 所示，若存在则表示 MySQL 集群的数据库同步成功。然后使用 SQL 语句"USE　mydb;"选择该数据库，如图 4-2-41 所示。

图 4-2-40　查看数据库信息

```
mysql> USE mydb;
Reading table information for completion of table and column names
You can turn off this feature to get a quicker startup with -A

Database changed
```

图 4-2-41　选择数据库

（6）进一步验证数据库的表及其中的数据是否同步成功。在 MySQL 数据库的控制台界面中使用 SQL 语句"SHOW　TABLES;"查看数据库的表信息中是否存在名为"user_info"的表，如图 4-2-42 所示。若存在则继续使用 SQL 语句

“SELECT * FROM user_info”查看表中的数据是否与之前添加的数据一致，如图 4-2-43 所示。

```
mysql> SHOW TABLES;
+-----------------+
| Tables_in_mydb |
+-----------------+
| user_info       |
+-----------------+
1 row in set (0.01 sec)
```

图 4-2-42　查看数据库表信息

```
mysql> SELECT * FROM user_info;
+----+------+
| id | name |
+----+------+
|  1 | Jack |
|  2 | Tom  |
|  3 | Adam |
+----+------+
3 rows in set (0.00 sec)
```

图 4-2-43　查看表中数据信息

项目 5　Hive 数据仓库的搭建和使用

任务 5.1　Hive 数据仓库的搭建

5.1.1　Hive 数据仓库

5.1.1.1　数据仓库

数据仓库（Data Warehouse）这一概念由比尔·恩门（Bill Inmon）于 1990 年提出，并将其定义为一个面向主题的（Subject Oriented）、集成的（Integrated）、不可更新的（Non-Volatile）、随时间不断变化的（Time Variant）数据集合，用于支持各种管理决策（Decision Making Support）。

数据仓库为企业所有级别的决策制定过程，并提供所有类型的数据支持。它将资讯系统经年累月所累积的大量资料，透过数据仓库理论所特有的资料储存架构，进行系统的分析和整理。同时利用联机分析处理（OLAP）、数据挖掘（Data Mining）等分析方法创建决策支持系统（DSS）、主管资讯系统（EIS）等平台，帮助决策者快速有效地从大量资料中分析出有价值的资讯，帮助决策的拟定和快速回应外在环境的变动，建构商业智能（BI）。

数据仓库实际上是一个抽象的概念，它以数据库为基础，其实现的载体就是我们所常见的各种数据库表，但在需求、客户、体系结构、运行机制等方面与数据库又有着很大的不同。数据仓库利用数据库为决策支持系统和联机分析应用提供数据源的结构化数据环境，并进一步研究和解决从数据库中获取有价值信息的方式和方法。

1. 面向主题（Subject Oriented）

与传统数据库的面向应用进行数据组织的特点不同，数据仓库中的数据是面向主题进行组织的。主题是一个抽象的概念，是在较高层次上对企业信息系统中的数据进行综合、归类、分析、利用的抽象。在逻辑意义上，它是对应企业中某个宏观分析领域所涉及的分析对象。面向主题的数据组织方式，就是在较高层次上对分析对象的数据的一个完整、一致的描述，能完整、统一地刻画各个分析对象所涉及的企业的各项数据,以及数据之间的联系。所谓较高层次是相对面向应用的数据组织方式而言的,是指按照主题进行数据组织的方式具有更高的数据抽象级别。

2. 集成（Integrated）

数据仓库的数据是从原有的分散的数据库数据中抽取而来的。操作型数据与 DSS 分析型数据之间差别很大。首先,数据仓库的每个主题所对应的原数据在原有的各个分散数据库中可能有许多重复和不一致的地方,并且来源于不同的联机系统的数据一般都和不同的应用逻辑捆绑在一起。其次,数据仓库中的综合数据一般也不能从原有的数据库系统直接得到。因此在数据进入数据仓库之前,必然要经过统一与综合,这一步是数据仓库建设中最关键、最复杂的一步。这一步所要完成的工作包括统一源数据中所有矛盾之处,如字段的同名异义、异名同义、单位不统一、字长不一致等。另外还要进行数据综合和计算。数据仓库中的数据综合工作可以在从原有数据库抽取数据时生成,但许多时候是在数据仓库内部生成的,即进入数据仓库之后再进行综合生成的。

3. 不可更新（Non-Volatile）

数据仓库的数据主要是供企业决策分析之用,所涉及的数据操作主要是数据查询,一般情况下并不进行修改操作。数据仓库的数据通常反映的是一段相当长的时间内历史数据的内容,是不同时间点的数据库快照的集合,以及基于这些快照进行统计、综合、重组的导出数据,而不是联机处理的数据。数据库中进行联机处理的数据经过集成输入到数据仓库中,一旦数据仓库存放的数据已经超过数据仓库的数据存储期限,这些数据将从当前的数据仓库中删除。因为数据仓库只进行数据查询操作,所以数据仓库管理系统相比数据库管理系统而言要简单得多。数据库管理系统中许多技术难点,如完整性保护、并发控制等,在数据仓库的管理中几乎完全可以省去。但是由于数据仓库的查询数据量往往很巨大,所以就对数据查询提出了更高的要求,它要求采用各种复杂的索引技术。同时由于数据仓库面向的是商业企业的高层管理者,他们会对数据查询的界面友好性和数据表示方式提出更高的要求。

4. 随时间不断变化（Time Variant）

数据仓库中的数据不可更新是针对应用而言的。也就是说,数据仓库的用户进行分析处理时是不进行数据更新操作的。但并不是说,在从数据集成输入数据仓库开始到最终被删除的整个数据生存周期中,所有的数据仓库数据都是永远不变的。数据仓库的数据会随时间的变化而不断变化,这一特征表现在三个方面。

第一,数据仓库随时间变化会不断增加新的数据内容。数据仓库系统必须不断捕捉 OLTP 数据库中变化的数据,并追加到数据仓库中,也就是要不断地生成 OLTP 数据库的快照,经统一集成后增加到数据仓库中。但对于确实不再变化的数据库快照,如果捕捉到新的变化数据,则只生成一个新的数据库快照增加进去,而不会对原有的数据库快照进行修改。

第二,数据仓库随时间变化会不断删去旧的数据内容。数据仓库的数据也有存储期限,一旦超过了这一期限,过期数据就要被删除。只是数据仓库内的数据时限要远远长于操作型环境中的数据时限。在操作型环境中一般只会保存 60~90 天的数据,而在数据仓库中则需要保存较长时限的数据(如 5~10 年),以适应 DSS 进行趋势分析的要求。

第三,数据仓库中包含有大量的综合数据,这些综合数据中很多都跟时间有关,如数据经常按照时间段进行综合,或间隔一定的时间片进行抽样等。这些数据要随着时间的变化不断地进行重新综合。因此,数据仓库的数据特征都包含时间项,以标明数据的历史时期。

数据仓库的发展大致经历了三个阶段。第一个是简单报表阶段。在这个阶段系统的主要目标是解决一些日常的工作中业务人员需要的报表,以及生成一些简单的能够帮助领导进行决策所需要的汇总数据,大部分表现形式为数据库和前端报表工具。第二个阶段是数据集市阶段。在这个阶段主要是根据某个业务部门的需要,进行一定的数据的采集和整理,并按照业务人员的需要,进行多维报表的展现,能够提供对特定业务指导的数据,并且能够提供特定的领导决策数据。第三个阶段是数据仓库阶段。这个阶段主要是按照一定的数据模型,对整个企业的数据进行采集和整理,并且能够按照各个业务部门的需要,提供跨部门的、完全一致的业务报表数据,能够通过数据仓库生成对业务具有指导性的数据,同时为领导决策提供全面的数据支持,其数据集市阶段的重要区别就在于对数据模型的支持。

5. ODS 层(临时存储层)

临时存储层是接口数据的临时存储区域,为后一步的数据处理做准备。一般来说 ODS 层的数据和数据源系统的数据是同构的,主要目的是简化后续数据加工处理的工作。从数据粒度上来说 ODS 层的数据粒度是最细的。ODS 层的表通常包括两类,一个用于存储当前需要加载的数据,一个用于存储处理完后的历史数据。历史数据一般保存3~6个月后需要清除，以节省空间。但对于不同的项目需要区别对待，如果数据源系统的数据量不大，可以保留更长的时间，甚至全量保存。

6. PDW 层（数据仓库层）

数据仓库层的数据应该是一致的、准确的、干净的数据，即对数据源系统数据进行了清洗（去除了杂质）后的数据。这一层的数据一般是遵循数据库第三范式的，其数据粒度通常和 ODS 的粒度相同。在 PDW 层会保存 BI 系统中所有的历史数据，例如保存10年的数据。

7. DM 层（数据集市层）

数据集市层的数据是面向主题来组织数据的，通常是星形或雪花形结构的数据。从数据粒度来说，这层的数据是轻度汇总级的数据，已经不存在明细数据了。从数据的时间跨度来说，通常只是 PDW 层数据的一部分，其主要的目的是为了满足用户分析的需求，因为从数据分析的角度来说，用户通常只需要分析近几年的（如近三年的）数据即可。而从数据的广度来说，仍然覆盖了所有业务数据。

8. APP 层（应用层）

应用层的数据是完全为了满足具体的分析需求而构建的数据，也是星形或雪花形结构的数据。从数据粒度来说是高度汇总的数据。从数据的广度来说，则并不一定会覆盖所有业务数据，而是 DM 层数据的一个真子集，从某种意义上来说是 DM 层数据的一个重复。从极端情况来说，可以为每一张报表在 APP 层构建一个模型来支持，达到以空间换时间的目的。

以上是数据仓库的标准分层，而这只是一个建议性质的标准，实际实施时需要根据实际情况确定数据仓库的分层，不同类型的数据也可能采取不同的分层方法。对数据仓库分层的目的主要是用空间换时间，通过大量的预处理来提升应用系统的用户体验（效率），因此数据仓库会存在大量冗余的数据。如果不进行分层的话，如果数据源所在的业务系统的业务规则发生变化，将会影响到整个数据清洗的过程，会产生巨大的工作量。另外，通过数据分层管理可以简化数据清洗的过程，因为把原来一步的工作分到了多个步骤去完成，相当于把一个复杂的工作拆成了多个简单的工作，把一个大的黑盒拆分成了多个小的白盒，每一层的处理逻辑都相对简单和容易理解，这样比较容易保证每一个步骤的正确性。当数据发生错误的时候，往往只需要局部调整某个步骤即可。

5.1.1.2 Hive

Hive 是基于 Hadoop 的一个数据仓库工具，是建立在 Hadoop 之上的数据仓库基础架构。它可以将结构化的数据文件映射成为一张数据库表，并提供简单的 SQL 查询功能，可以将 SQL 语句转换成为 MapReduce 任务运行。其优点是学习成本低，可以通过类 SQL 语句快速实现简单的 MapReduce 统计，不必开发专门的 MapReduce 应用，十分适合数据仓库的统计分析。

Hive 最早起源于 Facebook。Hadoop 作为一个开源的 MapReduce 实现，可以轻松地处理 Facebook 网站每天所产生的大量数据。但是 MapReduce 程序对于 Java 程序员来说比较容易写，而对于其他语言的使用者来说则不太方便。Facebook 最早开始研发 Hive 就是为了能够方便地在 Hadoop 上使用 SQL 进行数据的分析查询，这可以使得那些非 Java 程序员也可以方便地使用 Hadoop 来进行数据分析。而 Hive 最早的目的也就是为了分析处理海量的日志。

在某种程度上 Hive 可以看成是面向用户编程接口，其本身并不提供存储数据和处理数据的功能，而是依赖于 HDFS 存储数据，依赖于 MapReduce 处理数据。Hive 专门定义了一种类 SQL 查询语言 HiveQL，有类似于 SQL 的功能，但又不完全支持 SQL 标准，如不支持更新、索引、事务等功能，其子查询和连接操作也存在很多限制。Hive 使用 HiveQL 语句表述查询操作，并立刻将其自动转化成一个或多个 MapReduce 作业对海量数据进行处理，最后将结果反馈给用户。这免去了在使用 MapReduce 对存储在 HDFS 上的数据执行查询前编写 mappper 和 reducer 任务的过程，不再需要 Java 软件开发人员参与。当然，HiveQL 语言也允许熟悉 MapReduce 的开发人员开发自定义的 mapper 和 reducer 来处理内建的 mapper 和 reducer 无法完成的复杂查询和分析工作。另外 Hive 还提供了一系列对数据进行提取、转换、加载的工具，可以存储、查询、分析存储在 HDFS 上的数据。

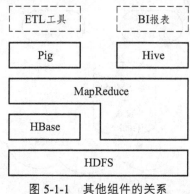

图 5-1-1　其他组件的关系

Hive 与 Hadoop 生态系统中的其他组件的关系如图 5-1-1 所示。

Hive 依赖于 HDFS 存储数据，依赖 MapReduce 处理数据。

Pig 可作为 Hive 的替代工具，是一种数据流语言和运行环境，适合用于在 Hadoop 平台上查询半结构化的数据集。也可以用于 ETL 过程的一部分，即将外部数据装载到 Hadoop 集群中，转换为用户需要的数据格式。

Hive 只能处理静态数据，主要是 BI 报表数据，其初衷是为减少复杂 MapReduce 应用程序的编写工作。而 HBase 是一个面向列的、分布式可伸缩的数据库，可提供数据的实时访问功能，是对 Hive 功能的一种补充。

5.1.1.3 Hive 的架构

Hive 的架构如图 5-1-2 所示。

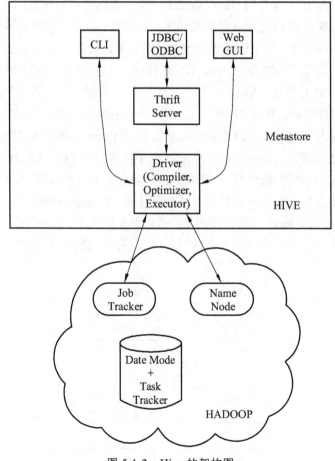

图 5-1-2 Hive 的架构图

1. 用户接口模块

用户接口模块包含了 CLI、HWI、JDBC、Thrift Server 等，用来实现对 Hive 的访问。CLI 是 Hive 自带的命令行界面；HWI 是 Hive 的一个简单网页界面；JDBC、ODBC 以及 Thrift Server 可向用户提供可编程接口，其中 Thrift Server 是基于 Thrift 软件框架开发的，提供 Hive 的 RPC 通信接口。

2. 驱动模块（Driver）

驱动模块包含了编译器、优化器、执行器等，它们负责把 HiveQL 语句转换成一系列 MapReduce 作业。所有命令和查询都会进入驱动模块，通过该模块的解析和变异，对计算过程进行优化，然后再按照指定的步骤执行。

3. 元数据存储模块（Metastore）

元数据模块是一个独立的关系型数据库，通常是与 MySQL 数据库进行连接后创建的一个 MySQL 实例，也可以是 Hive 自带的 Derby 数据库的一个实例。此模块主要是保存表的模式和其他系统元数据，如表的名称、表的列及其属性、表的分区及其属性、表的属性、表中数据所在位置信息等。

5.1.1.4 Hive 的工作原理

1. 用 MapReduce 实现连接操作

假设连接（join）的两个表分别是用户表 User（uid，name）和订单表 Order（uid，orderid），具体的 SQL 命令如下：

SELECT name，orderid FROM User u JOIN Order o ON u.uid=o.uid；

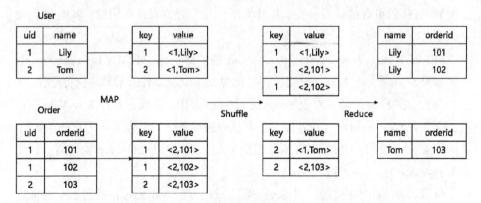

图 5-1-3　连接操作转换 MapReduce 操作

如图 5-1-3 所示，描述了连接操作转换为 MapReduce 操作任务的具体执行过程。

首先，在 Map 阶段，User 表以 uid 为 key，以 name 和表的标记位（这里 User 的标记位记为 1）为 value，进行 Map 操作，把表中记录转换生成一系列 Key-Value 对的形式。比如，User 表中记录（1，Lily）转换成为 Key-Value 对（1，<1，Lily>），其中第一个"1"是 uid 的值，第二个"1"是表 User 的标记位，用来标示这个键值对来自 User 表。同样，Order 表以 uid 为 key，以 orderid 和表的标记位（这里表 Order 的标记位记为 2）为 value 进行 Map 操作，把表中的记录转换生成一系列 Key-Value 对的形式。

接着，在 Shuffle 阶段，把 User 表和 Order 表生成的 Key-Value 对按 key 的值进行 Hash，然后传送给对应的 Reduce 机器执行。比如 Key-Value 对（1，<1，Lily>）、（1，<2，101>）、（1，<2，102>）传送到同一台 Reduce 机器上。当 Reduce 机器接收到这些 Key-Value 对时，还需要按表的标记位对这些 Key-Value 对进行排序，以优化连接操作。

最后，在 Reduce 阶段，对同一台 Reduce 机器上的 Key-Value 对，根据 value 中的表标记位，对来自表 User 和 Order 的数据进行笛卡尔积连接操作，以生成最终的结果。比如 Key-Value 对（1，<1，Lily>）与 Key-Value 对（1，<2，101>）、（1，<2，102>）的连接结果是（Lily，101）、（Lily，102）。

2. 用 MapReduce 实现分组操作

假设分数表 Score（rank，level）具有 rank（排名）和 level（级别）两个属性，需要进行一个分组（Group By）操作，功能是把表 Score 的不同片段按照 rank 和 level 的组合值进行合并，并计算不同的组合值有几条记录。SQL 语句命令如下：

SELECT rank, level, count(*) as value FROM score GROUP BY rank, level;

如图 5-1-4 所示，描述了分组操作转化为 MapReduce 任务的具体执行过程。

首先，在 Map 阶段，对表 Score 进行 Map 操作，生成一系列 Key-Value 对，其 key 为<rank，level>，value 为"拥有该<rank，level>组合值的记录的条数"。比如，Score 表的第一片段中有两条记录（A，1），所以进行 Map 操作后，转化为 Key-Value 对（<A，1>，2）。

接着，在 Shuffle 阶段，对 Score 表生成的 Key-Value 对，按照 key 的值进行 Hash，然后根据 Hash 结果传送给对应的 Reduce 机器去执行。比如，Key-Value

对（<A，1>，2）、（<A，1>，1）传送到同一台 Reduce 机器上，Key-Value 对（<B，2>，1）传送另一 Reduce 机器上。然后，Reduce 机器对接收到的这些 Key-Value 对，按 key 的值进行排序。

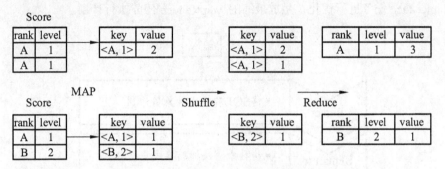

图 5-1-4　分组操作转化为 MapReduce 任务

最后，在 Reduce 阶段，把具有相同 key 的所有 Key-Value 对的 value 进行累加，生成分组的最终结果。比如，在同一台 Reduce 机器上的 Key-Value 对（<A，1>，2）和（<A，1>，1）在 Reduce 操作后的输出结果为（A，1，3）。

3. SQL 查询转换成 MapReduce 作业的过程

当 Hive 接收到一条 HQL 语句之后，需要与 Hadoop 交互工作来完成该操作。HQL 首先进入驱动模块，由驱动模块中的编译器解析编译，并由优化器对该操作进行优化计算，然后交给执行器去执行。执行器通常会启动一个或多个 MapReduce 任务，但有时也会有不启动 MapReduce 任务的情况（如执行全表扫描时）。

如图 5-1-5 所示，描述了 Hive 把 HQL 语句转化成 MapReduce 任务进行执行的详细过程。

首先由驱动模块中的编译器 Antlr 语言识别工具对用户输入的 SQL 语句进行词法和语法解析，将 HQL 语句转换成抽象语法树（AST Tree）的形式。因为 AST 结构复杂，不方便直接翻译成 MapReduce 算法程序，所以还会遍历抽象语法树，转化成 QueryBlock 查询单元。其中 QueryBlock 是一条最基本的 SQL 语法组成单元，包括输入源、计算过程、和输入三个部分。

接着遍历 QueryBlock，生成 OperatorTree（操作树），OperatorTree 由很多逻辑操作符组成，这些逻辑操作符可在 Map、Reduce 阶段完成某一特定操作。Hive 驱动模块中的逻辑优化器会对 OperatorTree 进行优化，变换 OperatorTree

的形式,合并多余的操作符,减少 MapReduce 任务数以及 Shuffle 阶段的数据量。

然后遍历优化之后的 OperatorTree,根据 OperatorTree 中的逻辑操作符生成需要执行的 MapReduce 任务。并启动 Hive 驱动模块中的物理优化器,对生成的 MapReduce 任务进行优化,生成最终的 MapReduce 任务执行计划。

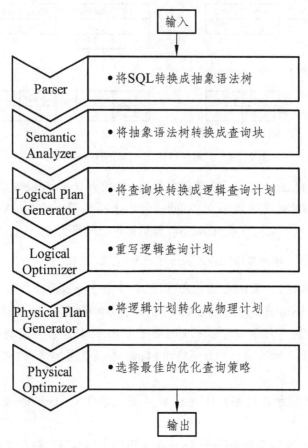

图 5-1-5 HQL 语句转化 MapReduce 任务过程

最后由 Hive 驱动模块中的执行器对最终的 MapReduce 任务执行输出。Hive 驱动模块中的执行器执行最终的 MapReduce 任务时,Hive 本身不会生成 MapReduce 算法程序。它通过一个表示"Job 执行计划"的 XML 文件来驱动内置的、原生的 Mapper 和 Reducer 模块。Hive 通过和 JobTracker 通信来初始化 MapReduce 任务,而不需要将 Hive 数据仓库工具直接部署在 JobTracker 所在管理节点上。

5.1.1.5 Hive 与传统数据库

表 5-1-1 Hive 与传统数据库比较

对比项	Hive	传统数据库
数据插入	支持批量的数据导入，但不支持单条数据的插入	支持批量导入和单挑数据的插入
数据更新	不支持	支持
索引	拥有有限的索引功能，并且不像 RDBMS 中那样有键的概念。可以在某些列上建立索引,以此来加速一些查询操作。创建的索引数据会被保存在另外的表中	支持
分区	支持，Hive 的表是以分区的形式进行组织的，并且根据"分区列"的值对表进行粗略划分，以加快数据的查询速度	支持，提供了分区功能来改善大型表以及具备各种访问模式的表的可伸缩性和可管理性，同时提高数据库的效率
执行延迟	高，由于构建在 HDFS 和 MapReduce 之上，比传统数据库延迟要高很多，HQL 语句的延迟可达分钟级	低，运行传统 SQL 语句的延迟一般少于 1 s
扩展性	好，可以随着 Hadoop 集群而扩展，有很好的横向扩展性	有限，RDBMS 一般为非分布式模式，横向扩展难以实现，而纵向扩展受限于硬件技术的发展和成本问题，其扩展性也很有限。

5.1.2 Hive 数据仓库的安装规划

1. 环境要求

本地磁盘剩余空间 600 M 以上，其中 Hive 数据仓库软件自身需要 200 M 空间，Hadoop 软件需要 400M 空间；

已安装 CentOS 7 1611 64 位操作系统；

已安装 JDK 的 1.8.0_131 版本；

已完成基础网络环境配置；

已完成 Hadoop 平台的安装和部署；

已完成 MySQL 数据库平台的安装和部署。

2. 软件版本

选用 Hive 的 2.1.1 版本，软件包名 apache-hive-2.1.1-bin.tar.gz，该软件包可以在 Hive 项目位于 Apache 的官方网站（http：//hive.apache.org）的 Downloads 页面（http：//hive.apache.org/downloads.html）获取，如图 5-1-6 所示。

图 5-1-6　获取 Hive 软件包

3. 依赖软件

Hive 数据仓库软件的使用需要连接外部的关系型数据库，这里使用的是 MySQL 数据库。而 Hive 数据仓库软件是用 Java 语言实现，所以使用的是 JDBC

方式来连接 MySQL 数据库，需要使用到对应的 MySQL 数据库的连接驱动软件包。选用该软件包的 5.1.42 版本，软件包名 mysql-connector-java-5.1.42-bin.jar，该软件包可以在 MySQL 的官方网站（https://www.mysql.com）的开发者社区 DEVELOPER ZONE 的 Downloads 页面（https://dev.mysql.com/downloads/connector/j）获取，如图 5-1-7 所示。

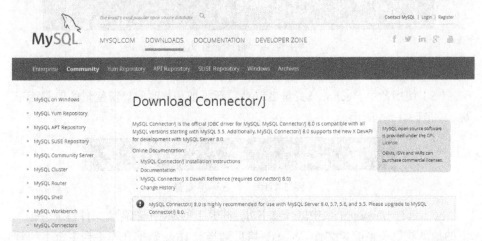

图 5-1-7　获取连接驱动软件包

5.1.3　Hive 数据仓库基础环境配置

（1）创建"hive"目录用于存放 Hive 相关文件，该目录可自行选择创建位置，创建完成后将当前工作目录切换到该目录。

（2）使用命令"tar -xzf Hive 安装包路径"将软件包解压解包到"hive"目录下，如图 5-1-8 所示，解压解包出来的目录名称为"apache-hive-2.1.1-bin"。

```
[hadoop@Hive hive]$ ls -l
total 0
drwxrwxr-x. 9 hadoop hadoop 171 Jul 11 11:59 apache-hive-2.1.1-bin
```

图 5-1-8　解压安装包

（3）在用户的配置文件".bash_profile"中配置 Hive 相关的环境变量，如图 5-1-9 所示，在文件末尾添加以下内容：

\# hive environment
HIVE_HOME=Hive 软件目录路径
PATH=$HIVE_HOME/bin：$PATH

export　HIVE_HOME　PATH

Hive 软件目录即 Hive 软件包解压解包出来的"apache-hive-2.1.1-bin"目录，这里需要书写该目录及其所在的绝对路径。

```
# hive environment
HIVE_HOME=/home/hadoop/hive/apache-hive-2.1.1-bin
PATH=$HIVE_HOME/bin:$PATH
export HIVE_HOME PATH
```

图 5-1-9　修改配置文件

（4）使用命令"source　~/.bash_profile"使新配置的环境变量立即生效。

（5）使用命令"echo　$变量名"查看新添加和修改的环境变量的值是否正确，如图 5-1-10 所示。

```
[hadoop@Hive ~]$ echo $HIVE_HOME
/home/hadoop/hive/apache-hive-2.1.1-bin
```

图 5-1-10　查看环境变量

（6）选择 Hadoop 平台中的任意主机，将其上的操作系统配置文件目录"etc"下的地址映射关系配置文件"hosts"拷贝到当前 Hive 数据仓库平台主机下的对应目录下，覆盖原有的文件。

若是直接在 Hadoop 平台中的某台主机上安装 Hive 数据仓库，可以省略掉该步骤以及之后的步骤（7）。

（7）创建"hadoop"目录用于存放 Hadoop 相关文件，该目录可自行选择创建位置。创建完成之后，选择 Hadoop 平台中的任意主机，将其上的 Hadoop 软件目录"hadoop-2.7.3"拷贝到该目录之中。然后如图 5-1-11 所示，在用户的配置文件".bash_profile"中配置 Hadoop 相关的环境变量"HADOOP_HOME"和"PATH"。

```
# hadoop environment
HADOOP_HOME=/home/hadoop/hadoop/hadoop-2.7.3
PATH=$HADOOP_HOME/bin:$HADOOP_HOME/sbin:$PATH
export HADOOP_HOME PATH
```

图 5-1-11　修改配置文件

（8）使用 Hadoop 客户端命令"hadoop　fs　-mkdir　hive"在 Hadoop 平台中创建 Hive 数据仓库的专用目录，接着再在该目录下分别创建 Hive 的临时文件的存放目录"tmp"、数据文件的存放目录"warehouse"、日志文件的存放目录"log"。然后使用 Hadoop 客户端命令"hadoop fs -chmod -R 777 hive"修改 Hive 数据仓库的专用目录及其所有子目录的权限为开放所有权限,如图 5-1-12 所示。

```
[hadoop@Hive ~]$ hadoop fs -ls hive
Found 3 items
drwxrwxrwx   - hadoop supergroup          0 2018-07-12 09:57 hive/log
drwxrwxrwx   - hadoop supergroup          0 2018-07-12 09:57 hive/tmp
drwxrwxrwx   - hadoop supergroup          0 2018-07-12 09:57 hive/warehouse
```

图 5-1-12　创建数据仓库专用目录

Hive 数据仓库的搭建虽然使用的是与 Hadoop 平台不同的主机,但依然使用的是名为"hadoop"的操作系统用户,所以使用上面的命令创建的 Hadoop 平台上的 Hive 专用目录在 HDFS 文件系统中的完整路径为"/user/hadoop/hive",这个完整路径会在后面的配置过程中使用。

(9)将 JDBC 连接 MySQL 数据库的驱动软件包"mysql-connector-java-5.1.42-bin.jar"拷贝到 Hive 数据仓库软件所在目录中的"lib"目录下,如图 5-1-13 所示。

```
[hadoop@Hive lib]$ ls -l mysql*
-rwxr-xr-x. 1 hadoop hadoop 996444 Jul 12 10:49 mysql-connector-java-5.1.42-bin.jar
```

图 5-1-13　拷贝驱动软件包

(10)登录到 Hive 数据仓库所使用的 MySQL 数据库,如图 5-1-14 所示,在其中创建一个名为"hive"的数据库用于存放 Hive 数据仓库的元数据,并设定该数据库的用户名和密码为"hive"。当然数据库名、用户名、密码都可以另行自定义,但需要注意的是自定义的数据库名、用户名、密码要与之后的 Hive 相关配置文件中的内容相对应。以下是需要在 MySQL 数据库中执行的相关 SQL 语句,依次输入并执行即可,括号中为对该 SQL 语句执行内容的说明:

CREATE DATABASE hive;(创建数据库"hive")

USE hive;(切换到新创建的"hive"数据库)

CREATE USER 'hive'@'%' IDENTIFIED BY 'hive';(创建数据库"hive"的用户"hive",同时设置该用户的登录密码为"hive"并且支持远程访问)

GRANT ALL ON hive.* TO 'hive'@'%';(设置"hive"数据库的"hive"用户的权限为拥有对该数据库的所有操作权限)

FLUSH PRIVILEGES;(刷新数据库的权限信息)

quit;(退出 MySQL 数据库的控制台界面

```
[hadoop@Hive ~]$ mysql -h 192.168.60.140 -u root -p
Enter password:
Welcome to the MySQL monitor.  Commands end with ; or \g.
Your MySQL connection id is 4
Server version: 5.7.18 MySQL Community Server (GPL)

Copyright (c) 2000, 2017, Oracle and/or its affiliates. All rights reserved.

Oracle is a registered trademark of Oracle Corporation and/or its
affiliates. Other names may be trademarks of their respective
owners.

Type 'help;' or '\h' for help. Type '\c' to clear the current input statement.

mysql> CREATE DATABASE hive;
Query OK, 1 row affected (0.00 sec)

mysql> USE hive;
Database changed
mysql> CREATE USER 'hive'@'%' IDENTIFIED BY 'hive';
Query OK, 0 rows affected (0.00 sec)

mysql> GRANT ALL ON hive.* TO 'hive'@'%';
Query OK, 0 rows affected (0.00 sec)

mysql> FLUSH PRIVILEGES;
Query OK, 0 rows affected (0.00 sec)

mysql> quit;
Bye
[hadoop@Hive ~]$
```

图 5-1-14 创建"hive"数据库

若是远程连接 MySQL 数据库服务器，则需要保证本机安装了 MySQL 数据库的客户端工具软件。Linux 操作系统中可以直接使用之前搭建 MySQL 数据库的单机模式平台的软件包 mysql-5.7.18-linux-glibc2.5-x86_64.tar.gz，将其解压解包之后，如图 5-1-15 所示，配置相应的环境变量，然后使用命令"mysql -h MySQL 数据库服务器的主机名或 IP 地址 -u 用户名 -p"便可以远程连接并登录到 MySQL 数据库。

```
# mysql environment
MYSQL_HOME=/home/hadoop/mysql/mysql-5.7.18-linux-glibc2.5-x86_64
PATH=$MYSQL_HOME/bin:$PATH
export MYSQL_HOME PATH
```

图 5-1-15 配置环境变量

5.1.4 Hive 数据仓库软件配置

（1）进入 Hive 的相关文件目录"hive"，如图 5-1-16 所示，创建 Hive 在本地文件系统中的临时文件的存放目录"tmp"。

```
[hadoop@Hive hive]$ ls -l
total 0
drwxrwxr-x. 9 hadoop hadoop 171 Jul 11 11:59 apache-hive-2.1.1-bin
drwxrwxr-x. 2 hadoop hadoop   6 Jul 12 12:21 tmp
```

图 5-1-16　创建目录"tmp"

（2）Hive 的配置文件位于其软件目录的"conf"目录下，如图 5-1-17 所示，进入该目录，对其中的配置文件进行编辑。

```
[hadoop@Hive conf]$ pwd
/home/hadoop/hive/apache-hive-2.1.1-bin/conf
```

图 5-1-17　进入"conf"目录

（3）依次使用下列配置文件的模板文件生成 Hive 的对应配置文件，如图 5-1-18 所示，生成方法为复制对应模板文件并去掉文件名中的".template"部分后缀。

hive-env.sh.template

hive-log4j2.properties.template

hive-exec-log4j2.properties.template

同时使用配置文件的模板文件"hive-default.xml.template"生成 Hive 的对应配置文件"hive-site.xml"。

```
[hadoop@Hive conf]$ ls -l
total 496
-rw-r--r--. 1 hadoop hadoop   1596 Nov 28  2016 beeline-log4j2.properties.template
-rw-r--r--. 1 hadoop hadoop 229198 Nov 29  2016 hive-default.xml.template
-rw-r--r--. 1 hadoop hadoop   2378 Jul 12 07:23 hive-env.sh
-rw-r--r--. 1 hadoop hadoop   2378 Nov 28  2016 hive-env.sh.template
-rw-r--r--. 1 hadoop hadoop   2274 Jul 12 07:24 hive-exec-log4j2.properties
-rw-r--r--. 1 hadoop hadoop   2274 Nov 28  2016 hive-exec-log4j2.properties.template
-rw-r--r--. 1 hadoop hadoop   2925 Jul 12 07:24 hive-log4j2.properties
-rw-r--r--. 1 hadoop hadoop   2925 Nov 28  2016 hive-log4j2.properties.template
-rw-r--r--. 1 hadoop hadoop 229198 Jul 12 07:25 hive-site.xml
-rw-r--r--. 1 hadoop hadoop   2060 Nov 28  2016 ivysettings.xml
-rw-r--r--. 1 hadoop hadoop   2719 Nov 28  2016 llap-cli-log4j2.properties.template
-rw-r--r--. 1 hadoop hadoop   4353 Nov 28  2016 llap-daemon-log4j2.properties.template
-rw-r--r--. 1 hadoop hadoop   2662 Nov 28  2016 parquet-logging.properties
```

图 5-1-18　模板文件

（4）编辑配置文件"hive-env.sh"，如图 5-1-19 所示，找到下列配置项的所在行并修改其对应内容：

HADOOP_HOME=Hadoop 软件目录路径

export　HIVE_CONF_DIR=Hive 软件目录路径/conf

export　HIVE_AUX_JARS_PATH=Hive 软件目录路径/lib

```
47 # Set HADOOP_HOME to point to a specific hadoop install directory
48 HADOOP_HOME=/home/hadoop/hadoop/hadoop-2.7.3

50 # Hive Configuration Directory can be controlled by:
51 export HIVE_CONF_DIR=/home/hadoop/hive/apache-hive-2.1.1-bin/conf

53 # Folder containing extra ibraries required for hive compilation/execution can be controlled
   by:
54 export HIVE_AUX_JARS_PATH=/home/hadoop/hive/apache-hive-2.1.1-bin/lib
```

图 5-1-19　编辑配置文件"hive-env.sh"

配置项说明：

HADOOP_HOME——指定本地文件系统中 Hadoop 软件的所在路径。

HIVE_CONF_DIR——指定本地文件系统中 Hive 软件的配置文件目录的所在路径。

HIVE_AUX_JARS_PATH——指定本地文件系统 Hive 软件的 lib 文件目录的所在路径。

（5）编辑配置文件"hive-site.xml"，配置 Hive 数据仓库所使用到在 HDFS 中的相关目录信息，如图 5-1-20 所示，在文件中找到下列所有标签"<name>"和"</name>"之间所标识的属性项名称的所在位置，然后修改其对应标签"<value>"和"</value>"之间所标识的属性对应值的内容：

<configuration>
　<property>
　　<name>hive.exec.scratchdir</name>
　　<value>/user/hadoop/hive/tmp</value>
　</property>
　<property>
　　<name>hive.metastore.warehouse.dir</name>
　　<value>/user/hadoop/hive/warehouse</value>
　</property>
　<property>
　　<name>hive.querylog.location</name>
　　<value>/user/hadoop/hive/log</value>
　</property>
</configuration>

配置项说明：

hive.exec.scratchdir——指定 Hive 数据仓库的临时文件的存放目录路径，需要是位于 Hadoop 平台中的目录路径。

hive.metastore.warehouse.dir——指定 Hive 的数据文件的存放目录路径，需要是位于 Hadoop 平台中的目录路径。

hive.querylog.location——指定 Hive 的日志文件的存放目录路径，需要是位于 Hadoop 平台中的目录路径。

```
37  <property>
38    <name>hive.exec.scratchdir</name>
39    <value>/user/hadoop/hive/tmp</value>
40    <description>HDFS root scratch dir for Hive jobs which gets created with write all (733)
   permission. For each connecting user, an HDFS scratch dir: ${hive.exec.scratchdir}/&lt;user
   name&gt; is created, with ${hive.scratch.dir.permission}.</description>

325 <property>
326   <name>hive.metastore.warehouse.dir</name>
327   <value>/user/hadoop/hive/warehouse</value>
328   <description>location of default database for the warehouse</description>
329 </property>

1514 <property>
1515   <name>hive.querylog.location</name>
1516   <value>/user/hadoop/hive/log</value>
1517   <description>Location of Hive run time structured log file</description>
1518 </property>
```

图 5-1-20　编辑配置文件 "hive-site.xml"

该配置文件中配置项较多，可以在 vi 编辑器内使用命令 "/关键字" 直接进行搜索，同时使用快捷键 "n" 可以切换到下一个关键字的所在位置。

（6）继续编辑配置文件 "hive-site.xml"，配置 MySQL 数据库连接的相关信息，如图 5-1-21 所示，在文件中找到下列所有标签 "<name>" 和 "</name>" 之间所标识的属性项名称的所在位置，然后修改其对应标签 "<value>" 和 "</value>" 之间所标识的属性对应值的内容：

<configuration>

　<property>

　　<name>javax.jdo.option.ConnectionURL</name>

　　<value>jdbc：mysql：//MySQL 数据库服务器的主机名或 IP 地址：3306/hive? createDatabaseIfNotExist=true& characterEncoding=UTF-8& useSSL= false</value>

　</property>

　<property>

　　<name>javax.jdo.option.ConnectionDriverName</name>

　　<value>com.mysql.jdbc.Driver</value>

```xml
    </property>
    <property>
        <name>javax.jdo.option.ConnectionUserName</name>
        <value>hive</value>
    </property>
    <property>
        <name>javax.jdo.option.ConnectionPassword</name>
        <value>hive</value>
    </property>
</configuration>
```

★ 配置项说明：

javax.jdo.option.ConnectionURL – 指定 Hive 数据仓库所使用的 MySQL 数据库的连接路径，这里的连接路径为 JDBC 连接方式的连接路径，包括了 MySQL 数据库服务器的地址以及数据库的名称。同时连接路径后面还可以带上一些连接相关的配置参数，其中参数 "createDatabaseIfNotExist" 表示若指定名称的数据库不存在则自动进行创建，参数 "characterEncoding" 表示所使用的字符编码格式，参数 "useSSL" 表示是否对数据包进行 SSL 方式的加密。

javax.jdo.option.ConnectionDriverName – 指定 Hive 数据仓库使用 JDBC 方式连接 MySQL 数据库所使用的驱动类的名称及路径。

javax.jdo.option.ConnectionUserName – 指定 Hive 数据仓库登录 MySQL 数据库所使用的用户。

javax.jdo.option.ConnectionPassword – 指定 Hive 数据仓库登录 MySQL 数据库所使用的用户对应的登录密码。

```
499     <property>
500         <name>javax.jdo.option.ConnectionURL</name>
501         <value>jdbc:mysql://192.168.60.140:3306/hive?createDatabaseIfNotExist=true&amp;characterEncoding=UTF-8&amp;useSSL=false</value>
502         <description>
503             JDBC connect string for a JDBC metastore.
504             To use SSL to encrypt/authenticate the connection, provide database-specific SSL flag in the connection URL.
505             For example, jdbc:postgresql://myhost/db?ssl=true for postgres database.
506         </description>
507     </property>
```

```
931     <property>
932         <name>javax.jdo.option.ConnectionDriverName</name>
933         <value>com.mysql.jdbc.Driver</value>
934         <description>Driver class name for a JDBC metastore</description>
935     </property>
```

```
956     <property>
957       <name>javax.jdo.option.ConnectionUserName</name>
958       <value>hive</value>
959       <description>Username to use against metastore database</description>
960     </property>

484     <property>
485       <name>javax.jdo.option.ConnectionPassword</name>
486       <value>hive</value>
487       <description>password to use against metastore database</description>
488     </property>
```

图 5-1-21　编辑配置文件 "hive-site.xml"

（7）继续编辑配置文件 "hive-site.xml"，在 vi 编辑器中使用快捷键 ":" 进入到编辑器的命令模式，也称为末行模式，然后依次使用下列命令替换掉配置文件中的原有内容：

%s#${system：java.io.tmpdir}#/home/hadoop/hive/tmp#g

%s#${system：user.name}#${user.name}#g

5.1.5　Hive 数据仓库初始化和启动和验证

（1）使用命令 "schematool -initSchema -dbType mysql" 对 Hive 进行初始化，如图 5-1-22 所示。

```
[hadoop@Hive ~]$ schematool -initSchema -dbType mysql
which: no hbase in (/home/hadoop/mysql/mysql-5.7.18-linux-glibc2.5-x86_64/bin:/home/hadoop/hadoop/ha
doop-2.7.3/bin:/home/hadoop/hadoop/hadoop-2.7.3/sbin:/home/hadoop/hive/apache-hive-2.1.1-bin/bin:/ho
me/hadoop/java/jdk1.8.0_131/bin:/home/hadoop/hadoop/hadoop-2.7.3/bin:/home/hadoop/hadoop/hadoop-2.7.
3/sbin:/home/hadoop/hive/apache-hive-2.1.1-bin/bin:/home/hadoop/java/jdk1.8.0_131/bin:/home/hadoop/h
ive/apache-hive-2.1.1-bin/bin:/usr/local/bin:/bin:/usr/bin:/usr/local/sbin:/usr/sbin:/home/hadoop/.l
ocal/bin:/home/hadoop/bin:/home/hadoop/.local/bin:/home/hadoop/bin:/home/hadoop/.local/bin:/home/had
oop/bin:/home/hadoop/.local/bin:/home/hadoop/bin)
SLF4J: Class path contains multiple SLF4J bindings.
SLF4J: Found binding in [jar:file:/home/hadoop/hive/apache-hive-2.1.1-bin/lib/log4j-slf4j-impl-2.4.1
.jar!/org/slf4j/impl/StaticLoggerBinder.class]
SLF4J: Found binding in [jar:file:/home/hadoop/hadoop/hadoop-2.7.3/share/hadoop/common/lib/slf4j-log
4j12-1.7.10.jar!/org/slf4j/impl/StaticLoggerBinder.class]
SLF4J: See http://www.slf4j.org/codes.html#multiple_bindings for an explanation.
SLF4J: Actual binding is of type [org.apache.logging.slf4j.Log4jLoggerFactory]
Metastore connection URL:        jdbc:mysql://192.168.60.140:3306/hive?createDatabaseIfNotExist=true
&characterEncoding=UTF-8&useSSL=false
Metastore Connection Driver :    com.mysql.jdbc.Driver
Metastore connection User:       hive
Starting metastore schema initialization to 2.1.0
Initialization script hive-schema-2.1.0.mysql.sql
Initialization script completed
schemaTool completed
```

图 5-1-22　对 Hive 初始化

（2）使用命令 "hive" 启动 Hive 数据仓库，启动成功后会进入 Hive 的控制台界面，如图 5-1-23 所示。

```
[hadoop@Hive ~]$ hive
which: no hbase in (/home/hadoop/mysql/mysql-5.7.18-linux-glibc2.5-x86_64/bin:/home/hadoop/hadoop/ha
doop-2.7.3/bin:/home/hadoop/hadoop/hadoop-2.7.3/sbin:/home/hadoop/hive/apache-hive-2.1.1-bin/bin:/ho
me/hadoop/java/jdk1.8.0_131/bin:/home/hadoop/hadoop/hadoop-2.7.3/bin:/home/hadoop/hadoop/hadoop-2.7.
3/sbin:/home/hadoop/hive/apache-hive-2.1.1-bin/bin:/home/hadoop/java/jdk1.8.0_131/bin:/home/hadoop/h
ive/apache-hive-2.1.1-bin/bin:/usr/local/bin:/bin:/usr/bin:/usr/local/sbin:/usr/sbin:/home/hadoop/.l
ocal/bin:/home/hadoop/bin:/home/hadoop/.local/bin:/home/hadoop/bin:/home/hadoop/.local/bin:/home/had
oop/bin:/home/hadoop/.local/bin:/home/hadoop/bin)
SLF4J: Class path contains multiple SLF4J bindings.
SLF4J: Found binding in [jar:file:/home/hadoop/hive/apache-hive-2.1.1-bin/lib/log4j-slf4j-impl-2.4.1
.jar!/org/slf4j/impl/StaticLoggerBinder.class]
SLF4J: Found binding in [jar:file:/home/hadoop/hadoop/hadoop-2.7.3/share/hadoop/common/lib/slf4j-log
4j12-1.7.10.jar!/org/slf4j/impl/StaticLoggerBinder.class]
SLF4J: See http://www.slf4j.org/codes.html#multiple_bindings for an explanation.
SLF4J: Actual binding is of type [org.apache.logging.slf4j.Log4jLoggerFactory]

Logging initialized using configuration in file:/home/hadoop/hive/apache-hive-2.1.1-bin/conf/hive-lo
g4j2.properties Async: true
Hive-on-MR is deprecated in Hive 2 and may not be available in the future versions. Consider using a
 different execution engine (i.e. spark, tez) or using Hive 1.X releases.
hive>
```

图 5-1-23　启动数据仓库

（3）在 Hive 的控制台界面中使用命令"CREATE DATABASE mydb;"新建一个测试用的数据库，如图 5-1-24 所示。

```
hive> create database mydb;
OK
Time taken: 0.46 seconds
```

图 5-1-24　新建测试用数据库 Mydb

（4）在 Hive 的控制台界面中使用命令"SHOW DATABASES;"可以查看当前数据仓库中的数据库列表，如图 5-1-25 所示。

```
hive> show databases;
OK
default
mydb
Time taken: 0.046 seconds, Fetched: 2 row(s)
```

图 5-1-25　查看数据库列表信息

（5）在 Hive 的控制台界面中使用命令"quit;"或"exit;"可以退出 Hive 的控制台返回到操作系统的命令行界面，如图 5-1-26 所示。

```
hive> quit;
[hadoop@Hive ~]$

hive> exit;
[hadoop@Hive ~]$
```

图 5-1-26　退出 Hive 控制台

任务 5.2　Hive 数据仓库的使用和管理

Hive 数据仓库中也是以类似于关系型数据库中的表的方式来存储数据，其中的很多命令和操作与 MySQL 数据库中的命令和操作相似设置相同，而且还可以直接使用 SQL 语句来对数据仓库中的数据进行查询。

在使用 Hive 之前，需要先了解 Hive 中的分区表的概念。在 Hive 中进行查询一般会扫描整个表的内容，若此时只需要查询其中的一部分数据，便会消耗掉很多时间做一些多余的工作。因此 Hive 在建立表时引入了分区（Partition）概念，是指在创建表时指定按照数据中的某个列来进行分区，而每个分区会在 HDFS 上创建一个对应的目录。比如按照月份进行分区，那么该表在 HDFS 上面便会按照××××年××月来创建目录，并将属于对应月份的数据存放在该目录中。分区虽然依然是以字段的形式在表结构中存在，但该字段不再存放实际的数据内容。

将数据组织成分区的模式主要是为了提高数据的查询速度。而用户所存储的每一条记录到底放到哪个分区是由用户决定的，即用户在加载数据的时候必须显示的指定该部分数据将要存放在哪个分区之中。

同时 Hive 中的分区还支持多级分区，即指定数据中的多个列来进行分区，Hive 会根据多个列的书写顺序来创建多级的分区目录。

这里使用和 HBase 的使用演示中一样的简单示例数据库来展示 Hive 的使用，该数据库只包含部门信息表（departments）和员工信息表（employees）这两个表。下面的表 5-2-1 和表 5-2-2 展示了这两个表的表结构，表 5-2-3 和表 5-2-4 展示了需要在这两个表中分别添加的数据。

表 5-2-1　部门信息表结构

部门信息表（departments）					
列名	数据类型	长度	允许空	主外键	说明
dept_no	CHAR	4	×	主	
dept_name	VARCHAR	40	×		

表 5-2-2　员工信息表结构

员工信息表（employees）					
列名	数据类型	长度	允许空	主外键	说明
emp_no	INT		×	主	

续表

员工信息表（employees）					
列名	数据类型	长度	允许空	主外键	说明
dept_no	CHAR	4	×	外	
birth_date	DATE		×		
first_name	VARCHAR	14	×		
last_name	VARCHAR	16	×		
gender	ENUM		×		枚举值：M、F
hire_date	DATE		×		

表 5-2-3　部门信息表数据

部门信息表数据	
dept_no	dept_name
D001	Marketing
D002	Finance
D003	Production
D004	Sales
D005	Customer Service

表 5-2-4　员工信息表数据

员工信息表数据						
emp_no	dept_no	birth_date	first_name	last_name	gender	hire_date
10001	D001	1953-09-02	Georgi	Facello	M	1986-06-26
10002	D001	1964-06-02	Bezalel	Simmel	F	1985-11-21
10003	D001	1959-12-03	Parto	Bamford	M	1986-08-28
10004	D001	1954-05-01	Chirstian	Koblick	M	1986-12-01
10005	D001	1955-01-21	Kyoichi	Maliniak	M	1989-09-12
10006	D002	1953-04-20	Anneke	Preusig	F	1989-06-02
10007	D002	1957-05-23	Tzvetan	Zielinski	F	1989-02-10
10008	D002	1958-02-19	Saniya	Kalloufi	M	1994-09-15
10009	D002	1952-04-19	Sumant	Peac	F	1985-02-18
10010	D002	1963-06-01	Duangkaew	Piveteau	F	1989-08-24

5.2.1　创建数据库和表

（1）进入 Hive 的控制台界面，本小节之后的所有步骤以及后面小节的所有步骤，若没有特殊说明，默认都是在 Hive 控制台界面之中进行操作。

（2）在 Hive 数据仓库中创建一个名为“testdb”的数据库，如图 5-2-1 所示，具体命令如下：

CREATE　DATABASE　testdb；

```
hive> CREATE DATABASE testdb;
OK
Time taken: 1.469 seconds
```

图 5-2-1　创建数据库

（3）查看当前 Hive 数据仓库中的数据库列表，检查“testdb”数据库是否创建成功，如图 5-2-2 所示，具体命令如下：

SHOW　DATABASES；

```
hive> SHOW DATABASES;
OK
default
mydb
testdb
Time taken: 0.29 seconds, Fetched: 3 row(s)
```

图 5-2-2　查看数据仓库

（4）将当前使用的数据库切换到新创建的“testdb”数据库，如图 5-2-3 所示，具体命令如下：

USE　testdb；

```
hive> USE testdb;
OK
Time taken: 1.204 seconds
```

图 5-2-3　切换数据库

（5）在该数据库中创建部门信息表（departments），如图 5-2-4 所示，具体命令如下：

CREATE　TABLE　departments　(

dept_no　CHAR（4），

dept_name VARCHAR（40）
)
ROW FORMAT DELIMITED FIELDS TERMINATED BY '\t'
STORED AS TEXTFILE；

Hive 中表的创建与传统的关系型数据库中表的创建有一些区别，即不需要指定主键和外键，因为 Hive 本身并不支持传统关系型数据库中的主键和外键。Hive 通过将 HDFS 上的文件进行表结构的映射，所以它需要通过特定的分隔符来区分数据与表中的列的对应关系，"ROW FORMAT DELIMITED FIELDS TERMINATED BY" 就是用于指定这个特定的分隔符，也就是 Hive 中的序列化和反序列化的规则，这里的 "\t" 代表的是使用 Tab 键作为分隔符。"STORED AS" 用于指定 Hive 的数据文件在 HDFS 存放时所使用的文件格式，这里的 "TEXTFILE" 就是最常用的文本文件格式。

```
hive> CREATE TABLE departments (
    > dept_no CHAR(4),
    > dept_name VARCHAR(40)
    > )
    > ROW FORMAT DELIMITED FIELDS TERMINATED BY '\t'
    > STORED AS TEXTFILE;
OK
Time taken: 1.0 seconds
```

图 5-2-4 创建部门信息表

（6）查看新建的部门信息表（departments）的表结构信息，如图 5-2-5 所示，具体命令如下：

DESCRIBE departments；

```
hive> DESCRIBE departments;
OK
dept_no                 char(4)
dept_name               varchar(40)
Time taken: 0.307 seconds, Fetched: 2 row(s)
```

图 5-2-5 查看部门信息表结构

（7）查看新建的部门信息表（departments）的完整信息，包括各种配置属性的数据，如图 5-2-6 所示，具体命令如下：

DESCRIBE FORMATTED departments；

```
hive> DESCRIBE FORMATTED departments;
OK
# col_name              data_type               comment

dept_no                 char(4)
dept_name               varchar(40)

# Detailed Table Information
Database:               testdb
Owner:                  hadoop
CreateTime:             Wed Aug 22 07:04:38 EDT 2018
LastAccessTime:         UNKNOWN
Retention:              0
Location:               hdfs://hadoop-ha/user/hadoop/hive/warehouse/testdb.db/departments
Table Type:             MANAGED_TABLE
Table Parameters:
        numFiles                1
        numRows                 0
        rawDataSize             0
        totalSize               82
        transient_lastDdlTime   1534940840

# Storage Information
SerDe Library:          org.apache.hadoop.hive.serde2.lazy.LazySimpleSerDe
InputFormat:            org.apache.hadoop.mapred.TextInputFormat
OutputFormat:           org.apache.hadoop.hive.ql.io.HiveIgnoreKeyTextOutputFormat
Compressed:             No
Num Buckets:            -1
Bucket Columns:         []
Sort Columns:           []
Storage Desc Params:
        field.delim             \t
        serialization.format    \t
Time taken: 0.241 seconds, Fetched: 31 row(s)
```

图 5-2-6　查看部门信息表完整信息

（8）在该数据库中使用分区表的方式创建员工信息表（employees），以员工所属的部门编号（dept_no）作为分区列，如图 5-2-7 所示，具体命令如下：

CREATE TABLE employees （
emp_no   INT，
birth_date   DATE，
first_name   VARCHAR（14），
last_name   VARCHAR（16），
gender   CHAR（1），
hire_date   DATE
）
PARTITIONED BY　（dept_no   CHAR（4））
ROW FORMAT DELIMITED FIELDS TERMINATED BY '\t'
STORED AS TEXTFILE；

在创建分区表时，用作分区的列将不再写入表的列属性列表中。如果使用多个列来创建多级分区，则在"PARTITION BY"后面的括号中依次书写用作分区的列的列属性列表，书写方式与表的列属性列表的书写方式相同。但需要注意的是一定要按照层级的顺序书写，Hive 会严格按照书写的顺序来逐级建立对应的目录。另外 Hive 中并没有 ENUM 类型的数据，所以"gender"属性的数据类型需要替换成字符或者字符串类型。

```
hive> CREATE TABLE employees (
    > emp_no INT,
    > birth_date DATE,
    > first_name VARCHAR(14),
    > last_name VARCHAR(16),
    > gender CHAR(1),
    > hire_date DATE
    > )
    > PARTITIONED BY (dept_no CHAR(4))
    > ROW FORMAT DELIMITED FIELDS TERMINATED BY '\t'
    > STORED AS TEXTFILE;
OK
Time taken: 0.218 seconds
```

图 5-2-7　创建员工信息表

（9）查看当前数据库中的所有表信息，如图 5-2-8 所示，具体命令如下：
SHOW　TABLES;

```
hive> SHOW TABLES;
OK
departments
employees
Time taken: 0.083 seconds, Fetched: 2 row(s)
```

图 5-2-8　查看所有表信息

5.2.2　向数据库导入数据

由于 Hive 数据仓库建立在 HDFS 文件系统之上，而 HDFS 文件系统并不支持对数据的修改操作。所以 Hive 也不支持 INSERT、UPDATE、DELETE 等传统数据库的修改操作，这样做不仅可以避免在 HDFS 文件系统上进行数据的修改，也可以不需要复杂的锁机制来进行数据的读写。Hive 最常用的添加数据手段是从文本文件来导入数据，文本文件中的一行对应数据库的表中的一行数据，一行数据中的每个属性列之间使用特定的符号进行分隔，这个分隔符可以在创建表时进行指定。

（1）在操作系统的控制台界面创建一个空文本文件"departments.txt"，将部门信息表（departments）的数据输入到该文本文件当中，列与列之间用分隔符"\t"（Tab 键）隔开，如图 5-2-9 所示。

```
D001    Marketing
D002    Finance
D003    Production
D004    Sales
D005    Customer Service
```

图 5-2-9　写部门信息表数据到文档

（2）在操作系统的控制台界面中使用 Hadoop 的 FS 命令在 HDFS 文件系统中的操作系统用户对应目录下创建一个"hive-data"目录，如图 5-2-10 所示。然后将之前创建的数据文件"departments.txt"上传到 HDFS 文件系统的该目录下，如图 5-2-11 所示。

```
[hadoop@Hive hive]$ hadoop fs -mkdir hive-data
[hadoop@Hive hive]$ hadoop fs -ls
Found 5 items
drwxr-xr-x   - hadoop supergroup          0 2018-08-14 12:13 data
drwxr-xr-x   - hadoop supergroup          0 2018-08-20 11:35 hbase
drwxrwxrwx   - hadoop supergroup          0 2018-07-12 09:57 hive
drwxr-xr-x   - hadoop supergroup          0 2018-08-22 07:33 hive-data
drwxr-xr-x   - hadoop supergroup          0 2018-08-17 11:57 sqoop-data
```

图 5-2-10　创建目录

```
[hadoop@Hive hive]$ hadoop fs -put departments.txt hive-data
[hadoop@Hive hive]$ hadoop fs -ls hive-data
Found 1 items
-rw-r--r--   3 hadoop supergroup         84 2018-08-22 07:34 hive-data/departments.txt
```

图 5-2-11　上传文件到目录

（3）从 HDFS 文件系统中将数据文件"departments.txt"中的数据导入到 Hive 数据仓库的部门信息表（departments）中，如图 5-2-12 所示，具体命令如下：

LOAD DATA INPATH '/user/hadoop/hive-data/departments.txt' OVERWRITE INTO TABLE departments；

"OVERWRITE"表示删除表中原有的所有数据，然后再进行加载，否则就是在原有数据的基础上进行添加。

在退出过 Hive 的控制台界面之后再次进入 Hive 的控制台界面时，需要再次使用命令"USE　testdb；"选择使用"testdb"数据库之后，才能对该数据库及其中的表进行操作，后面也是一样。

```
hive> LOAD DATA INPATH '/user/hadoop/hive-data/departments.txt' OVERWRITE INTO TABLE departments;
Loading data to table testdb.departments
OK
Time taken: 1.15 seconds
```

图 5-2-12　导入文件数据列表中

（4）在操作系统的控制台界面中使用 Hadoop 的 FS 命令在 HDFS 文件系统中查看之前导入到 Hive 数据仓库中的部门信息表（departments）的数据。之前在 Hive 数据仓库平台时，指定了 Hive 中数据库的数据在 HDFS 文件系统中的保存位置为 HDFS 文件系统中的 Hive 专用目录下的"warehouse"目录，如图 5-2-13 所示。在该目录中，每一个数据库都会有一个与数据库名对应的目录来保存该数据库的数据，如图 5-2-14 所示，其中"mydb.db"目录下是数据库"mydb"的数据，"testdb.db"目录下是数据库"testdb"的数据。然后在数据库的目录下，每个表也会有一个与其表名相同的目录来保存该表的数据，如图 5-2-15 所示。而如果该表是分区表，则在该表的目录下还会有各个分区的目录来保存该分区的数据，但部门信息表（departments）不是分区表，所以该表的目录下直接保存的就是该表的数据文件，如图 5-2-16 所示。数据文件的具体内容如图 5-2-17 所示，可以看到和用来导入数据的数据文件的名称和内容是一样的。

```
[hadoop@Hive hive]$ hadoop fs -ls hive
Found 3 items
drwxrwxrwx   - hadoop supergroup          0 2018-07-12 09:57 hive/log
drwxrwxrwx   - hadoop supergroup          0 2018-07-13 05:15 hive/tmp
drwxrwxrwx   - hadoop supergroup          0 2018-08-22 06:57 hive/warehouse
```

图 5-2-13　数据保存位置目录

```
[hadoop@Hive hive]$ hadoop fs -ls hive/warehouse
Found 2 items
drwxrwxrwx   - hadoop supergroup          0 2018-07-13 05:31 hive/warehouse/mydb.db
drwxrwxrwx   - hadoop supergroup          0 2018-08-22 07:13 hive/warehouse/testdb.db
```

图 5-2-14　每个数据库对应保存数据目录

```
[hadoop@Hive hive]$ hadoop fs -ls hive/warehouse/testdb.db
Found 2 items
drwxrwxrwx   - hadoop supergroup          0 2018-08-22 08:27 hive/warehouse/testdb.db/departments
drwxrwxrwx   - hadoop supergroup          0 2018-08-22 09:35 hive/warehouse/testdb.db/employees
```

图 5-2-15　每个表对应保存数据目录

```
[hadoop@Hive hive]$ hadoop fs -ls hive/warehouse/testdb.db/departments
Found 1 items
-rwxrwxrwx   3 hadoop supergroup         82 2018-08-22 08:27 hive/warehouse/testdb.db/departments/de
partments.txt
```

图 5-2-16　部门信息表数据文件

```
[hadoop@Hive hive]$ hadoop fs -cat hive/warehouse/testdb.db/departments/departments.txt
D001    Marketing
D002    Finance
D003    Production
D004    Sales
D005    Customer Service
```

图 5-2-17　部门信息表数据文件内容

(5)在操作系统的控制台界面中创建员工信息表(employees)的数据文件。这里需要注意的是，由于该表在 Hive 数据仓库中是一个以部门编号(dept_no)列作为分区列的分区表，作为分区列的部门编号(dept_no)虽然仍然存在于表结构中，但却已经不存在于表的数据部分的结构中，所以该表的数据文件不需要包含该列的内容。而表的数据中不再有部门编号(dept_no)之后，要区分属于不同部门的员工就必须通过将不同部门的员工信息存放在不同的分区之中的方式。所以在创建员工信息表(employees)的数据文件的时候，需要对应属于不同部门的员工信息创建不同的数据文件。当前给出的员工信息中员工所属的部门编号(dept_no)总共有"D001"和"D002"两个，对应这两个部门编号分别创建"employees-D001.txt"和"employees-D002.txt"两个数据文件，如图 5-2-18 所示。然后将员工信息表(employees)中属于部门编号"D001"的员工信息的数据输入到"employees-D001.txt"文件当中，将属于部门编号"D002"的员工信息的数据输入到"employees-D002.txt"文件当中，输入的数据中不包含部门编号(dept_no)信息，每行数据的列与列之间用分隔符"\t"(Tab 键)隔开，如图 5-2-19 和图 5-2-20 所示。

```
[hadoop@Hive hive]$ ls -l
total 12
drwxrwxr-x. 9 hadoop hadoop 171 Jul 11 11:59 apache-hive-2.1.1-bin
-rwxr-xr-x. 1 hadoop hadoop  84 Aug 22 07:29 departments.txt
-rwxr-xr-x. 1 hadoop hadoop 236 Aug 22 07:59 employees-D001.txt
-rwxr-xr-x. 1 hadoop hadoop 469 Aug 22 07:58 employees-D002.txt
drwxrwxr-x. 3 hadoop hadoop  20 Aug 22 07:57 tmp
```

图 5-2-18　创建两个部门数据文件

```
10001   1953-09-02   Georgi    Facello   M   1986-06-26
10002   1964-06-02   Bezalel   Simmel    F   1985-11-21
10003   1959-12-03   Parto     Bamford   M   1986-08-28
10004   1954-05-01   Christian Koblick   M   1986-12-01
10005   1955-01-21   Kyoichi   Maliniak  M   1989-09-12
```

图 5-2-19　输入部门编号"D001"员工信息数据

```
10006   1953-04-20   Anneke    Preusig    F   1989-06-02
10007   1957-05-23   Tzvetan   Zielinski  F   1989-02-10
10008   1958-02-19   Saniya    Kalloufi   M   1994-09-15
10009   1952-04-19   Sumant    Peac       F   1985-02-18
10010   1963-06-01   Duangkaew Piveteau   F   1989-08-24
```

图 5-2-20　输入部门编号"D002"员工信息数据

（6）向分区表中员工信息表（employees）中添加新的分区，当前给出的员工信息中员工所属的部门编号（dept_no）总共有"D001"和"D002"两个，所以需要在员工信息表（employees）中添加"D001"和"D002"两个分区，如图 5-2-21 所示，具体命令如下：

ALTER TABLE employees ADD PARTITION （dept_no='D001'）；
ALTER TABLE employees ADD PARTITION （dept_no='D002'）；

```
hive> ALTER TABLE employees ADD PARTITION (dept_no='D001');
OK
Time taken: 0.719 seconds
hive> ALTER TABLE employees ADD PARTITION (dept_no='D002');
OK
Time taken: 0.333 seconds
```

图 5-2-21　添加分区

（7）查看员工信息表（employees）的分区情况相关信息，如图 5-2-22 所示，具体命令如下：

SHOW PARTITIONS employees；

```
hive> SHOW PARTITIONS employees;
OK
dept_no=D001
dept_no=D002
Time taken: 0.524 seconds, Fetched: 2 row(s)
```

图 5-2-22　查看分区情况

（8）从本地文件系统中将数据文件"employees-D001.txt"和"employees-D002.txt"中的数据导入到 Hive 数据仓库的员工信息表（employees）中的指定数据分区之中，如图 5-2-23 和图 5-2-24 所示，具体命令如下

LOAD DATA LOCAL INPATH '/home/hadoop/hive/employees-D001.txt' INTO TABLE employees PARTITION （dept_no='D001'）；
LOAD DATA LOCAL INPATH '/home/hadoop/hive/employees-D002.txt' INTO TABLE employees PARTITION （dept_no='D002'）；

```
hive> LOAD DATA LOCAL INPATH '/home/hadoop/hive/employees-D001.txt' INTO TABLE employees PARTITION (dept_no='D001');
Loading data to table testdb.employees partition (dept_no=D001)
OK
Time taken: 2.748 seconds
```

图 5-2-23　导入文件"employees-D001.txt"

```
hive> LOAD DATA LOCAL INPATH '/home/hadoop/hive/employees-D002.txt' INTO TABLE employees PARTITION (
dept_no='D002');
Loading data to table testdb.employees partition (dept_no=D002)
OK
Time taken: 1.31 seconds
```

图 5-2-24　导入文件"employees-D002.txt"

（9）在操作系统的控制台界面中使用 Hadoop 的 FS 命令在 HDFS 文件系统中查看之前导入到 Hive 数据仓库中的员工信息表（employees）的数据。由于该表是个分区表，所以在 HDFS 文件系统中该表的数据存放目录下会有该表的各个分区的目录，如图 5-2-25 所示。在分区目录中保存着该分区数据的数据文件，如图 5-2-26 所示。数据文件的具体内容如图 5-2-27 所示，也是和用来导入数据的数据文件的名称和内容是一样的。

```
[hadoop@Hive hive]$ hadoop fs -ls hive/warehouse/testdb.db/employees
Found 2 items
drwxrwxrwx   - hadoop supergroup          0 2018-08-22 09:37 hive/warehouse/testdb.db/employees/dept
_no=D001
drwxrwxrwx   - hadoop supergroup          0 2018-08-22 09:43 hive/warehouse/testdb.db/employees/dept
_no=D002
```

图 5-2-25　查看导入数据

```
[hadoop@Hive hive]$ hadoop fs -ls hive/warehouse/testdb.db/employees/dept_no=D001
Found 1 items
-rwxrwxrwx   3 hadoop supergroup        234 2018-08-22 09:37 hive/warehouse/testdb.db/employees/dept
_no=D001/employees-D001.txt
```

图 5-2-26　查看分区数据文件

```
[hadoop@Hive hive]$ hadoop fs -cat hive/warehouse/testdb.db/employees/dept_no=D001/employees-D001.tx
t
10001   1953-09-02   Georgi    Facello    M    1986-06-26
10002   1964-06-02   Bezalel   Simmel     F    1985-11-21
10003   1959-12-03   Parto     Bamford    M    1986-08-28
10004   1954-05-01   Chirstian Koblick    M    1986-12-01
10005   1955-01-21   Kyoichi   Maliniak   M    1989-09-12
```

图 5-2-27　查看分区数据文件内容

5.2.3　从数据库查询数据

Hive 数据仓库中的数据查询可以直接使用关系型数据库中通用的 SQL 语言，标准 SQL 语言中的绝大部分功能 Hive 都能够提供完美的支持。

（1）查询部门信息表（departments）中的所有数据，如图 5-2-28 所示，具体命令如下：

SELECT ＊ FROM departments;

（2）查询员工信息表（employees）中的所有数据，如图 5-2-29 所示，具体命令如下：

SELECT ＊ FROM employees;

```
hive> SELECT * FROM departments;
OK
D001    Marketing
D002    Finance
D003    Production
D004    Sales
D005    Customer Service
Time taken: 1.652 seconds, Fetched: 5 row(s)
```

图 5-2-28　查询部门信息表数据

```
hive> SELECT * FROM employees;
OK
10001   1953-09-02   Georgi  Facello  M     1986-06-26            D001
10002   1964-06-02   Bezalel Simmel   F     1985-11-21            D001
10003   1959-12-03   Parto   Bamford  M     1986-08-28            D001
10004   1954-05-01   Chirstian        Koblick M    1986-12-01    D001
10005   1955-01-21   Kyoichi Maliniak         M    1989-09-12    D001
10006   1953-04-20   Anneke  Preusig  F     1989-06-02            D002
10007   1957-05-23   Tzvetan Zielinski        F    1989-02-10    D002
10008   1958-02-19   Saniya  Kalloufi         M    1994-09-15    D002
10009   1952-04-19   Sumant  Peac     1985-02-18            D002
10010   1963-06-01   Duangkaew        Piveteau     F    1989-08-24    D002
Time taken: 0.319 seconds, Fetched: 10 row(s)
```

图 5-2-29　查询员工信息表数据

（3）查询员工信息表（employees）中"D001"分区的所有数据，如图 5-2-30 所示，具体命令如下：

SELECT * FROM employees WHERE dept_no='D001';

查询指定分区的数据实际上就是条件查询，因为被用作分区的列也是存在于表结构当中的，是可以作为查询条件的属性列的。分区表一般都用于数据量较大的表，所以对于分区表不建议使用上面的全表扫描的方式进行查询，这样会浪费大量资源。

```
hive> SELECT * FROM employees WHERE dept_no='D001';
OK
10001   1953-09-02   Georgi  Facello  M     1986-06-26            D001
10002   1964-06-02   Bezalel Simmel   F     1985-11-21            D001
10003   1959-12-03   Parto   Bamford  M     1986-08-28            D001
10004   1954-05-01   Chirstian        Koblick M    1986-12-01    D001
10005   1955-01-21   Kyoichi Maliniak         M    1989-09-12    D001
Time taken: 0.97 seconds, Fetched: 5 row(s)
```

图 5-2-30　查询"D001"分区数据

（4）连表查询，从部门信息表（departments）和员工信息表（departments）中获取员工信息与部门名称的对应数据，如图 5-2-31 所示，具体命令如下：

SELECT　a.emp_no，a.first_name，a.last_name，b.dept_name　FROM

employees a，departments b WHERE a.dept_no=b.dept_no;

对于比较复杂的 SQL 查询操作，Hive 会将其自动转换成 MapReduce 任务来执行。

```
hive> SELECT a.emp_no,a.first_name,a.last_name,b.dept_name FROM employees a,departments b WHERE a.de
pt_no=b.dept_no;
WARNING: Hive-on-MR is deprecated in Hive 2 and may not be available in the future versions. Conside
r using a different execution engine (i.e. spark, tez) or using Hive 1.X releases.
Query ID = hadoop_20180822102252_9bdde007-46a6-46ba-8a41-ee65ce7916a7
Total jobs = 1
SLF4J: Class path contains multiple SLF4J bindings.
SLF4J: Found binding in [jar:file:/home/hadoop/hive/apache-hive-2.1.1-bin/lib/log4j-slf4j-impl-2.4.1
.jar!/org/slf4j/impl/StaticLoggerBinder.class]
SLF4J: Found binding in [jar:file:/home/hadoop/hadoop/hadoop-2.7.3/share/hadoop/common/lib/slf4j-log
4j12-1.7.10.jar!/org/slf4j/impl/StaticLoggerBinder.class]
SLF4J: See http://www.slf4j.org/codes.html#multiple_bindings for an explanation.
SLF4J: Actual binding is of type [org.apache.logging.slf4j.Log4jLoggerFactory]
2018-08-22 10:23:06     Starting to launch local task to process map join;      maximum memory = 518
979584
2018-08-22 10:23:08     Dump the side-table for tag: 1 with group count: 5 into file: file:/home/had
oop/hive/tmp/hadoop/b81ae347-6356-4763-9dae-934de82f830f/hive_2018-08-22_10-22-52_052_38909207182026
1897-1/-local-10004/HashTable-Stage-3/MapJoin-mapfile11--.hashtable
2018-08-22 10:23:08     Uploaded 1 File to: file:/home/hadoop/hive/tmp/hadoop/b81ae347-6356-4763-9da
e-934de82f830f/hive_2018-08-22_10-22-52_052_389092071820261897-1/-local-10004/HashTable-Stage-3/MapJ
oin-mapfile11--.hashtable (427 bytes)
2018-08-22 10:23:08     End of local task; Time Taken: 1.961 sec.
Execution completed successfully
MapredLocal task succeeded
Launching Job 1 out of 1
Number of reduce tasks is set to 0 since there's no reduce operator
Starting Job = job_1534518620990_0005, Tracking URL = http://Cluster-01:8088/proxy/application_15345
18620990_0005/
Kill Command = /home/hadoop/hadoop/hadoop-2.7.3/bin/hadoop job  -kill job_1534518620990_0005
Hadoop job information for Stage-3: number of mappers: 1; number of reducers: 0
2018-08-22 10:24:03,925 Stage-3 map = 0%,  reduce = 0%
2018-08-22 10:24:31,258 Stage-3 map = 100%,  reduce = 0%, Cumulative CPU 2.96 sec
```

```
MapReduce Total cumulative CPU time: 2 seconds 960 msec
Ended Job = job_1534518620990_0005
MapReduce Jobs Launched:
Stage-Stage-3: Map: 1   Cumulative CPU: 2.96 sec   HDFS Read: 8635 HDFS Write: 516 SUCCESS
Total MapReduce CPU Time Spent: 2 seconds 960 msec
OK
10001   Georgi  Facello Marketing
10002   Bezalel Simmel  Marketing
10003   Parto   Bamford Marketing
10004   Chirstian       Koblick Marketing
10005   Kyoichi Maliniak        Marketing
10006   Anneke  Preusig Finance
10007   Tzvetan Zielinski       Finance
10008   Saniya  Kalloufi        Finance
10009   Sumant  Peac    Finance
10010   Duangkaew       Piveteau        Finance
Time taken: 101.451 seconds, Fetched: 10 row(s)
```

图 5-2-31　连表查询

5.2.4　其他常用操作命令

（1）删除指定数据库——DROP DATABASE 数据库名;

（2）删除指定表——DROP TABLE 表名;

（3）清空指定表中的所有数据——TRUNCATE TABLE 表名;

（4）修改指定表的表名——ALTER TABLE 原表名 RENAME TO 新

表名；

（5）向指定表中添加新列——ALTER TABLE 表名 ADD COLUMNS（列名1 字段类型1，列名2 字段类型2，…）；

（6）删除指定分区表中的指定分区——ALTER TABLE 表名 DROP PARTION （分区列名='分区值'）；

（7）从其他表中查询数据并添加到指定表中——INSERT INTO|OVERWRITE TABLE 表名 查询语句；

INTO 表示在表中原有数据的基础上添加新的数据，OVERWRITE 表示删除表中原有的所有数据再加载新数据。

（8）查看 Hive 的功能函数——SHOW FUNCTIONS。

项目 6 使用 ETL 工具 Sqoop 转换数据

任务 6.1 Sqoop 工具的安装

6.1.1 Sqoop 工具介绍

6.1.1.1 ETL 工具

ETL，是英文 Extract Transform Load 的缩写，它是用来描述将数据从源端经过抽取（Extract）、转换（Transform）、加载（Load）之后到达目的端的整个过程。

ETL 是构建数据仓库的重要一环，是用户将业务系统中的数据经过抽取、清洗、转换之后，按照预先定义好的数据仓库模型加载到数据仓库中的过程。目的是将企业的业务系统中分散、零乱、标准不统一的数据整合到一起，为企业的决策提供分析的依据。ETL 是 BI 项目中的一个重要组成部分。通常情况下，BI 项目中 ETL 会花掉整个项目至少 1/3 的时间，ETL 设计的好坏直接关系到 BI 项目的成败。

ETL 的设计如它的名字所代表的含义一样，主要分为三部分，分别是数据抽取、数据清洗和转换、数据加载。数据加载一般是在数据清洗和转换完成之后将产生的结果数据直接写入到 DW（Data Warehousing，数据仓库）中。

1. 数据抽取（Extract）

数据抽取是指从各个不同的数据源中抽取数据到 ODS（Operational Data Store，操作型数据存储）中，这个过程中也可以做一些数据的清洗和转换工作。这一部分需要在调研阶段做大量的工作，要搞清楚数据是从哪几个业务系统中来，各个业务系统的数据库服务器运行的是什么样的 DBMS，是否存在手工数据，手工数据量有多大，是否存在非结构化的数据等。当收集完这些信息之后才可以进行数据抽取的设计。

2. 数据清洗和转换（Cleaning、Transform）

一般情况下数据仓库至少分为 ODS 和 DW 两个部分。通常的做法是数据从业务系统到 ODS 时做清洗操作，将脏数据和不完整数据过滤掉。然后再在数据从 ODS 到 DW 的过程中做转换操作，进行一些业务规则的计算和聚合。这部分是三个部分中花费时间最长的，一般情况下这部分的工作量会占据整个 ETL 过程的 2/3。

1）数据清洗

数据清洗的任务是过滤那些不符合要求的数据，主要包括不完整的数据、错误的数据、重复的数据三大类。不完整的数据主要是指一些应该有的信息缺失了，这一类数据一般会整理提交给客户，要求其在规定的时间内补全。错误的数据一般是由于业务系统不健全导致，如对输入的数据没有进行格式验证和有效性判断等，这一类数据也会整理提交给客户，要求其在规定的时间内修正。重复的数据在现在的数据库设计中比较常见，这一类数据会将重复数据记录的字段整理出来，让客户确认最终使用的字段。数据清洗是一个反复的过程，不可能在几天之内完成，需要不断地发现并解决问题。对于是否修正、是否过滤等问题一般要求客户进行确认。

2）数据转换

数据转换的任务主要是进行不一致数据的转换，数据粒度的转换，以及一些基于商务规则的计算。不一致数据的转换一个整合的过程，会将不同业务系统的相同类型的数据进行统一，比如同一个供应商在结算系统中的编码是 XX0001，而在 CRM 中的编码是 YY0001，这样在抽取过来之后需要统一转换成同一个编码。数据粒度的转换主要是针对业务系统和数据仓库不同的数据粒度要求，业务系统一般存储非常明细的数据，而数据仓库中用来分析的数据不需要非常明细。一般情况下会将业务系统中的数据按照数据仓库中的数据粒度进行聚合。商务规则的计算则是根据不同的 ETL 业务有着不同的业务规则和数据指标，这些计算很多时候不是简单的加加减减就能完成的，需要在 ETL 中将这些数据的规则和指标计算好之后存储在数据仓库中，以供后面的分析使用。

ETL 的常用实现方法有三种。一种是借助 ETL 工具实现，如 Oracle 的 OWB、SQL Server 2000 的 DTS、SQL Server2005 的 SSIS 服务、Informatic 等；另一种是 SQL 方式实现；最后一种是 ETL 工具和 SQL 相结合来实现。借助 ETL 工具的方法可以快速地建立起 ETL 流程，屏蔽了复杂的编码任务，提高了速度，降低了难度，但是缺少灵活性。SQL 的方法优点是灵活，提高了 ETL 的运行效率，但是编码复杂，对技术要求比较高。第三种综合了前面两种的优点，能够极大地提高 ETL 的开发速度和效率。

6.1.1.2　Sqoop

Sqoop 是一款的开源 ETL 工具，主要用于在 Hadoop 平台的组件（HDFS、HBase、Hive 等）和传统关系型数据库（MySQL、PostgreSQL 等）之间进行数据的传递。比如它可以将一个传统关系型数据库中的数据导入到 Hadoop 的 HDFS 中，也可以将 HDFS 中的数据导入到传统关系型数据库中。

Sqoop 项目开始于 2009 年，由 Cloudera 公司开发。最早是作为 Hadoop 的一个第三方模块存在，后来为了让使用者能够进行快速部署，同时也是为了让软件开发人员能够更快速地进行迭代开发，Sqoop 被捐给了 Apache 软件基金会，独立成为一个 Apache 项目。

随着 Hadoop 成为一个越来越通用的分布式计算环境，更多的用户需要将数据集在 Hadoop 的组件与传统关系型数据库之间进行转移，而 Sqoop 就是这样一款能够帮助进行数据传输的工具。而对于某些 NoSQL 数据库，Sqoop 也提供了对应连接器。

Sqoop 类似于其他 ETL 工具，使用元数据模型来判断数据类型并在数据从数据源转移到 Hadoop 时确保类型安全的数据处理。而 Sqoop 中的一大亮点就是通过 Hadoop 的 MapReduce 把数据在传统关系型数据库和 HDFS 之间进行转移。Sqoop 是专门为大数据批量传输设计，能够分割数据集并创建 Hadoop 任务来处理每个区块，可以在 Hadoop 和传统关系型数据库之间转移大量数据。

Sqoop 目前分为 Sqoop 1 和 Sqoop 2 这两个大的版本分支，其中 Sqoop 1 当前的版本系列编号为 1.4.X，而 Sqoop 2 当前的版本系列编号为 1.99.X。

6.1.1.3　Sqoop 1 的架构

Sqoop 1 在架构上使用的是 Sqoop 客户端直接提交的方式，如图 6-1-1 所示，访问方式采用的是 CLI 控制台方式进行访问。通过在命令或脚本中指定用户数据库的名称和密码来保证安全性。

6.1.1.4　Sqoop 2 的架构

Sqoop 2 在架构上引入了 Sqoop Server，并对 Connector 实现了集中的管理，如图 6-1-2 所示。访问方式也更加丰富，可以通过 REST API、JAVA API、WEB UI、CLI 控制台等多种方式进行访问。同时 Sqoop 2 还引入了基于角色的安全机制。

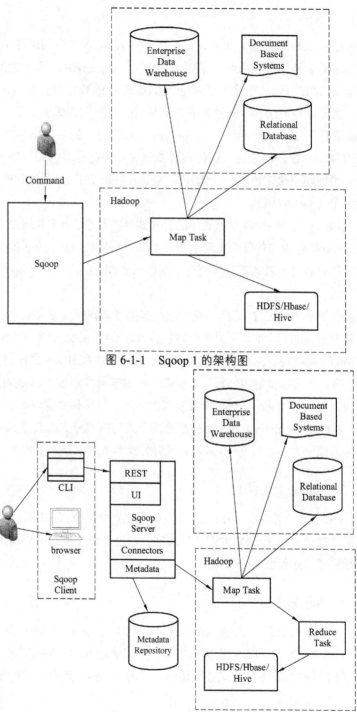

图 6-1-1　Sqoop 1 的架构图

图 6-1-2　Sqoop 2 的架构图

6.1.1.5 Sqoop 1 与 Sqoop 2 的对比

1. Sqoop 1 和 Sqoop 2 的功能对比

表 6-1-1　Sqoop 1 和 Sqoop 2 对比

功能	Sqoop 1	Sqoop 2
用于所有主要 RDBMS 的连接器	支持	不支持，但可以使用通用 JDBC 连接器，不过性能上可能无法与 Sqoop 1 中的专用连接器相比。还可以通过 API 编写连接程序
Kerberos 安全集成	支持	不支持
数据从 RDBMS 传输至 Hive 或 HBase	支持	不支持，但可以先将数据从 RDBMS 导入到 HDFS 之中，然后再在 Hive 中使用相应的工具和命令（如 LOAD DATA 语句）手动将数据载入到 Hive 或 HBase 之中。还可以通过 API 自己编写程序来执行连接和数据传输
数据从 Hive 或 HBase 传输至 RDBMS	不支持，但可以先从 Hive 或 HBase 将数据以文本或 Avro 文本的形式提取至 HDFS，然后再使用 Sqoop 将数据从 HDFS 导出到 RDBMS	不支持，但按照与 Sqoop 1 相同的解决方法操作。还可以通过 API 编写程序来执行连接和数据传输

2. Sqoop 1 和 Sqoop 2 的优缺点对比

Sqoop 1 的架构部署更为简单；

Sqoop 1 的命令行方式容易出错，命令格式紧耦合；

Sqoop 1 无法支持所有数据类型；

Sqoop 1 安全机制不够完善，例如显式密码等；

Sqoop 1 的安装需要 ROOT 权限；

Sqoop 1 的 Connector 必须符合 JDBC 模型；

Sqoop 2 拥有多种交互方式；

Sqoop 2 的 Conncetor 集中化管理，所有的链接都安装在 Sqoop Server 之上；

Sqoop 2 的 Connector 更加规范化，仅仅负责数据的读写；

Sqoop 2 具备完善的权限管理机制；

Sqoop 2 架构较为复杂，配置部署更加烦琐。

6.1.2 Sqoop 工具的安装规划

1. 环境要求

本地磁盘剩余空间 600 M 以上，其中 Sqoop 工具软件自身需要 200 M 空间，Hadoop 软件需要 400M 空间；

已安装 CentOS 7 1611 64 位操作系统；

已安装 JDK 的 1.8.0_131 版本；

已完成基础网络环境配置；

已完成 Hadoop 整合平台的安装和部署；

已完成 MySQL 数据库平台的安装和部署。

2. 软件版本

选用 Sqoop 的 1.99.7 版本，软件包名 sqoop-1.99.7-bin-hadoop200.tar.gz，该软件包可以在 Sqoop 项目位于 Apache 的官方网站（http://sqoop.apache.org）获取，如图 6-1-3 所示。

图 6-1-3　获取软件包

3. 依赖软件

Sqoop 工具使用的是 Java 语言实现，所以在进行 MySQL 数据库的数据转换时，是使用 JDBC 的方式连接 MySQL 数据库，需要使用到对应的 MySQL 数据库的连接驱动软件包。同 Hive 数据仓库平台搭建时一样，选用该软件包的 5.1.42 版本，软件包名 mysql-connector-java-5.1.42-bin.jar。

6.1.3 Hadoop 平台配置

Sqoop 工具在访问 Hadoop 平台的 MapReduce 组件时使用的是代理方式，所以必须在 Hadoop 平台中配置 Proxy 服务所能够接收的用户和组。另外，Sqoop 工具在使用 Hadoop 平台的 MapReduce 组件时还需要使用到该组件的 JobHistory 服务，这是用来记录 MapReduce 中的任务运行状态并对外提供访问的服务。所以这里需要先对 Hadoop 平台原有的配置进行一些修改。

（1）选择 Hadoop 平台中的任意一台主机，进入其上的 Hadoop 软件目录下的配置文件所在目录"etc/hadoop"。

（2）编辑 Hadoop 的配置文件"core-site.xml"，如图 6-1-4 所示，在其中配置以下内容：

```
<configuration>
  <property>
    <name>hadoop.proxyuser.$SERVER_USER.hosts</name>
    <value>*</value>
  </property>
  <property>
    <name>hadoop.proxyuser.$SERVER_USER.groups</name>
    <value>*</value>
  </property>
</configuration>
```

配置项说明：

hadoop.proxyuser.用户名.hosts——指定用户"$SERVER_USER"可以代理的主机的主机名，多个主机名之间用","分隔，"*"号表示可以代理所有主机。"$SERVER_USER"为安装 Sqoop 工具的主机上用于运行 Sqoop 工具的用户，这里使用的是 Hadoop 平台的专用用户"hadoop"。

hadoop.proxyuser.用户名.groups——指定用户"$SERVER_USER"可以代理的主机上的用户组，多个组名之间用","分隔，"*"号表示可以代理所有用户组。"$SERVER_USER"为安装 Sqoop 工具的主机上用于运行 Sqoop 工具的用户，这里使用的是 Hadoop 平台的专用用户"hadoop"。

```
<property>
  <name>hadoop.proxyuser.hadoop.hosts</name>
  <value>*</value>
</property>
<property>
  <name>hadoop.proxyuser.hadoop.groups</name>
  <value>*</value>
</property>
```

图 6-1-4　编辑配置文件

（3）编辑 Hadoop 的配置文件"mapred-site.xml"，如图 6-1-5 所示，在其中配置以下内容：

<configuration>

 <property>

  <name>mapreduce.jobhistory.address</name>

  <value>Cluster-01：10020</value>

 </property>

</configuration>

配置项说明：

mapreduce.jobhistory.address——指定集群中 JobHistory 服务节点对应的地址和端口。

```
<property>
  <name>mapreduce.jobhistory.address</name>
  <value>Cluster-01:10020</value>
</property>
```

图 6-1-5　编辑配置文件

（4）将修改后的配置文件"core-site.xml"和"mapred-site.xml"发送到 Hadoop 平台中的其他所有主机上的 Hadoop 软件配置文件目录下，覆盖原有文件。

（5）重新启动 Hadoop 平台，使新配置的 Hadoop 平台的代理服务生效。

（6）在 Hadoop 平台中提供 JobHistory 服务的主机上依次使用"mr-jobhistory-daemon.sh start historyserver"命令和"yarn-daemon.sh start timelineserver"命令来启动 JobHistory 服务，如图 6-1-6 所示。然后查看 Java 进程信息，如图 6-1-7 所示，若存在名为"JobHistoryServer"和"ApplicationHistoryServer"的两个进程，则表示 Hadoop 平台的 JobHistory 服务启动成功。

```
[hadoop@Cluster-01 hadoop]$ mr-jobhistory-daemon.sh start historyserver
starting historyserver, logging to /home/hadoop/hadoop/hadoop-2.7.3/logs/mapred-hadoop-historyserver
-Cluster-01.out
[hadoop@Cluster-01 hadoop]$ yarn-daemon.sh start timelineserver
starting timelineserver, logging to /home/hadoop/hadoop/hadoop-2.7.3/logs/yarn-hadoop-timelineserver
-Cluster-01.out
```

图 6-1-6　启动 JobHistory

```
[hadoop@Cluster-01 hadoop]$ jps
3874 QuorumPeerMain
117122 ApplicationHistoryServer
117060 JobHistoryServer
117220 Jps
116631 DFSZKFailoverController
116326 NameNode
116726 ResourceManager
```

图 6-1-7　查看 Java 进程信息

6.1.4　Sqoop 工具基础环境配置

（1）创建"sqoop"目录用于存放 Sqoop 相关文件，该目录可自行选择创建位置，创建完成后将当前工作目录切换到该目录。

（2）使用命令"tar -xzf Sqoop 安装包路径"将软件包解压解包到"sqoop"目录下，如图 6-1-8 所示，解压解包出来的目录名称为"sqoop-1.99.7-bin-hadoop200"。

```
[hadoop@Sqoop sqoop]$ ls -l
total 0
drwxrwxr-x. 8 hadoop hadoop 157 Jul 16 07:06 sqoop-1.99.7-bin-hadoop200
```

图 6-1-8　解压软件包

（3）在 Sqoop 的相关文件目录中创建 Sqoop 工具的第三方 Jar 包（即使用 Sqoop 连接数据库或存储系统时使用到的驱动软件包或其他工具软件包）的存放目录"extra"，如图 6-1-9 所示。

```
[hadoop@Sqoop sqoop]$ ls -l
total 0
drwxrwxr-x. 2 hadoop hadoop   6 Jul 16 07:07 extra
drwxrwxr-x. 8 hadoop hadoop 157 Jul 16 07:06 sqoop-1.99.7-bin-hadoop200
```

图 6-1-9　创建存放目录"extra"

（4）将 JDBC 连接 MySQL 数据库的驱动软件包"mysql-connector-java-5.1.42-bin.jar"拷贝到 Sqoop 工具的第三方 Jar 包的存放目录"extra"下，如图

6-1-10 所示。

```
[hadoop@Sqoop extra]$ ls -l
total 976
-rwxr-xr-x. 1 hadoop hadoop 996444 Jul 16 07:08 mysql-connector-java-5.1.42-bin.jar
```

图 6-1-10　拷贝软件包

（5）在用户的配置文件".bash_profile"中配置 Sqoop 相关的环境变量，如图 6-1-11 所示，在文件末尾添加以下内容：

sqoop environment

SQOOP_HOME=Sqoop 软件目录路径

SQOOP_SERVER_EXTRA_LIB=Sqoop 相关文件目录/extra

PATH=$SQOOP_HOME/bin：$PATH

export　SQOOP_HOME　SQOOP_SERVER_EXTRA_LIB　PATH

Sqoop 软件目录即 Sqoop 软件包解压解包出来的"sqoop-1.99.7-bin-hadoop200"目录，这里需要书写该目录及其所在的绝对路径。

```
# sqoop environment
SQOOP_HOME=/home/hadoop/sqoop/sqoop-1.99.7-bin-hadoop200
SQOOP_SERVER_EXTRA_LIB=/home/hadoop/sqoop/extra
PATH=$SQOOP_HOME/bin:$PATH
export SQOOP_HOME SQOOP_SERVER_EXTRA_LIB PATH
```

图 6-1-11　修改配置文件

（6）使用命令"source ~/.bash_profile"使新配置的环境变量立即生效。

（7）使用命令"echo $变量名"查看新添加和修改的环境变量的值是否正确，如图 6-1-12 所示。

```
[hadoop@Sqoop ~]$ echo $SQOOP_SERVER_EXTRA_LIB
/home/hadoop/sqoop/extra
```

图 6-1-12　查看环境变量

（8）选择 Hadoop 平台中的任意主机，将其上的操作系统配置文件目录"etc"下的地址映射关系配置文件"hosts"拷贝到当前用于安装 sqoop 工具的主机下的对应目录中，覆盖原有的文件。

若是直接在 Hadoop 平台中的某台主机上安装 Sqoop 工具，可以省略掉该步骤以及之后的步骤（10）。

（9）编辑操作系统配置文件目录"etc"下的地址映射关系配置文件"hosts"，在其中添加当前主机的 IP 地址与主机名的映射关系，如图 6-1-13 所示。

```
192.168.60.130    Cluster-01
192.168.60.131    Cluster-02
192.168.60.132    Cluster-03
192.168.60.133    Cluster-04
192.168.60.134    Cluster-05
192.168.60.201    Sqoop
```

图 6-1-13　编辑配置文件

（10）创建"hadoop"目录用于存放 Hadoop 相关文件，该目录可自行选择创建位置。创建完成之后，选择 Hadoop 平台中的任意主机，将其上的 Hadoop 软件目录"hadoop-2.7.3"拷贝到该目录之中。然后如图 6-1-14 所示，在用户的配置文件".bash_profile"中配置 Hadoop 相关的环境变量"HADOOP_HOME"和"PATH"。

```
# hadoop environment
HADOOP_HOME=/home/hadoop/hadoop/hadoop-2.7.3
PATH=$HADOOP_HOME/bin:$HADOOP_HOME/sbin:$PATH
export HADOOP_HOME PATH
```

图 6-1-14　配置相关环境变量

6.1.5　Sqoop 工具安装与配置

（1）Sqoop 的配置文件位于其软件目录的"conf"目录下，如图 6-1-15 所示，进入该目录并对其中的配置文件进行编辑。

```
[hadoop@Sqoop conf]$ pwd
/home/hadoop/sqoop/sqoop-1.99.7-bin-hadoop200/conf
```

图 6-1-15　进入"conf"目录

（2）配置文件"sqoop_bootstrap.properties"里可以对 Sqoop 软件的 config 支持类进行配置，一般使用默认值即可，不需要进行更改。

（3）编辑配置文件"sqoop.properties"，如图 6-1-16 所示，找到下列配置项的所在行并修改其对应内容：

org.apache.sqoop.submission.engine.mapreduce.configuration.directory=Hadoop 软件目录路径/etc/hadoop

org.apache.sqoop.security.authentication.type=SIMPLE

org.apache.sqoop.security.authentication.handler=org.apache.sqoop.security.authentication.SimpleAuthenticationHandler org.apache.sqoop.security.authentication.anonymous=true

配置项说明：

org.apache.sqoop.submission.engine.mapreduce.configuration.directory——指定本地文件系统中 Hadoop 软件的配置文件目录的所在路径。

org.apache.sqoop.security.authentication.type——指定 Sqoop 软件的安全验证机制类型，支持 Hadoop 平台的"SIMPLE"和"KERBEROS"两种验证机制。原配置文件中已有对应的配置项值，只需要去掉前面的注释符号即可。

org.apache.sqoop.security.authentication.handler——指定用于进行 Sqoop 软件的安全验证的 Java 实现类。原配置文件中已有对应的配置项值，只需要去掉前面的注释符号即可。

org.apache.sqoop.security.authentication.anonymous——指定 Sqoop 软件是否允许使用匿名用户。原配置文件中已有对应的配置项值，只需要去掉前面的注释符号即可。

该配置文件中配置项较多，可以在 vi 编辑器内使用命令"/关键字"直接进行搜索，同时使用快捷键"n"可以切换到下一个关键字的所在位置。

```
139 #
140 # Configuration for Mapreduce submission engine (applicable if it's configured)
141 #
142
143 # Hadoop configuration directory
144 org.apache.sqoop.submission.engine.mapreduce.configuration.directory=/home/hadoop/hadoop/hadoop-2.7.3/etc/hadoop

153
154 #
155 # Authentication configuration
156 #
157 org.apache.sqoop.security.authentication.type=SIMPLE
158 org.apache.sqoop.security.authentication.handler=org.apache.sqoop.security.authentication.SimpleAuthenticationHandler
159 org.apache.sqoop.security.authentication.anonymous=true
```

图 6-1-16　编辑配置文件"sqoop.properties"

6.1.6　Sqoop 工具初始化和启动

（1）进入用于存放 Sqoop 相关文件的目录"sqoop"，使用命令"sqoop2-tool upgrade"对 Sqoop 的元数据进行初始化，如图 6-1-17 所示。初始化操作完成之后会在当前工作目录下生成"@BASEDIR@"和"@LOGDIR@"两个目录，如图 6-1-18 所示，其中"@BASEDIR@"目录存放的是 Sqoop 的元数据信息，而"@LOGDIR@"目录存放的是 Sqoop 的日志信息。

（2）使用命令"sqoop2-tool　verify"验证 Sqoop 软件的配置是否正确，如图 6-1-19 所示。

项目 6 使用 ETL 工具 Sqoop 转换数据

图 6-1-17 对 Sqoop 无数据初始化

图 6-1-18 查看目录

图 6-1-19 验证 Sqoop 软件配置

（3）使用命令"sqoop2-server start"启动 Sqoop 工具的服务，如图 6-1-20 所示。

图 6-1-20 启动服务

（4）查看 Java 进程信息，如图 6-1-21 所示，若有名为"SqoopJettyServer"的进程，则表示 Sqoop 服务启动成功。

图 6-1-21 查看 Java 进程信息

（5）使用命令"sqoop2-shell"进入 Sqoop 的控制台界面，如图 6-1-22 所示，若能够正常进入则表示 Sqoop 工具能够正常使用。

```
[hadoop@Sqoop ~]$ sqoop2-shell
Setting conf dir: /home/hadoop/sqoop/sqoop-1.99.7-bin-hadoop200/bin/../conf
Sqoop home directory: /home/hadoop/sqoop/sqoop-1.99.7-bin-hadoop200
Sqoop Shell: Type 'help' or '\h' for help.

sqoop:000>
```

图 6-1-22　进入 Sqoop 控制台界面

（6）在 Sqoop 的控制台界面中使用命令"：exit"可以退出 Sqoop 的控制台界面返回到操作系统的命令行界面，如图 6-1-23 所示。

```
sqoop:000> :exit
[hadoop@Sqoop ~]$
```

图 6-1-23　退出控制台

6.1.7　初始化 Sqoop 服务器连接参数

（1）使用命令"sqoop2-shell"进入 Sqoop 的控制台界面，本小节之后的所有步骤都是在 Sqoop 的控制台界面之中进行操作。

（2）使用命令"set server --host Sqoop 服务器的主机名或 IP 地址 --port 12000 --webapp sqoop"配置 Sqoop 服务器的连接参数。"如图 6-1-24 所示，--port"选项设定的是 Sqoop 服务所使用的端口号，这里使用的是默认的端口号"12000"。而"--webapp"选项是指定 Sqoop 工具软件的 Jetty 服务器的标识名称，可以自己定义。

```
sqoop:000> set server --host Sqoop --port 12000 --webapp sqoop
Server is set successfully
```

图 6-1-24　配置连接参数

（3）使用命令"show version --all"验证 Sqoop 服务器的连接是否正常，如图 6-1-25 所示，该命令能够显示"client"和"server"的版本信息，若显示的版本信息正确，则表示服务器连接没有问题。

```
sqoop:000> show version --all
client version:
  Sqoop 1.99.7 source revision 435d5e61b922a32d7bce567fe5fb1a9c0d9b1bbb
  Compiled by abefine on Tue Jul 19 16:08:27 PDT 2016
0   [main] WARN  org.apache.hadoop.util.NativeCodeLoader  - Unable to load native-hadoop library fo
r your platform... using builtin-java classes where applicable
server version:
  Sqoop 1.99.7 source revision 435d5e61b922a32d7bce567fe5fb1a9c0d9b1bbb
  Compiled by abefine on Tue Jul 19 16:08:27 PDT 2016
API versions:
  [v1]
```

图 6-1-25　验证服务器连接

任务 6.2 使用 Sqoop 工具进行数据转换

6.2.1 准备 MySQL 数据库的数据

这里使用一个在网上公开发布的员工与部门信息的示例数据库作为示例数据来演示 Sqoop 工具的数据转换操作,该数据库可以从一个名为 Launchpad 的网站中获取,具体的下载地址为"https://launchpad.net/test-db/+download",选择其中对应完整数据库的文件包"employees_db-full-1.0.6.tar.bz2"进行下载,如图 6-2-1 所示。

图 6-2-1 获取文件包

该示例数据库的结构如图 6-2-2 所示。

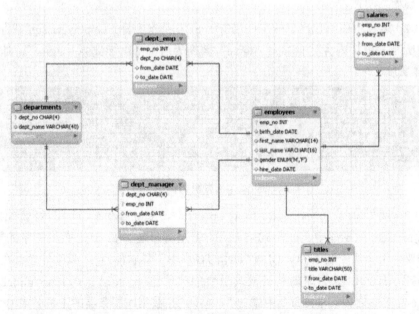

图 6-2-2 数据库结构

（1）使用命令"tar -xjf 示例数据库文件包路径"将之前下载下来的的示例数据库的文件包"employees_db-full-1.0.6.tar.bz2"解压解包，解压解包出来的目录名称为"employees_db"，该目录的存放路径可以自行选择，该文件包中的内容如图 6-2-3 所示。

```
[hadoop@Sqoop employees_db]$ ls -l
total 164492
-rwxr-xr-x. 1 hadoop hadoop       752 Aug 16 11:10 Changelog
-rwxr-xr-x. 1 hadoop hadoop      6460 Aug 16 11:10 employees_partitioned2.sql
-rwxr-xr-x. 1 hadoop hadoop      7624 Aug 16 11:10 employees_partitioned3.sql
-rwxr-xr-x. 1 hadoop hadoop      5660 Aug 16 11:10 employees_partitioned.sql
-rwxr-xr-x. 1 hadoop hadoop      3861 Aug 16 11:10 employees.sql
-rwxr-xr-x. 1 hadoop hadoop       241 Aug 16 11:10 load_departments.dump
-rwxr-xr-x. 1 hadoop hadoop  13828291 Aug 16 11:10 load_dept_emp.dump
-rwxr-xr-x. 1 hadoop hadoop      1043 Aug 16 11:10 load_dept_manager.dump
-rwxr-xr-x. 1 hadoop hadoop  17422825 Aug 16 11:10 load_employees.dump
-rwxr-xr-x. 1 hadoop hadoop 115848997 Aug 16 11:10 load_salaries.dump
-rwxr-xr-x. 1 hadoop hadoop  21265549 Aug 16 11:11 load_titles.dump
-rwxr-xr-x. 1 hadoop hadoop      3889 Aug 16 11:11 objects.sql
-rwxr-xr-x. 1 hadoop hadoop      2211 Aug 16 11:11 README
-rwxr-xr-x. 1 hadoop hadoop      4455 Aug 16 11:11 test_employees_md5.sql
-rwxr-xr-x. 1 hadoop hadoop      4450 Aug 16 11:11 test_employees_sha.sql
```

图 6-2-3　文件包内容

若是无法正常的进行解包解压操作，可能是因为当前 Linux 操作系统的发行版或是安装模式中没有默认安装 bzip2 工具，需要安装该工具之后再进行解包解压操作。

（2）编辑解压解包出来的文件包中的 SQL 脚本文件"employees.sql"，将其中的所有关键字内容"storage_engine"替换为"default_storage_engine"，如图 6-2-4 所示。

```
38    set default_storage_engine = InnoDB;
39 -- set default_storage_engine = MyISAM;
40 -- set default_storage_engine = Falcon;
41 -- set default_storage_engine = PBXT;
42 -- set default_storage_engine = Maria;
43
44 select CONCAT('default_storage_engine: ', @@default_storage_engine) as INFO;
```

图 6-2-4　替换关键字

（3）使用命令"mysql -u root -p -t < employees.sql"将示例数据导入到 MySQL 数据库中，如图 6-2-5 所示。若是远程连接 MySQL 数据库服务器，则需要保证本机安装了 MySQL 数据库的客户端工具软件，该工具软件包含在之前搭建 MySQL 数据库的单机模式平台的软件包 mysql-5.7.18-linux-glibc2.5-x86_64.tar.gz 之中，并在命令中添加"-h MySQL 数据库服务器的主机名或 IP 地址"来指定远程服务器。

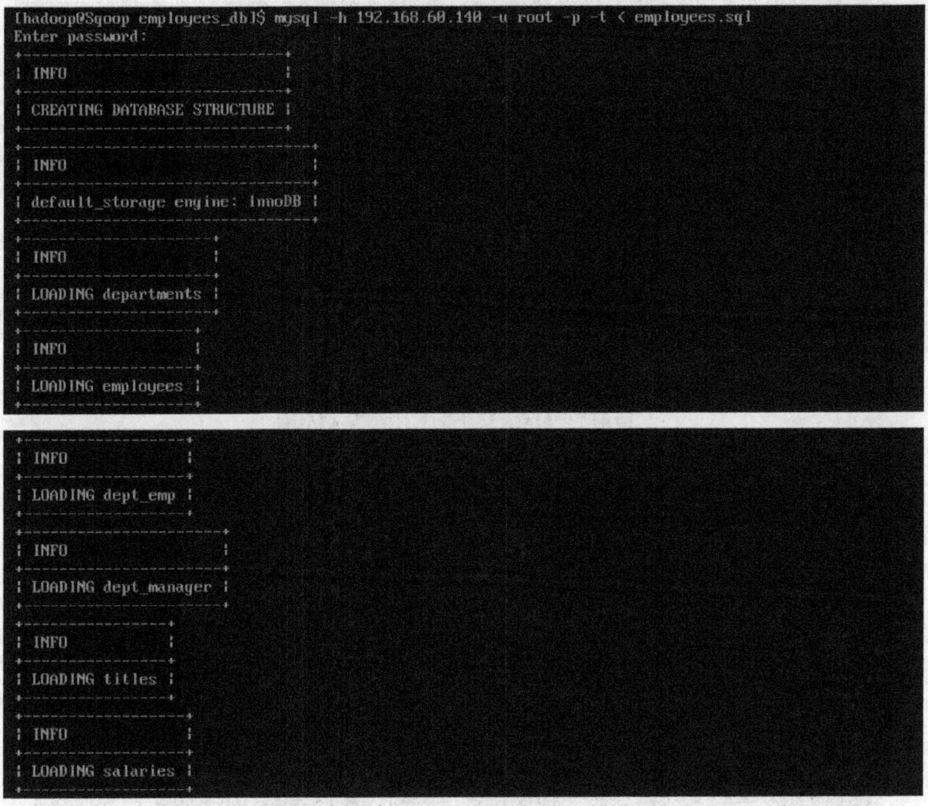

图 6-2-5　连接数据库服务器

（4）登录到 MySQL 数据库，通过查看导入的示例数据库的数据库、表、数据等内容检查导入操作是否成功完成。相关 SQL 语句如下：

SHOW　DATABASES;（查看数据库信息，见图 6-2-6）

图 6-2-6　查看数据库信息

USE　employees;（切换到导入的"employees"数据库，见图 6-2-7）
SHOW　TABLES;（查看表信息，见图 6-2-8）

```
mysql> USE employees;
Reading table information for completion of table and column names
You can turn off this feature to get a quicker startup with -A

Database changed
```

图 6-2-7　切换数据库

```
mysql> SHOW TABLES;
+----------------------+
| Tables_in_employees  |
+----------------------+
| departments          |
| dept_emp             |
| dept_manager         |
| employees            |
| salaries             |
| titles               |
+----------------------+
6 rows in set (0.00 sec)
```

图 6-2-8　查看表信息

SELECT * FROM employees LIMIT 20;（查看表"employees"中的数据，如图 6-2-9 所示，由于该表中的数据较多，而检查导入操作是否成功不需要查看所有数据，所以这里只查看了其中的 20 行数据）

```
mysql> SELECT * FROM employees LIMIT 20;
+--------+------------+------------+-----------+--------+------------+
| emp_no | birth_date | first_name | last_name | gender | hire_date  |
+--------+------------+------------+-----------+--------+------------+
| 10001  | 1953-09-02 | Georgi     | Facello   | M      | 1986-06-26 |
| 10002  | 1964-06-02 | Bezalel    | Simmel    | F      | 1985-11-21 |
| 10003  | 1959-12-03 | Parto      | Bamford   | M      | 1986-08-28 |
| 10004  | 1954-05-01 | Chirstian  | Koblick   | M      | 1986-12-01 |
| 10005  | 1955-01-21 | Kyoichi    | Maliniak  | M      | 1989-09-12 |
| 10006  | 1953-04-20 | Anneke     | Preusig   | F      | 1989-06-02 |
| 10007  | 1957-05-23 | Tzvetan    | Zielinski | F      | 1989-02-10 |
| 10008  | 1958-02-19 | Saniya     | Kalloufi  | M      | 1994-09-15 |
| 10009  | 1952-04-19 | Sumant     | Peac      | F      | 1985-02-18 |
| 10010  | 1963-06-01 | Duangkaew  | Piveteau  | F      | 1989-08-24 |
| 10011  | 1953-11-07 | Mary       | Sluis     | F      | 1990-01-22 |
| 10012  | 1960-10-04 | Patricio   | Bridgland | M      | 1992-12-18 |
| 10013  | 1963-06-07 | Eberhardt  | Terkki    | M      | 1985-10-20 |
| 10014  | 1956-02-12 | Berni      | Genin     | M      | 1987-03-11 |
| 10015  | 1959-08-19 | Guoxiang   | Nooteboom | M      | 1987-07-02 |
| 10016  | 1961-05-02 | Kazuhito   | Cappelletti| M     | 1995-01-27 |
| 10017  | 1958-07-06 | Cristinel  | Bouloucos | F      | 1993-08-03 |
| 10018  | 1954-06-19 | Kazuhide   | Peha      | F      | 1987-04-03 |
| 10019  | 1953-01-23 | Lillian    | Haddadi   | M      | 1999-04-30 |
| 10020  | 1952-12-24 | Mayuko     | Warwick   | M      | 1991-01-26 |
+--------+------------+------------+-----------+--------+------------+
20 rows in set (0.00 sec)
```

图 6-2-9　查看 employee 表数据

SELECT * FROM departments;（查看表 "departments" 中的数据，见图 6-2-10，这是后面将要把数据从 MySQL 数据库导入到 HDFS 文件系统中的表）

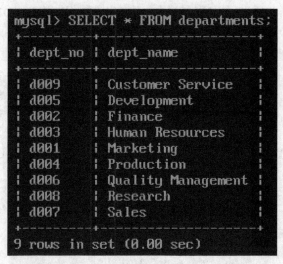

图 6-2-10　查看 departments 中的数据

6.2.2　建立 MySQL 数据库的连接

（1）使用命令 "sqoop2-shell" 进入 Sqoop 的控制台界面，本小节之后的所有步骤以及后面小节的所有步骤，若没有特殊说明，默认都是在 Sqoop 控制台界面之中进行操作。

（2）使用命令 "show connector" 可以查看当前所有的连接模板信息，如图 6-2-11 所示。"Name" 列是连接模板的名称，"Class" 列是连接模板的连接操作所对应的 Java 实现类，"Supported Directions" 列是连接模板所支持的数据流向，"FROM" 表示可以读取，"TO" 表示可以写入。Sqoop 的 1.99.7 版本默认提供了七种连接的模板：普通 JDBC 连接、Kite 数据集连接、Oracle 数据库的 JDBC 连接、FTP 连接、HDFS 文件系统连接、Kafka 消息系统连接、SFTP 连接。其中 FTP、Kafka 消息系统、SFTP 这三种连接只能执行写入的单向数据转换，而其他连接都能执行读取和写入的双向数据转换。

Sqoop 的 1.99.7 版本中并没有提供 HBase 和 Hive 的连接模板，如果需要使用该版本的 Sqoop 来进行 HBase 和 Hive 与其他文件系统或数据库之间的数据转换，需要使用 Sqoop 所提供的 API 来自行开发相应的连接程序。理论上，使用 Sqoop 所提供的 API 自行开发连接和转换程序，可以实现所有文件系统或数据库之间的数据转换。

Hive 数据仓库由于其本身就是使用 HDFS 文件系统或是本地文件系统之中的文件来导入数据，所以可以通过使用 Sqoop 先将数据转换到 HDFS 文件系统，然后再从 HDFS 文件系统中的数据文件将数据导入到 Hive 数据仓库中这种间接的方式来实现其他文件系统或数据库到 Hive 数据仓库的数据转换。

```
sqoop:000> show connector
+------------------------+---------+------------------------------------------------------+----+
|          Name          | Version |                        Class                         | Su
| pported Directions     |         |                                                      |
+------------------------+---------+------------------------------------------------------+----+
| generic-jdbc-connector | 1.99.7  | org.apache.sqoop.connector.jdbc.GenericJdbcConnector | FR
| OM/TO                  |         |                                                      |
| kite-connector         | 1.99.7  | org.apache.sqoop.connector.kite.KiteConnector        | FR
| OM/TO                  |         |                                                      |
| oracle-jdbc-connector  | 1.99.7  | org.apache.sqoop.connector.jdbc.oracle.OracleJdbcConnector | FR
| OM/TO                  |         |                                                      |
| ftp-connector          | 1.99.7  | org.apache.sqoop.connector.ftp.FtpConnector          | TO
| hdfs-connector         | 1.99.7  | org.apache.sqoop.connector.hdfs.HdfsConnector        | FR
| OM/TO                  |         |                                                      |
| kafka-connector        | 1.99.7  | org.apache.sqoop.connector.kafka.KafkaConnector      | TO
| sftp-connector         | 1.99.7  | org.apache.sqoop.connector.sftp.SftpConnector        | TO
+------------------------+---------+------------------------------------------------------+----+
```

图 6-2-11　查看所有连接模板信息

（3）使用命令"create link -connector generic-jdbc-connector"创建一个 JDBC 类型的连接，用于连接 MySQL 数据库。此时会出现交互式操作，如图 6-2-12 所示，需要依次设定下列各个参数项所对应的值，每次显示并输入一个参数项，每个参数项输入完成之后按回车键可以进行下一个参数项的输入。下面是每个参数项所对应的配置内容：

Name——指定该 JDBC 类型连接的名称，可自行定义。

Driver Class——指定该 JDBC 类型连接启动时需要加载的驱动类，当前需要连接的是 MySQL 数据库，所以对应的 JDBC 连接驱动类为 "com.mysql.jdbc.Driver"。

Connection String——指定该 JDBC 类型连接的连接字符串，当前需要连接的是 MySQL 数据库，所以连接字符串为"jdbc：mysql：//MySQL 服务器的主机名或 IP 地址：3306/数据库名称"。

Username——指定该 JDBC 类型连接在连接 MySQL 数据库时所使用的登录用户的名称，这里使用 MySQL 数据库的 "root" 用户来连接数据库。

Password——指定该 JDBC 类型连接在连接 MySQL 数据库时所使用的登录用户的密码，这里为 MySQL 数据库的 "root" 用户所对应的密码。

项目6 使用ETL工具Sqoop转换数据

```
sqoop:000> create link -connector generic-jdbc-connector
Creating link for connector with name generic-jdbc-connector
Please fill following values to create new link object
Name: MySQL-Connector

Database connection

Driver class: com.mysql.jdbc.Driver
Connection String: jdbc:mysql://192.168.60.140:3306/employees
Username: root
Password: ******
Fetch Size:
Connection Properties:
There are currently 0 values in the map:
entry# protocol=tcp
There are currently 1 values in the map:
protocol = tcp
entry# useSSL=false
There are currently 2 values in the map:
protocol = tcp
useSSL = false
entry#

SQL Dialect

Identifier enclose:
Fri Aug 17 08:53:23 EDT 2018 WARN: Establishing SSL connection without server's identity verificatio
n is not recommended. According to MySQL 5.5.45+, 5.6.26+ and 5.7.6+ requirements SSL connection mus
t be established by default if explicit option isn't set. For compliance with existing applications
not using SSL the verifyServerCertificate property is set to 'false'. You need either to explicitly
disable SSL by setting useSSL=false, or set useSSL=true and provide truststore for server certificat
e verification.
New link was successfully created with validation status OK and name MySQL-Connector
```

图 6-2-12　创建 JDBC 类型的连接

Fetch Size——指定查询操作的数据缓冲区大小，该项参数使用默认值即可。使用默认值的方式为不输入任何内容直接按回车键进入下一个参数项的输入。

entry#——指定该 JDBC 类型连接的其他相关连接属性，在这里配置的连接属性需要同时书写配置项名称和对应的值，中间用 "=" 号连接。可以配置多项连接属性，书写完一个配置项及其对应的值之后，按回车键会提示输入下一个配置项及其对应的值。在不输入任何内容的情况下按回车键可以结束该阶段配置的输入。这里为该 JDBC 类型连接配置连接属性 "protocol=tcp" 和 "useSSL=false"。

Identifier enclose——指定 SQL 语句界定符，默认是双引号 """，但这个界定符会导致在 MySQL 中执行 SQL 语句时出错，需要改为空格。即在这里输入一个空格字符，然后按回车键确认即可。

各个参数项配置完成之后若出现包含有 "successful" 内容的提示信息，则表示该连接创建成功。

（4）使用命令 "show link" 查看新创建的连接的信息，如图 6-2-13 所示。

```
sqoop:000> show link
+----------------+------------------------+---------+
|      Name      |     Connector Name     | Enabled |
+----------------+------------------------+---------+
| MySQL-Connector| generic-jdbc-connector |  true   |
+----------------+------------------------+---------+
```

图 6-2-13　查看连接信息

6.2.3 建立 HDFS 文件系统连接

（1）使用命令"create link -connector hdfs-connector"创建一个 HDFS 类型的连接，用于连接 Hadoop 平台的 HDFS 文件系统。此时会出现交互式操作，如图 6-2-14 所示，需要依次设定下列各个参数项所对应的值，每次显示并输入一个参数项，每个参数项输入完成之后按回车键可以进行下一个参数项的输入。下面是每个参数项所对应的配置内容：

Name——指定该 HDFS 类型连接的名称，可自行定义。

URI——指定 HDFS 文件系统的访问路径，这个访问路径需要与 Hadoop 的配置文件"core-site.xml"中配置的属性"fs.defaultFS"中的值一致。

Conf directory——指定 Hadoop 平台的配置文件在本地文件系统中的存放路径，Hadoop 的配置文件位于其软件目录下的"etc/hadoop"目录下，Hadoop 的软件目录以拷贝到搭建 Sqoop 平台的主机上的路径为准。

entry#——指定该 HDFS 类型连接的其他相关连接属性，和创建 JDBC 类型连接时一样，在这里配置的连接属性需要同时书写配置项名称和对应的值，中间用"="号连接。同时可以配置多项连接属性，书写完一个配置项及其对应的值之后，按回车键会提示输入下一个配置项及其对应的值。在不输入任何内容的情况下按回车键可以结束该阶段配置的输入。这里不需要为该 HDFS 类型连接配置其他相关连接属性，直接按回车键跳过即可。

和创建 JDBC 类型连接时一样，各个参数项配置完成之后若出现包含有"successful"内容的提示信息，则表示该连接创建成功。

```
sqoop:000> create link -connector hdfs-connector
Creating link for connector with name hdfs-connector
Please fill following values to create new link object
Name: HDFS-Connector

HDFS cluster

URI: hdfs://hadoop-ha
Conf directory: /home/hadoop/hadoop/hadoop-2.7.3/etc/hadoop
Additional configs::
There are currently 0 values in the map:
entry#
New link was successfully created with validation status OK and name HDFS-Connector
```

图 6-2-14 创建 HDFS 类型的连接

（2）使用命令"show link"可以查看新创建的连接的信息，如图 6-2-15 所示。

```
sqoop:000> show link
+-----------------+------------------------+---------+
|      Name       |     Connector Name     | Enabled |
+-----------------+------------------------+---------+
| MySQL-Connector | generic-jdbc-connector |  true   |
| HDFS-Connector  | hdfs-connector         |  true   |
+-----------------+------------------------+---------+
```

图 6-2-15　查看连接信息

6.2.4　建立数据传输事务

（1）使用命令 "create job -f 连接名1 -t 连接名2" 创建数据传输事务，其中选项 "-f" 指定数据的来源位置，选项 "-t" 指定数据的目标位置。当前将要操作的内容为从 MySQL 数据库将数据导入 HDFS 文件系统，即 "连接名1" 为前面创建的 JDBC 类型连接的连接名，而 "连接名2" 为前面创建的 HDFS 类型连接的连接名。运行命令时会出现交互式操作，如图 6-2-16 所示，需要依次设定下列各个参数项所对应的值，每次显示并输入一个参数项，每个参数项输入完成之后按回车键可以进行下一个参数项的输入。下面是每个参数项所对应的配置内容：

Name——指定该数据传输事务的名称，可自行定义。

首先是设置与数据的来源相关的参数项：

Schema Name——指定数据源所使用的 Database 或 Schema 的名称，在 MySQL 中 Database 与 Schema 类似，这里直接指定为作为数据源的数据库的名称 "employees" 即可。

Table Name——指定需要导出数据的数据库中的表的表名，可以指定多个表。这里只导出 "departments" 表的数据。

SQL Statement——指定 SQL 查询语句的查询条件，该参数项为可选项，这里选择导出指定表的所有数据，所以不需要输入任何内容，直接按回车键跳过该参数项的设置即可。

element#——设置数据导出的其他相关参数，该项为可选项，这里直接按回车键跳过不进行设置。

Partition column——指定用于分区的列，该参数项为可选项，这里直接按回车键跳过不进行设置。

Partition column nullable——设置用于分区的列中的数据是否允许空值，该参数项为可选项，这里直接按回车键跳过不进行设置。

Boundary query——指定查询操作的边界，该参数项为可选项，这里直接按回车键跳过不进行设置。

Check column——指定用于检查的列，该参数项为可选项，这里直接按回车键跳过不进行设置。

Last value——指定导入数据时的结束值，该参数项为可选项，这里直接按回车键跳过不进行设置。

然后是设置数据的目标相关的参数项：

Override null value——设置是否对数据来源中的空值进行替换，该参数项为可选项，这里直接按回车键跳过不进行设置。

Null value——指定用于替换数据来源中空值的值，该参数项为可选项，这里直接按回车键跳过不进行设置。

File format——指定在 HDFS 文件系统中用于存放导出数据的数据文件的文件格式，这里会给出多个选项来进行选择，选择的方式为输入对应文件格式的编号。这里使用文本文件来保存导出的数据，即对应选项中的"TEXT_FILE"文件格式，输入该文件格式对应的编号"0"即可。

Compression codec——指定用于存放导出数据的数据文件所使用的压缩格式，这里会给出多个选项来进行选择，选择的方式为输入对应压缩方式的编号。这里不使用任何压缩方式来压缩数据文件，即对应选项中的"NONE"方式，输入该方式对应的编号"0"即可。

Custom codec——指定客户端的字符编码格式，这里使用默认的编码格式，直接按回车键跳过即可。

Output directory——指定导出的数据文件在 HDFS 文件系统中的存储目录的路径，该目录的名称及其路径可以自行制定，但需要是一个在 HDFS 文件系统中不存在的目录，否则数据传输事务在运行过程中会报错，不过该错误不会影响数据传输事务的继续执行。这里使用"/user/hadoop/sqoop-data"目录来保存从 MySQL 数据库中导入到 HDFS 文件系统中的数据。

Append mode——设置在导出的数据文件已存在的情况下，是否可以将新数据追加到该数据文件的末尾，该参数项为可选项，这里直接按回车键跳过不进行设置。

Extractors——指定在 MapReduce 任务中该数据传输事务所使用的 Mapper 的数量，这个数量需要根据当前 Hadoop 平台的实际情况来设置，这里设置为"2"。

Loaders——指定在 MapReduce 任务中该数据传输事务所使用的 Reducer 的数量，这个数量同样需要根据当前 Hadoop 平台的实际情况来设置，这里设置为"2"。

element#——指定该数据传输事务所对应的 MapReduce 任务一些其他相关参数,该参数项为可选项,这里直接按回车键跳过不进行设置。

和创建连接时一样,各个参数配置完成之后若出现包含有"successful"内容的提示信息,则表示该数据传输事务创建成功。

从关系型数据库中导出数据到文件系统之中时,只能实现一个表、一个视图、一个查询结果集的导出,因为文件系统中的单个文件并不能像数据库一样同时分别存储多个表的数据。

```
sqoop:000> create job -f MySQL-Connector -t HDFS-Connector
Creating job for links with from name MySQL-Connector and to name HDFS-Connector
Please fill following values to create new job object
Name: MySQL-to-HDFS

Database source

Schema name: employees
Table name: departments
SQL statement:
Column names:
There are currently 0 values in the list:
element#
Partition column:
Partition column nullable:
Boundary query:

Incremental read

Check column:
Last value:

Target configuration

Override null value:
Null value:
File format:
  0 : TEXT_FILE
  1 : SEQUENCE_FILE
  2 : PARQUET_FILE
Choose: 0
Compression codec:
  0 : NONE
  1 : DEFAULT
  2 : DEFLATE
  3 : GZIP
  4 : BZIP2
  5 : LZO
  6 : LZ4
  7 : SNAPPY
  8 : CUSTOM
Choose: 0
Custom codec:
Output directory: /user/hadoop/sqoop_data
Append mode:

Throttling resources

Extractors: 2
Loaders: 2

Classpath configuration

Extra mapper jars:
There are currently 0 values in the list:
element#
New job was successfully created with validation status OK  and name MySQL-to-HDFS
```

图 6-2-16 创建数据传输事务

(2)使用命令"show job"可以查看新创建的数据传输事务的信息,如图 6-2-17 所示。

```
sqoop:000> show job
+----+---------------+-------------------------------------------+-----------------------------------+---------+
| Id | Name          | From Connector                            | To Connector                      | Enabled |
+----+---------------+-------------------------------------------+-----------------------------------+---------+
| 1  | MySQL-to-HDFS | MySQL-Connector (generic-jdbc-connector)  | HDFS-Connector (hdfs-connector)   | true    |
+----+---------------+-------------------------------------------+-----------------------------------+---------+
```

图 6-2-17　查看数据传输事务信息

6.2.5　将 MySQL 数据导入 HDFS

（1）使用命令"start job -n 数据传输事务名"启动并运行指定的数据传输事务，如图 6-2-18 所示。

```
sqoop:000> start job -n MySQL-to-HDFS
Submission details
Job Name: MySQL-to-HDFS
Server URL: http://localhost:12000/sqoop/
Created by: hadoop
Creation date: 2018-08-17 11:56:18 EDT
Lastly updated by: hadoop
External ID: job_1534518620990_0003
        http://Cluster-01:8088/proxy/application_1534518620990_0003/
2018-08-17 11:56:18 EDT: BOOTING  - Progress is not available
```

图 6-2-18　启动并运行事务

（2）使用命令"status job -n 事务名"可以查看指定的数据传输事务的状态信息。若该数据传输事务正在运行过程中，可以看到该数据传输事务当前的运行进度，如图 6-2-19 所示。若该数据传输事务已经运行完成，则可以看到该数据传输事务的所有运行统计数据，如图 6-2-20 所示。

```
sqoop:000> status job -n MySQL-to-HDFS
Submission details
Job Name: MySQL-to-HDFS
Server URL: http://localhost:12000/sqoop/
Created by: hadoop
Creation date: 2018-08-17 11:56:18 EDT
Lastly updated by: hadoop
External ID: job_1534518620990_0003
        http://Cluster-01:8088/proxy/application_1534518620990_0003/
2018-08-17 11:56:33 EDT: BOOTING  - 0.00 %

sqoop:000> status job -n MySQL-to-HDFS
Submission details
Job Name: MySQL-to-HDFS
Server URL: http://localhost:12000/sqoop/
Created by: hadoop
Creation date: 2018-08-17 11:56:18 EDT
Lastly updated by: hadoop
External ID: job_1534518620990_0003
        http://Cluster-01:8088/proxy/application_1534518620990_0003/
2018-08-17 11:56:54 EDT: RUNNING  - 50.00 %
```

项目6 使用ETL工具Sqoop转换数据

```
sqoop:000> status job -n MySQL-to-HDFS
Submission details
Job Name: MySQL-to-HDFS
Server URL: http://localhost:12000/sqoop/
Created by: hadoop
Creation date: 2018-08-17 11:56:18 EDT
Lastly updated by: hadoop
External ID: job_1534518620990_0003
        http://Cluster-01:8088/proxy/application_1534518620990_0003/
2018-08-17 11:57:01 EDT: RUNNING  - 75.00 %
```

图 6-2-19　查看数据传输事务状态信息

```
sqoop:000> status job -n MySQL-to-HDFS
Submission details
Job Name: MySQL-to-HDFS
Server URL: http://localhost:12000/sqoop/
Created by: hadoop
Creation date: 2018-08-17 11:56:18 EDT
Lastly updated by: hadoop
External ID: job_1534518620990_0003
        http://Cluster-01:8088/proxy/application_1534518620990_0003/
2018-08-17 11:57:41 EDT: SUCCEEDED
Counters:
        org.apache.hadoop.mapreduce.FileSystemCounter
                FILE_LARGE_READ_OPS: 0
                FILE_WRITE_OPS: 0
                HDFS_READ_OPS: 2
                HDFS_BYTES_READ: 295
                HDFS_LARGE_READ_OPS: 0
                FILE_READ_OPS: 0
                FILE_BYTES_WRITTEN: 1169486
                FILE_BYTES_READ: 228
                HDFS_WRITE_OPS: 2
                HDFS_BYTES_WRITTEN: 189
        org.apache.hadoop.mapreduce.lib.output.FileOutputFormatCounter
                BYTES_WRITTEN: 0
        org.apache.hadoop.mapreduce.lib.input.FileInputFormatCounter
                BYTES_READ: 0
        org.apache.hadoop.mapreduce.JobCounter
                TOTAL_LAUNCHED_MAPS: 2
                VCORES_MILLIS_REDUCES: 20239
                MB_MILLIS_MAPS: 21905408
                TOTAL_LAUNCHED_REDUCES: 2
                SLOTS_MILLIS_REDUCES: 20239
                VCORES_MILLIS_MAPS: 21392
                MB_MILLIS_REDUCES: 20724736
                SLOTS_MILLIS_MAPS: 21392
                MILLIS_REDUCES: 20239
                OTHER_LOCAL_MAPS: 2
                MILLIS_MAPS: 21392
        org.apache.sqoop.submission.counter.SqoopCounters
                ROWS_READ: 9
                ROWS_WRITTEN: 9
        org.apache.hadoop.mapreduce.TaskCounter
                MAP_OUTPUT_MATERIALIZED_BYTES: 240
                REDUCE_INPUT_RECORDS: 9
                SPILLED_RECORDS: 18
                MERGED_MAP_OUTPUTS: 4
                VIRTUAL_MEMORY_BYTES: 8408850432
                MAP_INPUT_RECORDS: 0
                SPLIT_RAW_BYTES: 295
                FAILED_SHUFFLE: 0
                MAP_OUTPUT_BYTES: 198
                REDUCE_SHUFFLE_BYTES: 240
                PHYSICAL_MEMORY_BYTES: 1112887296
                GC_TIME_MILLIS: 2776
                REDUCE_INPUT_GROUPS: 9
                COMBINE_OUTPUT_RECORDS: 0
                SHUFFLED_MAPS: 4
                REDUCE_OUTPUT_RECORDS: 9
                MAP_OUTPUT_RECORDS: 9
                COMBINE_INPUT_RECORDS: 0
                CPU_MILLISECONDS: 16000
                COMMITTED_HEAP_BYTES: 722427904
        Shuffle Errors
                CONNECTION: 0
                WRONG_LENGTH: 0
                BAD_ID: 0
                WRONG_MAP: 0
                WRONG_REDUCE: 0
                IO_ERROR: 0
Job executed successfully
```

图 6-2-20　传输事务统计数据

（3）返回到操作系统的控制台界面，查看 HDFS 文件系统中的文件与目录信息，可以看到在 HDFS 文件系统的操作系统用户对应目录中多出了一个名为“sqoop-data”的目录，如图 6-2-21 所示。而目录中有两个文件，如图 6-2-22 所示。

```
[hadoop@Sqoop ~]$ hadoop fs -ls
Found 4 items
drwxr-xr-x   - hadoop supergroup          0 2018-08-14 12:13 data
drwxr-xr-x   - hadoop supergroup          0 2018-07-23 12:48 hbase
drwxrwxrwx   - hadoop supergroup          0 2018-07-12 09:57 hive
drwxr-xr-x   - hadoop supergroup          0 2018-08-17 11:57 sqoop-data
```

图 6-2-21　查看“sqoop-data”目录

```
[hadoop@Sqoop ~]$ hadoop fs -ls sqoop-data
Found 2 items
-rw-r--r--   3 hadoop supergroup         92 2018-08-17 11:57 sqoop-data/53a2b2d0-405a-4086-960d-8a86
b7cfba5d.txt
-rw-r--r--   3 hadoop supergroup         97 2018-08-17 11:57 sqoop-data/ac33892a-2a9d-4476-94eb-48dd
d2f74bc4.txt
```

图 6-2-22　“sqoop-data”目录包含文件

（4）在操作系统的控制台界面使用命令“hadoop fs -cat sqoop-data/*”查看导出数据目录“sqoop-data”中的所有文件的内容，如图 6-2-23 所示。这些数据与 MySQL 数据库中“employees”数据库的“departments”表中的数据是一样的，如图 6-2-24 所示，说明数据转换操作成功完成。

```
[hadoop@Sqoop ~]$ hadoop fs -cat sqoop-data/*
'd002','Finance'
'd005','Development'
'd006','Quality Management'
'd009','Customer Service'
'd001','Marketing'
'd003','Human Resources'
'd004','Production'
'd007','Sales'
'd008','Research'
```

图 6-2-23　查看所有文件内容

```
mysql> SELECT * FROM departments;
+---------+--------------------+
| dept_no | dept_name          |
+---------+--------------------+
| d009    | Customer Service   |
| d005    | Development        |
| d002    | Finance            |
| d003    | Human Resources    |
| d001    | Marketing          |
| d004    | Production         |
| d006    | Quality Management |
| d008    | Research           |
| d007    | Sales              |
+---------+--------------------+
9 rows in set (0.00 sec)
```

图 6-2-24　查看 departments 表数据

参考文献

[1] White T. Hadoop 权威指南[M]. 北京：清华大学出版社，2015.
[2] Lam C. Hadoop 实战[M]. 北京：人民邮电出版社，2011.
[3] 陆嘉恒. Hadoop 实战[M]. 北京：机械工业出版社，2012.
[4] Turkington G. Hadoop 基础教程[M]. 北京：人民邮电出版社，2014.
[5] 唐汉明，翟振兴，关宝军，王洪权，黄潇. 深入浅出 MySQL[M]. 北京：人民邮电出版社，2014.